General Household Survey

Results for 2006

General Household Survey 2006

© Crown copyright 2008
Published with the permission of the Controller of Her Majesty's Stationary Office (HMSO) under the Click-Use License by Dandy Booksellers Ltd.

ISBN

ISSN 1754-4777

Applications for reproduction should be submitted to HMSO under HMSO's Class Licence: www.clickanduse.hmso.gov.uk

Alternatively applications can be made in writing to:

HMSO
Licensing Division
St. Clement's House
2-16 Colegate
Norwich
NR3 1BQ

Contact points:
To order this publication, contact
Dandy Booksellers Ltd:
Units 3 & 4
31-33 Priory Park Road
London NW6 7UP

Tel: +44 (0) 20 7624 2993
Fax: +44 (0) 20 7624 5049
Email: dandybooksellers@btconnect.com
Website: www.dandybooksellers.com

For enquiries about content, contact
National Statistics Customer Contact Centre
Room 1.015
Office for National Statistics
Cardiff Road
Newport
NP10 8XG

Tel: +44 (0) 845 601 3034
Fax: +44 (0) 1633 652747
ONS Minicom: +44 (0) 1633 812399
Email: info@statistics.gsi.gov.uk

You can also find National Statistics on the internet – go to www.statistics.gov.uk.

The report is compiled taking content from the National Statistics website
www.statistics.gov.uk/lib

About the Office for National Statistics
The Office for National Statistics (ONS) is the government agency responsible for compiling, analysing and disseminating many of the United Kingdom's economic, social and demographic statistics, including the retail prices index, trade figures and labour market data, as well as the periodic census of the population and heath statistics. The Director of ONS is also the National Statistician and the Registrar General for England and Wales, and the agency administers, the registry of birth, marriages and deaths there.

CONTENTS

General Household Survey Overview 1

Notes to Tables 6

Appendix A Definitions and terms..7

Appendix B Sample design and response 26

Appendix C Sampling errors.. 40

Appendix D Weighting and grossing...56

Appendix E Household and individual questionnaires 65

Appendix F Summary of main topics included in the GHS questionnaires 1971 –2005…………………………………………………………………………..... 180

Appendix G List of tables.. 193

Smoking and drinking among adults, 2006..198

Datasets……………………………………………………………………….215 - 368

General Household Survey, 2006
Overview Report

General Household Survey 2006

Overview Report

Riaz Ali

Julia Greer

David Matthews

Liam Murray

Simon Robinson

Ghazala Sattar

Office for National Statistics

Government Buildings

Cardiff Rd

Newport

NP10 8XG

Tel: 01633 455877

Email: ghs@ons.gsi.gov.uk

© Crown copyright

Office for National Statistics: November 2006

General Household Survey, 2006

Overview Report

Results for the 2006 General Household Survey (GHS) are now available. The GHS is a multi-purpose survey carried out by the Office for National Statistics (ONS).

From 1994-5 to 2004-5, the GHS was conducted on a financial year basis, with fieldwork spread evenly across the year April-March. However, in 2005 the survey period reverted to a calendar year and the whole of the annual sample (which was increased to 16,560) was dealt with in the nine months April to December 2005. From January 2006, the survey runs from January to December each year.

Since the 2005 survey did not cover the January-March quarter, this affected annual estimates for topics which are subject to seasonal variation. To rectify this, where the questions were the same in 2005 as in 2004-05, the final quarter of the 2004-05 survey was added (weighted in the correct proportion) to the nine months of the 2005 survey.

Another change in 2005 was that, in line with European requirements, the GHS adopted a longitudinal sample design, in which households remain in the sample for four years (waves) with one quarter of the sample being replaced each year. Thus approximately three quarters of the 2005 sample were re-interviewed in 2006. More details are given in Appendix B.

A major advantage of the longitudinal component of the design is that it is more efficient at detecting statistically significant estimates of change over time than the previous cross-sectional design. This is because an individual's responses to the same question at different points in time tend to be positively correlated, and this reduces the standard errors of estimates of change.

The GHS collects information on a range of topics from people living in private households in Great Britain. These are:

- households, families and people;
- housing and consumer durables;
- marriage and cohabitation;
- general health and use of health services;
- smoking and drinking;
- occupational and personal pension schemes.

A report on smoking and drinking can be found at www.statistics.gov.uk/ghs. Selected findings for other topics are summarised in this overview. GHS results will be combined with those from other sources in *Social Trends* and other reports due to be published in 2008.

Households, families and people

Between 1971 and 1991 the average size of household in Great Britain declined, from 2.91 persons to 2.48. It continued to decline, though at a slower rate, throughout the 1990s, falling to 2.32 by 1998, since when it has changed little. In 2006 the average number of persons per household was 2.34.

In 2006, 75 per cent of families with dependent children in Great Britain were headed by a married or cohabiting couple. This proportion fell markedly in the 1970s, 1980s and 1990s (92 per cent of families were of this type in 1971) but has changed little since the late 1990s. The percentage of families that are headed by a lone mother rose from 7 per cent in 1971 to 22 per cent in 1998 and has remained at about the same level since then. The proportion headed by a lone father has risen slightly since the early 1970s, but since the mid-1990s has remained at about 2-3 per cent. In total, the proportion of families headed by a lone parent has increased from 8 per cent in 1971 to 25 per cent in 2006.

In 2006, about one half of families headed by a lone parent had a usual weekly gross income of £300 or less, compared with only about one in ten families headed by a married couple, and one in six of those headed by a cohabiting couple.

Housing and consumer durables

Between 1971 and 2006, the proportion of households owning their home rose from 49 per cent to 70 per cent. Most of the increase occurred during the 1980s, and was due to a marked increase in the proportion of households owning with a mortgage.

The percentage of households renting council homes increased from 31 per cent in 1971 to a peak of 34 per cent in 1981, but since then has declined steadily to 12 per cent in 2006.

The percentage of households renting from a housing association rose from 1 per cent in 1971 to 3 per cent in 1991, the increase continuing since then to 8 per cent in 2006.

The percentage of households renting privately fell from 20 per cent in 1971 to 10 per cent in 1995, since when it has changed little.

Since the early 1970s, the GHS has recorded a steady rise in the ownership of many consumer durables and household amenities - some that were available only to a minority of households in the early 1970s were in much more widespread use by 2006. For example, 95 per cent of households had central heating in 2006, compared with only 37 per cent in 1972. Access to a car or van has also risen since the survey began (at least one car or van was available to 52 per cent of households in 1972, rising to 77 per cent in 2006).

Just over half (54 per cent) of all households had a telephone in 1975. In 2006, 99 per cent had a phone (either fixed or mobile). In 2000, when the GHS first asked about mobile phones, the proportion of households in which at least one person had a mobile phone was 58 per cent: this had risen to 83 per cent in 2006.

The proportion of households with a home computer increased from 13 per cent in 1985 to 69 per cent in 2006.

Marriage and cohabitation

In the 1970s, the GHS asked women aged 18 to 49 only about their current marriage: since then, the questions have been developed and extended from time to time to reflect changes taking place in society. The survey now obtains information about marital history and periods of cohabitation from all adults aged 16 and over.

In 2006, 13 per cent of men and women aged 16 to 59 were cohabiting. Among men, those aged 25 to 29 were more likely to cohabit than any other group (33 per cent of men aged 25 to 29 compared with fewer than 22 per cent in all other age groups). Among women, those in their twenties were more likely to be cohabiting than other age groups (27 per cent of women aged 20 to 24 and 29 per cent of women aged 25 to 29 compared with 21 per cent or fewer in the other age groups).

Among women aged 18 to 49, the group for which the longest time series is available, the proportion who were married at the time of interview declined from 74 per cent in 1979 to 47 per cent in 2005 but then increased to 50 per cent in 2006. The proportion of women who were single (i.e. who had never been married) more than doubled from 18 per cent in 1979 to 38 per cent in 2002, since when it has remained at about the same level.

The proportion of women aged 18 to 49 who were cohabiting at the time of interview also increased, from 11 per cent in 1979 to about 30 per cent in the late 1990s, and since then has stayed within the range 28 to 32 percent.

Occupational and personal pension schemes

The GHS has included questions on occupational pensions on a regular basis since 1981 and on personal pensions since 1987.

Since 1989 trends in participation in employer pension schemes have differed for men and women, and for those working part-time and full-time.

The proportion of men working full time who were members of their current employer's occupational pension scheme decreased from 64 per cent in 1989 to 53 per cent in 2006. The percentage of women working full time who were members of an occupational pension scheme showed a different pattern, rising from 55 per cent in 1989 to 60 per cent in 2002 and then falling to 58 per cent in 2006.

Among women working part time, the proportion who were members of an occupational pension scheme has more than doubled from 15 per cent in 1989 to 36 per cent in 2006.

Since 1991 the GHS has provided trend data on personal pension arrangements among self-employed men. The possession of a current personal pension among self-employed men working full-time remained fairly stable between 1991 and 1998 at around two-thirds. Between 1998 and 2006 the proportion with a current personal pension decreased from 64 per cent to 45 per cent.

Self-employed men were more likely than self-employed women to have personal pension arrangements (40 per cent of self-employed men compared with 27 per cent of self-employed women were currently in a personal pension scheme). Over a third (37 per cent) of self-employed men had never had a personal pension scheme compared with over a half (54 per cent) of women.

General Household Survey, 2006
Overview Report

General health and use of health services

The GHS provides information about the self-reported health of adults and children, and about their use of health services.

In 2006, 62 per cent of adults said they had good health, 26 per cent that they had fairly good health, and 12 per cent said their health was not good. In the past decade self-assessed general health has remained largely unchanged.

Thirty three per cent of people (all age groups) reported a long standing illness, that is one that had troubled them over a period of time, or was likely to affect them over a period of time. Just over half of those with a longstanding illness, 19 per cent of all respondents, said that it limited their activities.

In 2006, 11 per cent of males and 15 per cent of females had consulted an NHS GP in the 14 days before interview. This proportion has changed little since the survey began. Females had an average of five NHS GP consultations per year whereas males had four. Among adults, the likelihood of seeing a GP increased with age. For example, 12 per cent of those aged 16 to 44 had consulted a GP in the 14 days before interview, but 22 per cent of those aged 75 and over had done so.

The proportion of respondents attending a hospital outpatient or casualty department in the three months before interview rose from 10 per cent in 1972 to 16 per cent in 1998 and then fell back to 14 per cent in 2001, since when it has remained at that level. The proportion of people attending hospital as a day patient in the twelve months before interview increased from 4 per cent in 1992, when the question was first asked, to 8 per cent in 2005, but fell back to 6 per cent in 2006, the same level as ten years earlier. A similar proportion of respondents in 2006, 7 per cent, reported an inpatient stay in the twelve months before interview: this is a slightly lower proportion than in the early 1980s, but it has changed little since 2000.

Source: General Household Survey, 2006

Notes to tables

1. **Harmonised outputs**: where appropriate, tables including marital status, living arrangements, ethnic groups, tenure, economic activity, accommodation type, length of residence and general health have adopted the harmonised output categories described on the National Statistics website. However, where long established time series are shown, harmonised outputs may not have been used.

2. **Classification variables:** variables such as age and income, are not presented in a standard form throughout the report partly because the groupings of interest depend on the subject matter of the chapter, and partly because many of the trend series were started when the results used in the report had to be extracted from tabulations prepared to meet different departmental requirements.

3. **Nonresponse and missing information:** the information from a household which co-operates in the survey may be incomplete, either because of a partial refusal (eg to income), or because information was collected by proxy and certain questions omitted because considered inappropriate for proxy interviews (eg marriage and income data), or because a particular item was missed because of lack of understanding or an error.

Households who did not co-operate at all are omitted from all the analyses; those who omitted whole sections (eg marriages) because they were partial refusals or interviewed by proxy are omitted from the analyses of that section. The 'no answers' arising from omission of particular items have been excluded from the base numbers shown in the tables and from the bases used in percentaging. The number of 'no answers' is generally less than 0.5% of the total and at the level of precision used on GHS the percentages for valid answers are not materially affected by the treatment of 'no answers'. Socio-economic classification and income variables are the most common variables which have too many missing answers to ignore.

4. **Base numbers:** The reliability of estimates with a small base were investigated. Shaded figures indicate the estimates are unreliable and any analysis using these figures may be invalid. Any use of these shaded figures must be accompanied by this disclaimer.

5. **Percentages:** A percentage may be quoted in the text for a single category that is identifiable in the tables only by summing two or more component percentages. In order to avoid rounding errors, the percentage has been recalculated for the single category and therefore may differ by one percentage point from the sum of the percentages derived from the tables.

The row or column percentages may add to 99% or 101% because of rounding.

6. **Conventions:** The following conventions have been used within tables:
   .. data not available
   - category not applicable
   0 less than 0.5% or no observations

7. **Statistical significance:** Unless otherwise stated, changes and differences mentioned in the text have been found to be statistically significant at the 95% confidence level.

8. **Mean:** Throughout the report the arithmetic term 'mean' is used rather than 'average'. The mean is a measure of the central tendency for continuous variables, calculated as the sum of all scores in a distribution, divided by the total number of scores.

9. **Weighting:** All percentages and means presented in the tables in the substantive chapters are based on data weighted to compensate for differential nonresponse. Both the unweighted and weighted bases are given. The unweighted base represents the number of people/households interviewed in the specified group. The weighted base gives a grossed up population estimate in thousands. Trend tables show unweighted and weighted figures for 1998 to give an indication of the effect of the weighting.

Missing answers are excluded from the tables and in some cases this is reflected in the weighted bases, ie these numbers vary between tables. For this reason, the bases themselves are not recommended as a source for population estimates. Recommended data sources for population estimates for most socio-demographic groups are: ONS mid-year estimates, the Labour Force Survey, or Housing Statistics from the Office of the Deputy Prime Minister. See Appendix D for details of weighting.

General Household Survey, 2006
Definitions and terms

General Household Survey 2006

Definitions and terms

Appendix A

Riaz Ali   Office for National Statistics

Julia Greer   Government Buildings

David Matthews   Cardiff Rd

Liam Murray   Newport

Simon Robinson   NP10 8XG

Ghazala Sattar   Tel: 01633 455877

                            Email: ghs@ons.gsi.gov.uk

© Crown copyright

Office for National Statistics: January 2008

General Household Survey, 2006
Definitions and terms

Definitions and terms used are listed in alphabetical order

Acute sickness

See **Sickness**

Adults

Adults are defined as persons aged 16 or over in all tables except those showing dependent children where single persons aged 16 to 18 who are in full-time education are counted as dependent children. The GHS interviews all people aged 16 and over in private households.

Bedroom standard

This concept is used to estimate occupation density by allocating a standard number of bedrooms to each household in accordance with its age/sex/marital status composition and the relationship of the members to one another. A separate bedroom is allocated to each married couple, any other person aged 21 or over, each pair of adolescents aged 10 to 20 of the same sex, and each pair of children under 10. Any unpaired person aged 10 to 20 is paired if possible with a child under 10 of the same sex, or, if that is not possible, is given a separate bedroom, as is any unpaired child under 10. This standard is then compared with the actual number of bedrooms (including bedsitters) available for the sole use of the household, and deficiencies or excesses are tabulated. Bedrooms converted to other uses are not counted as available unless they have been denoted as bedrooms by the informants; bedrooms not actually in use are counted unless uninhabitable.

Central Heating

Central heating is defined as any system whereby two or more rooms (including kitchens, halls, landings, bathrooms and WCs) are heated from a central source, such as a boiler, a back boiler to an open fire, or the electricity supply. This definition includes a system where the boiler or back boiler heats one room and also supplies the power to heat at least one other room.

Under-floor heating systems, electric air systems, and night storage heaters are included.

Where a household has only one room in the accommodation, it is treated as having central heating if that room is heated from a central source along with other rooms in the house or building.

Office for National Statistics: January 2008

Chronic sickness

See **Sickness**

Cohabitation

See **Marital status**

Co-ownership or equity sharing schemes

Co-ownership or equity sharing schemes are those where a share in the property is bought by the occupier under an agreement with the housing association. The monthly charges paid for the accommodation include an amount towards the repayment of the collective mortgage on the scheme. The co-owner never becomes the sole owner of the property, but on leaving the scheme usually receives a cash sum.

See also **Tenure**

Dependent children

Dependent children are persons aged under 16, or single persons aged 16 to 18 and in full-time education, in the family unit and living in the household.

Doctor consultations

Data on doctor consultations presented in this report relate to consultations with National Health Service general medical practitioners during the two weeks before interview. Visits to the surgery, home visits, and telephone conversations are included, but contacts only with a receptionist are excluded. Consultations with practice nurses were excluded prior to 2000, but since then are identified separately. The GHS also collects information about consultations paid for privately.

The average number of consultations per person per year is calculated by multiplying the total number of consultations within the reference period, for any particular group, by 26 (the number of two-week periods in a year) and dividing the product by the total number of persons in the sample in that group.

Drinking

Questions about drinking alcohol were included in the General Household Survey every two years from 1978 to 1998. Following the review of the GHS, they have been included every year from 2000 onwards.

General Household Survey, 2006
Definitions and terms

Since 1998, the GHS has measured the maximum daily amount drunk last week. This is in line with the government's advice on sensible drinking which is now based on daily benchmarks rather than weekly consumption. Regular consumption of between three and four units a day for men and two to three units a day for women does not carry a significant health risk, but consistently drinking above these levels is not advised.

The questions are asked of all people aged 16 and over in the household with a self-completion form offered to those aged 16 or 17. Respondents are asked on how many days they drank alcohol during the previous week. They are then asked how much of each of six different types of drink they drank on their heaviest drinking day during the previous week. These amounts are added to give an estimate of the maximum the respondent had drunk on any one day.

Economic activity

Economically active

People over the minimum school-leaving age of 16, who were working or unemployed (as defined below) in the week before the week of interview. These persons constitute the labour force.

Working persons

This category includes persons aged 16 and over who, in the week before the week of interview, worked for wages, salary or other form of cash payment such as commission or tips, for any number of hours. It covers persons absent from work in the reference week because of holiday, sickness, strike, or temporary lay-off, provided they had a job to return to with the same employer. It also includes persons attending an educational establishment during the specified week if they were paid by their employer while attending it, people on Government training schemes and unpaid family workers.

Persons are excluded if they worked in a voluntary capacity for expenses only, or only for payment in kind, unless they worked for a business, firm or professional practice owned by a relative.

Full-time students are classified as 'working', 'unemployed' or 'inactive' according to their own reports of what they were doing during the reference week.

Unemployed persons

The GHS uses the International Labour Organisation (ILO) definition of unemployment. This classifies anyone as unemployed if he or she was out of work and had looked for work in the four weeks before interview, or would have but for temporary sickness or injury, and was available to start work in the two weeks after interview.

The treatment of all categories on the GHS is in line with that used on the Labour Force Survey (LFS).

Office for National Statistics: January 2008

Definitions and terms

Economically inactive

People who are neither working nor unemployed by the ILO measure. For example, this would include those who were looking after a home or retired.

Ethnic group

The GHS introduced the current National Statistics ethnic classification in 2001. The classification has a separate category for people from mixed ethnic backgrounds. In the previous system, people with these backgrounds had to select a specific ethnic group or categorise themselves as 'other'.

Household members are classified by the person answering the Household Schedule as:

- British, other White background;
- White and Black Caribbean, White and Black African, White and Asian, Other Mixed background;
- Indian, Pakistani, Bangladeshi, Other Asian background;
- Black Caribbean, Black African, Other Black background;
- Chinese; or
- Other ethnic group.

Family

A GHS family unit is defined as:

a) a married or opposite sex cohabiting couple on their own; or

b) a married or opposite sex cohabiting couple, or a lone parent, and their never-married children (who may be adult), provided these children have no children of their own.; or

c) a same sex couple in a legally recognised civil partnership.

Persons who cannot be allocated to a family as defined above are said to be persons not in the family – i.e. as 'non-family units'. In general, GHS family units cannot span more than two generations, i.e. grandparents and grandchildren cannot belong to the same family unit. The exception to this is where it is established that the grandparents are responsible for looking after the grandchildren (e.g. while the parents are abroad). Adopted and stepchildren belong to the same family unit as their adoptive/step-parents. Foster-children, however, are not part of their foster-parents' family (since they are not related to their foster-parents) and are counted as separate non-family units. See also **Lone-parent family.**

Full-time working

Full-time working is defined as more than 30 hours a week with the exception of occupations in education where more than 26 hours a week was included as full time.

Government Office Region (GOR)

Government Office Regions came into force in 1998. They replaced the Standard Statistical Regions as the primary classification for the presentation of English regional statistics. Standard Statistical Region was retained for some long term trend tables up to 2000. See also NHS Regional Office.

GP consultations

See Doctor consultations

Hospital visits

Inpatient stays

Inpatient data relate to stays overnight or longer (in a twelve month reference period) in NHS or private hospitals. All types of cases are counted, including psychiatric and maternity, except babies born in hospital who are included only if they remained in hospital after their mother was discharged.

Outpatient attendances

Outpatient data relate to attendances (in a reference period of three calendar months) at NHS or private hospitals, other than as an inpatient. Consultative outpatient attendances, casualty attendances, and attendances at ancillary departments are all included and a separate count is made of attendances at a casualty department.

Day patient

Day patients are defined as patients admitted to a hospital bed during the course of a day or to a day ward where a bed, couch or trolley is available for the patient's use. They are admitted with the intention of receiving care or treatment which can be completed in a few hours so that they do not require a stay in hospital overnight. If a patient admitted as a day patient then stays overnight they are counted as an inpatient.

Household

A household is defined as:

a single person or a group of people who have the address as their only or main residence and who either share one meal a day or share the living accommodation. (See L McCrossan, *A Handbook for Interviewers*. HMSO, London 1991.)

A group of people is not counted as a household solely on the basis of a shared kitchen or bathroom.

A person is in general regarded as living at the address if he or she (or the informant) considers the address to be his or her main residence. There are, however, certain rules which take priority over this criterion.

Children aged 16 or over who live away from home for purposes of either work or study and come home only for holidays are *not* included at the parental address under any circumstances.

Children of any age away from home in a temporary job and children under 16 at boarding school are *always* included in the parental household.

Anyone who has been away from the address *continuously* for six months or longer is excluded.

Anyone who has been living continuously at the address for six months or longer is included even if he or she has his or her main residence elsewhere.

Addresses used only as second homes are never counted as a main residence.

Household Reference Person (HRP)

For some topics it is necessary to select one person in the household to indicate the characteristics of the household more generally. In common with other government surveys, in 2000, the GHS replaced the Head of Household with the Household Reference Person for this purpose.

- The household reference person is defined as follows:
- in households with a *sole* householder that person is the household reference person;
- in households with *joint* householders the person with the *highest income* is taken as the household reference person;
- if both householders have exactly the same income, the *older* is taken as the household reference person.

Note that this definition does not require a question about people's actual incomes; only a question about who has the highest income. The main changes from the HOH definition are described in Appendix A in 'Living in Britain 2000'.

General Household Survey, 2006
Definitions and terms

Household type

There are many ways of grouping or classifying households into household types; most are based on the age, sex and number of household members.

The main classification of household type as used in Chapter 4 tables uses the following categories:

- 1 adult aged 16-59,
- 2 adults aged 16-59,
- small family - 1 or 2 persons aged 16 or over and 1 or 2 persons aged under 16,
- large family - 1 or more persons aged 16 or over and 3 or more persons aged under 16, or 3 or more persons aged 16 or over and 2 persons aged under 16,
- large adult household - 3 or more persons aged 16 or over, with or without 1 person aged under 16,
- 2 adults, 1 or both aged 60 or over,
- 1 adult aged 60 or over.

The term 'family' in this context does not necessarily imply any relationship.

GHS tables covering Chapter 3 also use a modified version of household type which takes account of the age of the youngest household member. 'Small family', 'large family' and 'large adult household' are replaced by the following:

- youngest person aged 0-4,
- 1 or more persons aged 16 or over and 1 or more persons aged under 5,
- youngest person aged 5-15,
- 1 or more persons aged 16 or over and 1 or more persons aged 5-15,
- 3 or more adults,
- 3 or more persons aged 16 or over and no-one aged under 16.

The first two categories above are combined in some tables.

Households are also classified according to the family units they contain (see Family for definition), into the following categories:

- One family households* containing:
    - married couple with dependent children,
    - married couple with non-dependent children only,
    - married couple with no children,
    - cohabiting couple with dependent children,
    - cohabiting couple with non-dependent children only,

- cohabiting couple with no children,
- lone parent with dependent children,
- lone parent with non-dependent children only.
- Households containing two or more families,
- Non-family households containing,
    - lone parent with dependent children,
    - lone parent with non-dependent children only,
    - 1 person only,
    - 2 or more non-family† adults.

Some of the above categories are combined for certain tables and figures.

* *Other individuals who were not family members may also have been present*

† *Individuals may, of course, be related without constituting a GHS family unit. A household consisting of a brother and sister, for example, is a non-family household of two or more non-family adults.*

Income

Usual gross weekly income

Total income for an individual refers to income at the time of the interview, and is obtained by summing the components of earnings, benefits, pensions, dividends, interest and other regular payments. Gross weekly income of employees and those on benefits is calculated if interest and dividends are the only components missing.

If the last pay packet/cheque was unusual, for example in including holiday pay in advance or a tax refund, the respondent is asked for usual pay. No account is taken of whether a job is temporary or permanent. Payments made less than weekly are divided by the number of weeks covered to obtain a weekly figure.

Usual gross weekly household income is the sum of usual gross weekly income for all adults in the household. Those interviewed by proxy are also included.

Lone-parent family

A lone-parent family consists of one parent, irrespective of sex, living with his or her never-married dependent children, provided these children have no children of their own.

Married or cohabiting women with dependent children, whose partners are not defined as resident in the household, are not classified as one-parent families because it is known that the majority of them are only temporarily separated from their husbands for a reason that does not imply the breakdown of the marriage (for example, because the husband usually works away from home). (See the GHS 1980 Report p.9 for further details.)

Longstanding conditions and complaints

See **Sickness**

Marriage and cohabitation

From 1971 to 1978 the Family Information section was addressed only to married women aged under 45 who were asked questions on their present marriage and birth expectations. In 1979 the section was expanded to include questions on cohabitation prior to marriage, previous marriages and all live births, and was addressed to all women aged 16 to 49 except non-married women aged 16 and 17. In 1986 the section was extended to cover all women and men aged 16 to 59. In 1998 all adults aged 16 to 59 were asked about any periods of cohabitation not leading to marriage. This section was extended in 2000 to include the length of past cohabitations not ending in marriage and what people perceived to be the end of the cohabitation (the end of the relationship, the end of sharing the accommodation or both).

Marital status

Since 1996 separate questions have been asked at the beginning of the questionnaire to identify the legal marital status and living arrangements of respondents in the household. The latter includes a category for cohabiting.

Before 1996, unrelated adults of the opposite sex were classified as cohabiting if they considered themselves to be living together as a couple. From 1996, this has included a small number of same sex couples, unless otherwise stated in the table. From 2006, those in a legally recognised civil partnerships have been presented separately from those in non-legal same sex relationships.

Married/non-married

In this dichotomy 'married ' generally includes cohabiting and 'non- married' covers those who are single, widowed, separated or divorced and not cohabiting.

Living arrangements (de facto marital status)

Before 1996, additional information from the Family Information section of the individuals' questionnaire was used to determine living arrangements (previously known as 'defacto marital status') and the classification only applied to those aged 16 to 59 who answer the marital history questions. For this population it only differed from the main marital status for those who revealed in the Family Information section that they were cohabiting rather than having the marital status given at the beginning of the interview. 'Cohabiting' took priority over other categories. Since 1996, information on legal marital status and living arrangements has been taken from the beginning of the interview where both are now asked.

Legal marital status

This classification applies to persons aged 16 to 59 who answer the marital history questions. Cohabiting people are categorised according to formal marital status. The classification differs from strict legal marital status in accepting the respondents' opinion of whether their marriage has terminated in separation rather than applying the criterion of legal separation.

Median

See **Quantiles**

National Statistics Socio-economic classification (NS-SEC)

From April 2001 the National Statistics Socio-economic Classification (NS-SEC) was introduced for all official statistics and surveys. It replaced Social Class based on occupation and Socio-economic Groups (SEG). Full details can be found in *'The National Statistics Socio-economic Classification User Manual 2002'* ONS 2002.

Descriptive definition	NS-SEC categories
Large employers and higher managerial occupations	L1, L2
Higher professional occupations	L3
Lower managerial and professional occupations	L4, L5, L6
Intermediate occupations	L7
Small employers and own account workers	L8, L9
Lower supervisory and technical occupations	L10, L11

General Household Survey, 2006
Definitions and terms

Semi-routine occupations	L12
Routine occupations	L13
Never worked and long-term unemployed	L14

The three residual categories: L15 (full time students); L16 (occupation not stated or inadequately described) and L17 (not classifiable for other reasons) are excluded when the classification is collapsed into its analytical classes.

The categories can be further grouped into:

Managerial and professional occupations	L1-L6
Intermediate occupations	L7-L9
Routine + manual occupations	L10-L13

This results in the exclusion of those who have never worked and the long term unemployed, in addition to the groups mentioned above.

The main differences users need to be aware of are:

- the introduction of SOC2000 which includes various new technology occupations not previously defined in SOC90,
- definitional variations in employment status in particular with reference to the term 'supervisor',
- the inclusion of armed forces personnel in the appropriate occupation group,
- the separate classification of full-time students, whether or not they have been or are presently in paid employment, and
- the separate classification of long term unemployed who previously were classified by their most recent occupation.

This change has resulted in a discontinuity in time series data. The operational categories of NS-SEC can be aggregated to produce an approximated version of the previous Socio-economic Group. These approximations have been shown to achieve an overall continuity level of 87%. Some tables on smoking have used this approximation.

Pensions

The GHS asks questions about any pension scheme, either occupational or personal, that the respondent belonged to on the date of interview. It is quite possible that some respondents may have held entitlement in the occupational pension scheme of a previous employer or a personal pension scheme in the past. The GHS measures current membership and not the percentage of respondents who will get an occupational or personal pension when they retire.

In April 2002 the State Second Pension (S2P) was introduced. This new pension reformed the State Earnings Related Pension Scheme (SERPS) to provide a more generous additional pension for low and moderate earners, certain groups of carers and people with a longstanding illness or disability. Since 1988, individual employees have had the option of contracting out of the S2P (formerly SERPS) by starting their own personal pension plan. Some respondents may be contributing to both an occupational and personal pension scheme.

From 2001 to 2002, the GHS asked employees whether or not they had arranged their own personal/stakeholder pension. If they answered 'yes', they were asked a supplementary question to establish whether it was a personal or stakeholder pension. In 2003, these two questions were replaced with a multiple response question that allowed the respondent to select one or more of the following answers:

- A personal or private pension or retirement annuity
- A Group Personal Pension (this is a collection of personal pensions arranged by an employer for a group of employees)
- A Stakeholder Pension arranged through your employer (who may or may not contribute to such a pension)
- None of these
- Don't know

Some of the change in the proportion of employees with a personal pension in 2002 and 2003 may be due to the different ways respondents were asked about personal pension arrangements.

Questions on personal pension arrangements for self-employed persons did not change.

The majority of tables show data for men working full time. This is because the sample sizes for male employees working part time and self-employed men working part time are too small to give reliable estimates.

Qualification levels

Degree or Degree equivalent, and above

Higher degree and postgraduate qualifications

First degree (including B.Ed.)

Postgraduate Diplomas and Certificates (including PGCE)

Professional qualifications at degree level e.g. graduate member of professional institute, chartered accountant or surveyor

NVQ or SVQ level 4 or 5

Other Higher Education below degree level

Diplomas in higher education & other higher education qualifications

HNC, HND, Higher level BTEC

Teaching qualifications for schools or further education (below Degree level standard)

Nursing, or other medical qualifications not covered above (below Degree level standard)

RSA higher diploma

A levels, vocational level 3 & equivalents

A level or equivalent

AS level

SCE Higher, Scottish Certificate Sixth Year Studies or equivalent

NVQ or SVQ level 3

GNVQ Advanced or GSVQ level 3

OND, ONC, BTEC National, SCOTVEC National Certificate

City & Guilds advanced craft, Part III (& other names)

RSA advanced diploma

Trade Apprenticeships

GCSE/O Level grade A*-C, vocational level 2 & equivalents

NVQ or SVQ level 2

GNVQ intermediate or GSVQ level 2

RSA Diploma

City & Guilds Craft or Part II (& other names)

BTEC, SCOTVEC first or general diploma et

O level or GCSE grade A-C, SCE Standard or Ordinary grades 1-3

General Household Survey, 2006
Definitions and terms

Qualifications at level 1 and below

NVQ or SVQ level 1

GNVQ Foundation level, GSVQ level 1

GCSE or O level below grade C, SCE Standard or Ordinary below grade 3

CSE below grade 1

BTEC, SCOTVEC first or general certificate

SCOTVEC modules

RSA Stage I, II, or III

City and Guilds part 1

Junior certificate

YT Certificate/ YTP

Other qualifications: level unknown

Other vocational or professional or foreign qualifications

No qualifications

Excludes those who never went to school (omitted from the classification altogether).

This is not a complete listing of all qualifications. In particular, it does not give all the names which have been used by BTEC or City and Guilds. Neither does it give names for vocational qualifications from other awarding bodies besides BTEC, City and Guilds, RSA and SCOTVEC, although it should cover the majority of vocational qualifications awarded.

The qualification levels do not in all cases correspond to those used in statistics published by the Department for Education and Skills.

Quantiles

The quantiles of a distribution, eg of household income, divide it into equal parts.

Median: the median of a distribution divides it into two equal parts. Thus half the households in a distribution of household income have an income higher than the median, and the other half have an income lower than the median.

Quartiles: the quartiles of a distribution divide it into quarters. Thus the upper quartile of a distribution of household income is the level of income that is expected by 25% of the households in the distribution; and 25% of the households have an income less than the lower quartile. It follows that 50% of the households have an income between the upper and lower quartiles.

Quintiles: the quintiles of a distribution divide it into fifths. Thus the upper quintile of a distribution of household income is the level of income that is expected by 20% of the households in the distribution; and 20% of the households have an income less than the lower quintile. It follows that 60% of the households have an income between the upper and lower quintiles.

Relatives in the household

The term 'relative' includes any household member related to the head of household by blood, marriage, or adoption. Foster-children are therefore not regarded as relatives.

Rooms

These are defined as habitable rooms, including (unless otherwise specified) kitchens, whether eaten in or not, but excluding rooms used solely for business purposes, those not usable throughout the year (eg conservatories), and those not normally used for living purposes such as toilets, cloakrooms, store rooms, pantries, cellars and garages.

Sickness

Acute sickness

Acute sickness is defined as restriction of the level of normal activity, because of illness or injury, at any time during the two weeks before interview. Since the two-week reference period covers weekends, normal activities include leisure activities as well as school attendance, going to work, or doing housework. Anyone with a chronic condition that caused additional restriction during the reference period is counted among those with acute sickness.

The average number of restricted activity days per person per year is calculated in the same way as the average number of doctor consultations.

Chronic sickness

Information on chronic sickness was obtained from the following two-part question:

'Do you have any longstanding illness, disability or infirmity? By longstanding I mean anything that has troubled you over a period of time or that is likely to affect you over a period of time.

General Household Survey, 2006

Definitions and terms

IF YES

Does this illness or disability limit your activities in any way?'

'Longstanding illness' is defined as a positive answer to the first part of the question, and 'limiting longstanding illness' as a positive answer to both parts of the question.

The data collected are based on people's subjective assessment of their health, and therefore changes over time may reflect changes in people's expectations of their health as well as changes in incidence or duration of chronic sickness. In addition, different sub-groups of the population may have varying expectations, activities and capacities of adaptation.

Longstanding conditions and complaints

The GHS collects information about the nature of longstanding illness. Respondents who report a longstanding illness are asked 'What is the matter with you?' and details of the illness or disability are recorded by the interviewers and coded into a number of broad categories. Interviewers are instructed to focus on the symptoms of the illness, rather than the cause, and code what the respondent said was currently the matter without probing for cause. This approach has been used in 1988, 1989, 1994 to 1996, 1998 and 2000 to 2006.

The categories used when coding the conditions correspond broadly to the chapter headings of the International Classification of Diseases (ICD). However, the ICD is used mostly for coding conditions and diseases according to cause whereas the GHS coding is based only on the symptoms reported. This gives rise to discrepancies in some areas between the two classifications.

Smoking

Questions about smoking behaviour have been included on the GHS in alternate years from 1974 to 1998, and every year from 2000 onwards. The questions are asked of all people aged 16 and over in the household with a self-completion form offered to those aged 16 or 17, where appropriate.

Information on tar yields is only collected for manufactured cigarettes. Tar yields are provided by the laboratory of the Government Chemist.

Socio-economic classification

See National Statistics Socio-economic classification

Office for National Statistics: January 2008

Step-family

See **Family**

Tenure

From 1981, households who were buying a share in the property from a housing association or co-operative through a shared ownership (equity sharing) or co-ownership scheme are included in the category of owner-occupiers. In earlier years such households were included with those renting from a housing association or co-operative.

Renting from a council includes renting from a local authority or New Town corporation or commissions or Scottish Homes (formerly the Scottish Special Housing Association).

Renting from a housing association also includes co-operatives and charitable trusts. It also covers fair rent schemes. Since 1996, housing associations are more correctly described as Registered Social Landlords (RSLs). RSLs are not-for-profit organisations which include: charitable housing associations, industrial and provident societies and companies registered under the Companies Act 1985.

Social sector renters includes households renting from a local authority or New Town corporation or commission or Scottish Homes and those renting from housing associations, cooperatives and charitable trusts.

Private renters include those who rent from a private individual or organisation and those whose accommodation is tied to their job even if the landlord is a local authority, housing association or Housing Action Trust, or if the accommodation is rent free. Squatters are also included in this category.

Unemployed

See **Economic activity**

Weighting

All percentages and means presented in the tables are based on data weighted to compensate for differential nonresponse. Both the unweighted and weighted bases are given. The unweighted base represents the number of people/households interviewed in the specified group. The weighted base gives a grossed up population estimate in thousands. Trend tables show unweighted and weighted figures for 1998 to give an indication of the effect of the weighting.

A full description of the method of weighting and the effects on data are in Appendix D.

General Household Survey, 2006

Definitions and terms

Working

See **Economic activity**

General Household Survey 2006

Sample Design and Response

Appendix B

Riaz Ali

Julia Greer

David Matthews

Liam Murray

Simon Robinson

Ghazala Sattar

© Crown copyright

Office for National Statistics

Government Buildings

Cardiff Rd

Newport

NP10 8XG

Tel: 01633 455877

Email: ghs@ons.gsi.gov.uk

Sample design and response

In 2006, 13,585 addresses were sampled. The GHS aims to interview all adults aged 16 or over at every household at the sampled address[1]. It uses a probability, stratified two-stage sample design. The Primary Sampling Units (PSUs) are postcode sectors, which are similar in size to wards and the secondary sampling units are addresses within those sectors.

Sample design

The revised 2000 survey design introduced new stratifiers[2]. Stratification involves the division of the population into sub-groups, or strata, from which independent samples are taken. This ensures that a representative sample will be drawn with respect to the stratifiers (i.e. the proportion of units sampled from any particular stratum will equal the proportion in the population with that characteristic). Stratification of a sample can lead to substantial improvements in the precision of survey estimates. Optimal precision is achieved where the factors used as strata are those that correlate most highly with the survey variables. From 2000, the stratification factors were based on an area classifier and selected indicators from the 1991 census. Details of how these were selected were reported in the January 2000 edition of the ONS Survey Methodology bulletin[3].

Initially, postcode sectors were allocated to 30 major strata. These were based on the 10 Government Office Regions in England, 5 subdivisions in Scotland and 2 in Wales. The English regions were divided between the former Metropolitan and non-Metropolitan counties. In addition London was subdivided into quadrants (Northwest, Northeast, Southwest and Southeast) with each quadrant being divided into inner and outer areas[4]. Using a finer division of London in the regional stratifier had a large effect on the increase in precision.

Within each major stratum, postcode sectors were then stratified according to the selected indicators taken from the 1991 Census. Sectors were initially ranked according to the proportion of households with no car, then divided into three bands containing approximately the same number of households. Within each band, sectors were re-ranked according to the proportion of households with household reference person in socio-economic groups 1 to 5 and 13, and these bands were then sub-divided into three further bands of approximately equal size. Finally, within each of these bands, sectors were re-ranked according to the proportion of people who were pensioners. In order to minimise the difference between one band and the next, the ranking by the pensioners and socio-economic group criteria were in the reverse order in consecutive bands, as shown in Figure B.A.

Figure B.A

Major strata were then divided into minor strata with equal numbers of addresses, the number of minor strata per major strata being proportionate to the size of the major stratum. Since 1984, the frame has been divided into 576 minor strata and one PSU has been selected from each per year. Of the 576 PSUs selected, 48 are randomly allocated to each month of the year. Each PSU forms a quota of work for an interviewer. Within each PSU, 23 addresses are randomly selected. In 2005 the frame was divided into 720 strata. In 2006, 588 of these were rolled forward to the next wave in the longitudinal design.

There were 132 pseudo wave 4 strata which were replaced and an additional 96 strata added, giving 816 for 2006. The number of PSUs has increased over time to counteract the attrition in the longitudinal sample.

Figure **B.A**

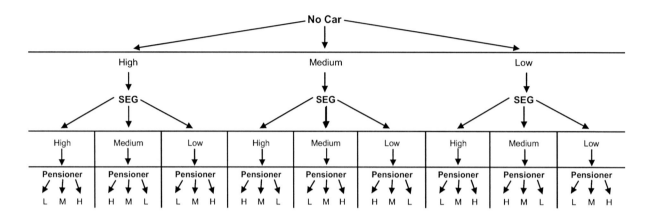

Sample design on the longitudinal General Household Survey

In 2005, the GHS adopted a new sample design in line with European requirements, changing from a cross-sectional to a longitudinal format.

The GHS sample design follows a four-year sample rotation in which households remain in the sample for four years (waves) with one quarter of the sample being replaced each year. Each quarter of the sample is known as a replication, and each replication is representative of the target population. Figure B.B illustrates how the design operates. Once the system is fully established (from year 4 onwards) the sample for any one year consists of four replications which have been in the survey for 1, 2, 3 or 4 years. Each year one of the four replications is dropped and a new one added, giving an overlap of 75 per cent between successive years. This has been implemented to avoid high attrition rates due to repeated interviewing. Because 2006 is the second year of this longitudinal design, this sample contains approximately 75 per cent follow-up interviews.

A major advantage of the longitudinal component of the design is that it is more efficient at detecting statistically significant estimates of change over time than is a design utilising independently selected cross-sections. This is because peoples' responses to the same question at different points in time tend to be positively correlated and this has the effect of reducing the estimated standard errors of change. Consequently, year on year comparisons will give the most efficient method of detecting changes over time, i.e. when overlap between the two time points is at its highest. This approach was adopted on the Labour Force Survey to give more efficient measures of un/employment.

The longitudinal design (as illustrated in figure B.B) means that GHS datasets from 2006 are not independent, with approximately three quarters of the sample from the previous year being re-interviewed during the current year.

Figure B.B

Figure **B.B**

Sample replication	Year 1	Year 2	Year 3	Year 4	Year 5	Year 6	
1	1st						
2	1st	2nd					
3	1st	2nd	3rd				
4	1st	2nd	3rd	4th			
5		1st	2nd	3rd	4th		
6			1st	2nd	3rd	4th	
7				1st	2nd	3rd	etc.
8					1st	2nd	etc.
9						1st	etc.

	New sample
	Follow-up sample

Conversion of multi-occupancy addresses to households

Most addresses contain just one private household, a few - such as institutions and purely business addresses[5] - contain no private households, while others contain more than one private household. For addresses containing more than one household, set procedures are laid down in order to give each household one and only one chance of selection.

As the Postcode Address File (PAF) does not give names of occupants of addresses, it is not possible to use the number of different surnames at an address as an indicator of the number of households living there. A rough guide to the number of households at an address is provided on the PAF by the multi-occupancy (MO) count. The MO count is a fairly accurate indicator in Scotland but is less accurate in England and Wales, so it is used only when sampling at addresses in Scotland.

All addresses in England and Wales, and those in Scotland with an MO count of two or less, are given only one chance of selection for the sample. At such addresses, interviewers interview all the households they find up to a maximum of three. If there are more than three households at the address, the interviewer selects the households for interview by listing all households at the address systematically then making a random choice by referring to a household selection table.

Addresses in Scotland with an MO count of three or more, where the probability that there is more than one household is fairly high, are given as many chances of selection as the value of the MO count. When the interviewer arrives at such an address, he or she checks the actual number of households and interviews a proportion of them according to instructions. The proportion is set originally by the MO count and adjusted according to the number of households actually found, with a maximum of three households being interviewed at any address. The interviewer selects the households for interview by listing all households at the address systematically and making a random choice, as above, by means of a table.

No addresses are deleted from the sample to compensate for the extra interviews that may result from these multi-household addresses, but a maximum of four extra interviews per quota of addresses is allowed. Once four extra interviews have been carried out in an interviewer's quota, only the first household selected at each multi-occupancy address is included. As a result of the limits on additional interviews, households in concealed multi-occupied addresses may be slightly under-represented in the GHS sample.

Data collection

Information for the GHS is collected week by week throughout the year[6] by personal interview. In 2006, interviews took place from January to December 2006. In 2005, the field period changed from financial year to calendar year. As the 2004/5 survey finished in March 2005, interviewing for the 2005 survey started in April 2005 and ended in December 2005. However, data from the 2005 January to March period of the 2004/5 survey was included in the analyses of the 2005 data in order to prevent seasonal bias. Since 1994 the survey has been carried out using Computer Assisted Personal Interviewing (CAPI) on laptop computers and Blaise software by face-to-face interviewers. Since 2000, telephone interviewers have converted GHS proxy interviews to full interviews using Computer Assisted Telephone Interviewing (CATI) from a central unit. Interviews are sought with all adult members (aged 16 or over) of the sample of private households and some information about children in the household is also collected.

A letter is sent in advance of an interviewer calling at an address. The letter briefly describes the purpose and nature of the survey and prepares the recipient for a visit by an interviewer. Since 2001, postage stamps have been included in the advance letter (see 'Improving response').

Data quality

The face-to-face and telephone interviewers who work on the GHS are recruited only after careful selection procedures after which they take part in an initial training course. Before working on the GHS they attend a briefing and new recruits are always supervised either by being accompanied in the field by a Field Manager or monitored by a Telephone Interviewing Unit (TIU) supervisor. All interviewers who continue to work on the GHS are observed regularly in their work.

Proxy interviews and the proxy conversion exercise

On occasion it may prove impossible, despite repeated calls, to contact a particular member of a household in person and, in strictly controlled circumstances, interviewers are permitted to conduct a proxy interview with a close household member. In these cases opinion-type questions and questions on smoking and drinking behaviour, qualifications, health, family information and income are omitted.

During the review of the GHS[7] the conversion of proxy interviews to full interviews was examined in order to improve the quality of data (a full interview is one in which the respondent has answered all sections of the questionnaire in person, either face-to-face or by telephone). This was achieved by re-contacting the household member, who was unavailable during the initial face-to-face interview, to answer the questions that were not asked of the proxy respondent on his/her behalf. The most efficient way of re-contacting these respondents was by employing Telephone Interviewing Unit (TIU) interviewers who could contact a widely dispersed population more efficiently than would be possible by conducting face-to-face interviews. Table B.1 shows the percentages of the types of interview taken for all persons in the co-operating households since 2000 before and after the proxy conversion exercise. In 2006 the TIU increased the percentage of full interviews conducted on the GHS from 71 per cent before proxy conversion to 73 per cent. The process of proxy conversion allows the GHS to provide more information on the topics not asked in detail of proxy respondents.

Table B.1 - Table B.1a

Table B1 Type of interview taken by proxy conversion status

All persons   *Great Britain*

	Proxy conversion status									
	2002		2003		2004		2005[1]		2006[2]	
	Before	After	Before	After	Before	After	Before	After	Before	After
	%	%	%	%	%	%	%	%	%	%
Full interview	71	74	70	71	71	73	70	72	71	73
Proxy	7	5	7	6	7	5	8	6	8	6
Child/Other	22	22	23	23	21	21	22	22	22	22
Unweighted sample	20149	20149	24489	24489	20421	20421	30069	30069	22924	22924

1 2005 data includes last quarter of 2004/5 data due to survey change from financial year to calendar year.
2 Results for 2006 include longitudinal data.

Table B1a Type of interview taken by wave and by proxy conversion status

All persons   *Great Britain 2006*

	Proxy conversion status			
	Wave1		Wave 2+	
	Before	After	Before	After
	%	%	%	%
Full interview	69	72	71	74
Proxy	9	6	7	5
Child/Other	22	22	22	22
Unweighted sample	7409	7409	15515	15515

Response

The GHS is conducted with people who volunteer their time to answer questions about themselves. The voluntary nature of the survey means that people who do not wish to take part in the survey can refuse to do so. Reasons for not participating in the survey vary from a dislike of surveys to poor health that prevents them from taking part. The sample is designed to ensure that the results of the survey represent the population of Great Britain. The risk of the survey not being representative is likely to increase with every refusal or non-contact with a sampled household (survey non-response). One measure of the quality of survey results is therefore the response rate.

Harmonised outcome codes and survey response rates

Harmonised outcome codes and definitions of response rates[8] were introduced for the first time on the 2002/3 GHS and other large household surveys, following recommendations from National Statistics and the National Centre for Social Research joint working group on standard survey outcomes. The harmonised outcome codes are categorised as complete and partial interviews, non-contact, refusals and other non-responders, and unknown and known ineligibility.

The joint working group also recommended that surveys include an estimate of the proportion of cases of unknown eligibility that are eligible from 2002 onwards. It is assumed that the proportion of eligible cases amongst those cases where eligibility is unknown is the same as that amongst cases where eligibility has been established.

Four response rates are now calculated for the GHS on the basis of these outcome codes (full interviews also include cases where the complete interview has been partly completed by the selected person, and partly by proxy):

Overall response rate

This indicates how many full and partial interviews were achieved as a proportion of those eligible for the survey. In order to obtain the most conservative response rate measures, the denominator includes an estimate of the proportion of cases of unknown eligibility that would in fact be eligible for interview. In 2006 the overall response rate was 74 per cent.

Full response rate

This is similar to the overall response rate calculated above, but only full interviews are included in the numerator. The full response rate for the GHS in 2006 was 64 per cent.

Co-operation rate

This indicates the number of achieved interviews as a proportion of those contacted during the fieldwork period. The co-operation rate for the GHS in 2006 was 75 per cent.

Contact rate

The contact rate measures the proportion of cases in which some household members were reached by the interviewer even though they might then have refused or been unable to give further information about the household or to participate in the survey. In 2006 the contact rate for the GHS was 96 per cent.

General Household Survey, 2006

Sample Design and Response

Table B.2 shows the outcome of visits to the addresses selected for the 2006 sample and the resultant number of households interviewed. Out of the 13,316 addresses that were selected, 12,455 were eligible and this yielded a sample of 12,562 eligible households. In 9,524 households, interviews (including proxies) were achieved with every member of the household. In a further 207 households interviews were achieved with some but not all members of the household. This produced a total of 9,731 full or partial interviews.

In total, 25 per cent of households selected for interview in 2006 were lost to the sample altogether, because they did not wish to take part (21 per cent) or because they could not be contacted (4 per cent) (table not shown). Table B.3 shows annual response by interview outcome category.

Tables B.2-B.3

Table B2 The sample of addresses and households

Great Britain: 2006

Selected addresses	13316
Ineligible addresses:	
Demolished or derelict	
Used wholly for business purposes	
Empty	
Institutions	861
Other ineligible	
No sample selected at address	
Address not traced	
Eligible addresses	12455
Number of households at eligible addresses	12562
Number of households where all individual interviews achieved (including proxies)	9524
Number of households where some but not all individual interviews achieved	207
Total full or partial interviews	9731

Table B3 Annual Response

Households *Great Britain: 2006*

Outcome category		No.	%
1	Complete household co-operation	8560	68.1
2	Non-interview of one of more household members, proxy taken	964	7.7
3	Non-contact with one or more household members	77	0.6
4	Refusal by at least one household member	130	1.0
5	Non-contact with household/resident	1871	14.9
6	HQ refusal	356	2.8
7	Other refusal	597	4.8
8	Other non-response	7	0.1
Unweighted sample base = 100%		*12562*	

Trends in response

In order to continue to measure the response to the survey over time, a 'middle' response rate has been calculated since 1971. The middle response rate includes full interviews and accepts some of the partial household interviews as response – that is, it includes households where information has been collected by proxy and is therefore missing certain sections (category 2 in Table B.3), but does not include those where information is missing altogether for one or more household members (categories 3 and 4 in Table B.3). In other words, this middle rate can be thought of as the proportion of the sample of households known to be eligible from whom all or nearly all the information was obtained.

For the purposes of comparison, the middle response rate has been calculated, although it is not in itself a classification of the harmonised response rates. In 2006, full interviews also included complete interviews completed partly by the selected person and partly by proxy.

In 2006 the middle response rate was 76 per cent. Table B.4 shows middle response rates by Government Office Region. Trends in the middle response rate are shown in Table B.5. Since 1971, the middle response rate has shown some fluctuation. The decline in response rate since the early 1990s is due to an increase in the proportion of households refusing to participate (12 per cent in 1991 rising to 21 per cent in 2006) rather than failure to contact people. This decline reflects a general trend in decreasing response experienced by all survey organisations.

Tables B.4-B.5

Figure B.4 Middle response rates by Government Office Region

Households *Great Britain: 2006*

Government Office Region	%	Rank
North East	77.2	6
North West	76.5	8
Yorkshire and the Humber	78.6	3
East Midlands	79.3	1
West Midlands	74.4	10
East of England	77.4	5
London	66.5	11
South East (excluding Greater London)	77.1	7
South West	78.9	2
Wales	75.5	9
Scotland	78.4	4
Great Britain	75.6	

Full interviews include complete interviews partly completed by selected person and partly by proxy.

Table B5 Trends in the middle response rate: 1971 to 2006

Households	Great Britain
Year	Response rate
	%
1971	83
1972	81
1973	81
1974	83
1975	84
1976	84
1977	83
1978	82
1979	83
1980	82
1981	84
1982	84
1983	82
1984	81
1985	82
1986	84
1987	85
1988	85
1989	84
1990	81
1991	84
1992	83
1993	82
1994	80
1995	80
1996	76
1998	72
2000	67
2001	72
2002	69
2003	70
2004	69
2005	72
2006	76

2005 data includes last quarter of 2004/5 data due to survey change from financial year to calendar year.

From 2005 full interviews include complete interviews partly completed by selected person and partly by proxy.

Improving response

The GHS uses a number of methods to try to improve response. One ongoing method of improving response has been to reissue addresses to interviewers where there is a possibility of obtaining a better outcome, for example if there was initially a non-contact or a circumstantial refusal.

During the 2001/2 survey the advance letter was changed to mention that the survey had been running for 30 years. This change shows respondents how important the survey is

General Household Survey, 2006

Sample Design and Response

and how many people take part. For subsequent years a similar advance letter has been produced informing respondents that the survey has been running for over 30 years.

Another method was introduced in 2002 following the findings of the Response Working Group of the Office for National Statistics. The Group found that response was higher among households who received postage stamps with their advance letters. Since August 2002, stamps have been included in all advance letters.

Respondents receive a £10 gift voucher for each full follow up interview they complete. Where a follow up interview is taken by proxy, a voucher is only issued if the TIU convert the proxy interview to a full interview.

Sample sizes

Tables B.6 and B.7 show the numbers of households and individuals interviewed on the 2006 GHS by age, sex, region and country.

Tables B.6-B.7

Table B6 Unweighted bases: number of household reference persons in GHS 2006 by age, sex, region and country

Household reference persons *Great Britain: 2006*

	Age							All
	16-24	25-34	35-44	45-54	55-64	65-74	75+	
Sex								
Male	122	764	1262	1140	1227	933	691	6139
Female	149	495	678	630	540	476	624	3592
Government Office Region								
North East	28	43	79	72	67	76	58	423
North West	35	167	245	197	197	193	170	1204
Yorkshire and the Humber	43	113	192	186	151	130	115	930
East Midlands	22	106	146	137	167	115	107	800
West Midlands	10	117	161	157	159	132	106	842
East of England	25	116	194	189	162	138	114	938
London	19	161	211	185	129	91	102	898
South East	31	169	270	227	269	198	201	1365
South West	13	103	187	162	176	134	148	923
Country								
England	226	1095	1685	1512	1477	1207	1121	8323
Wales	12	57	76	105	123	71	71	515
Scotland	33	107	179	153	167	131	123	893
Total	271	1259	1940	1770	1767	1409	1315	9731

Shaded figures also show the number of households in each region and country.

Office for National Statistics: January 2008

Table B7 Unweighted bases: number of people in GHS by age, sex, region and country

All persons
Great Britain: 2006

	Age									All
	0-4	5-15	16-24	25-34	35-44	45-54	55-64	65-74	75+	
Sex										
Male	685	1694	1048	1241	1632	1427	1516	1041	776	11060
Female	679	1652	1112	1416	1770	1582	1545	1153	955	11864
Government Office Region										
North East	47	119	94	100	137	118	116	109	76	916
North West	172	443	237	344	417	332	352	276	213	2786
Yorkshire and the Humber	116	309	237	235	337	296	253	188	153	2124
East Midlands	105	216	194	229	259	246	287	184	150	1870
West Midlands	136	317	188	250	265	285	283	203	131	2058
East of England	125	341	227	268	345	320	291	220	158	2295
London	175	405	208	343	366	288	206	143	133	2267
South East	178	476	285	357	489	417	455	327	265	3249
South West	133	309	180	218	343	268	317	207	198	2173
Country										
England	1187	2935	1850	2344	2958	2570	2560	1857	1477	19738
Wales	61	158	113	116	128	179	217	119	99	1190
Scotland	116	253	197	197	316	260	284	218	155	1996
Total	1364	3346	2160	2657	3402	3009	3061	2194	1731	22924

Notes and references

[1] A limit is put on the number of households that are contacted per address. This is explained in detail at the 'Conversion of multi-occupancy addresses to households' section of Appendix B.

[2] From 1984 to 1998, the stratifiers used were a regional variable (based on the standard statistical region until 1996 and on the Government Office Region in 1998) and variables that measured the prevalence of privately rented accommodation, local authority accommodation and people in professional and managerial socio-economic groups.

[3] Insalaco F Choosing stratifiers for the General Household Survey *ONS Social Survey Division, Survey Methodology Bulletin*, No. 46, January 2000.

[4] The GOR regional stratifier:
1. North East Met
2. North East Non Met
3. North West Met
4. North West Non Met
5. Merseyside

Office for National Statistics: January 2008

6. Yorks and Humberside Met
7. Yorks and Humberside Non Met
8. East Midlands
9. West Midlands Met
10. West Midlands Non Met
11. Eastern Outer Met
12. Eastern Other
13. Inner London North-East
14. Inner London North-West
15. Inner London South-East
16. Inner London South-West
17. Outer London North-East
18. Outer London North-West
19. Outer London South-East
20. Outer London South-West
21. South East Outer Met
22. South East Other
23. South West
24. Wales 1 – Glamorgan, Gwent
25. Wales 2 – Clwydd, Gwenneyd, Dyfed, Powys
26. Highlands, Grampian, Tayside
27. Fife, Central, Lothian
28. Glasgow Met
29. Strathclyde (excl. Glasgow)
30. Borders, Dumfries, Galloway

5 Most institutions and business addresses are not listed on the small-user PAF. If an address was found in the field to be non-private (e.g. boarding house containing four or more boarders at the time the interviewer calls), the interviewer was instructed not to take an interview. However, a household member in hospital at the time of interview was included in the sample provided that he or she had not been away from home for more than six months and was expected to return. In this case a proxy interview was taken.

6 In 1988, the GHS interviewing year was changed from a calendar year to a financial year basis. From 2005, the GHS interviewing year was changed back from financial year to calendar year.

7 Walker A et al *Living in Britain Results from the 2000 General Household Survey: Appendix E.* TSO London 2002. Also available on the web: www.statistics.gov.uk/lib

8 Lynn P, Beerten R, Laiho J and Martin J. Recommended Standard Final Outcome Categories and Standard Definition of Response Rate for Social Surveys. ISER Working Papers. Number 2001 – 23. http://www.iser.essex.ac.uk/pubs/workpaps/pdf/2001-23.pdf

General Household Survey 2006

Sampling Errors

Appendix C

Riaz Ali

Julia Greer

David Matthews

Liam Murray

Simon Robinson

Ghazala Sattar

Office for National Statistics

Government Buildings

Cardiff Rd

Newport

NP10 8XG

Tel: 01633 455877

Email: ghs@ons.gsi.gov.uk

© Crown copyright

General Household Survey, 2006
Sampling Errors

Tables in this appendix present estimates of sampling errors for some of the main variables used in this report, taking into account the complex sample design of the survey.

Sources of error in surveys

Survey results are subject to various sources of error. The total error in a survey estimate is the difference between the estimate derived from the sample data collected and the true value for the population. The total error is made up of two main types: systematic and random error.

Systematic error

Systematic error occurs when data are consistently biased in a certain way, such that the variation from the true values for the population will not average to zero over repeats of the survey. For example, if a certain section of the population is excluded from the sampling frame, estimates may be biased because non-respondents to the survey have different characteristics to respondents. Another cause of bias may be that interviewers systematically influence responses in one way or another. Substantial efforts have been made to avoid systematic errors, for example, through extensive interviewer training and by weighting the data collected for non-response.

Random error

Random error, or bias, is the variation in sample data from the true values for the population, which occurs by chance. This type of error is expected to average to zero over a number of repeats of the survey. Random error may result from sources such as variation in respondents' interpretation of the survey questions, or interviewer variation. Efforts are made to minimise these effects through pilot work and consistent interviewer training.

Sampling errors

An important component of random error is sampling error, which arises because the variable estimates are based on a sample rather than a full census of the population. The results obtained for any single sample would be likely to vary slightly from the true values for the population. The difference between the estimates derived from the sample and the true population values is referred to as the sampling error. The amount of variation can generally be reduced by increasing the size of the sample, and by improving the sample design. Sampling errors have been measured for estimates derived from the General Household Survey (GHS), and these may be used to assess the accuracy of the estimates presented in this report.

Office for National Statistics: January 2008

Calculating standard errors

Unlike non-sampling errors, it is possible to estimate the size of sampling error, by calculating the standard error of the survey estimates. The standard error (se) of a percentage p, based on a simple random sample of size n is calculated by the formula,

$$se(p)_{srs} = \sqrt{(p(100-p)/n)}.$$

The GHS uses a multi-stage sample design, which involves both clustering and stratification (see Appendix B). The complexity of the design means that sampling errors calculated on the basis of a simple random sample design will not reflect the true variance in the survey estimates. Clustering can lead to a substantial increase in sampling error if the households or individuals within the primary sampling units (PSUs) are relatively homogenous but the PSUs differ from one another. By contrast, stratification tends to reduce sampling error and is particularly effective when the stratification factor is related to the characteristics of interest on the survey.

Because of the complexity of the GHS sample design, the size of the standard error depends on how the characteristic of interest is spread within and between the PSUs and strata. The method used to calculate the standard errors for the survey takes this into account. It explicitly allows for the fact that the estimated values (percentages and means) are ratios of two survey estimates: the number with the characteristic of interest is the numerator (y) and the sample size is the denominator (x), both of which are subject to random error.

The standard error of a survey estimate is found by calculating the positive square root of the estimated variance of the ratio. The formula used to estimate the variance of a ratio estimator r (where $r = y/x$) is shown below.

$$var(r) = \frac{1}{x^2} [var(y) + r^2 var(x) - 2r\, cov(y,x)]$$

$Var(r)$ is the estimate of the variance of the ratio, r, expressed in terms of $var(y)$ and $var(x)$ which are the estimated variances of y and x, and $cov(y,x)$ which is their estimated covariance. The resulting estimate is only valid if the denominator (x) does not vary too greatly. The method compares the differences between totals for adjacent PSUs (postal sectors) in the characteristic of interest. The ordering of PSUs reflects the ranking of postal sectors on the stratifiers used in the sample design.

Design factors

The design factor, or deft, of an estimate p is the ratio of the complex standard error of p to the standard error of p that would have resulted had the survey design been a simple random sample of the same size.

$$deft(p) = \frac{se(p)}{se_{srs}(p)}$$

This is often used to give a broad indication of the effect of the clustering on the reliability of estimates. The size of the design factor varies between survey variables reflecting the degree to which a characteristic of interest is clustered within PSUs, or is distributed between strata. For a single variable the size of the design factor also varies according to the size of the subgroup on which the estimate is based, and on the distribution of that subgroup between PSUs and strata. Design factors below 1.0 show that the complex sample design improved on the estimate that we would have expected from a simple random sample, probably due to the benefits of stratification. Design factors greater than 1.0 show less reliable estimates than might be gained from a simple random sample, due to the effects of clustering. Design factors equal to 1.0 indicate no difference in the survey design on the reliability of the estimate.

The formula to calculate the standard error of the difference between two percentages for a complex sample design is:

$$se(p_1-p_2)=\sqrt{[deft^2_1(p_1(100-p_1)/n_1)+ deft^2_2 (p_2(100-p_2)/n_2)]}.$$

where p_1 and p_2 are observed percentages for the two sub-samples and n_1 and n_2 are the sub-sample sizes.

Confidence intervals

The estimate produced from a sample survey will rarely be identical to the population value, but statistical theory allows us to measure its accuracy. A confidence interval can be calculated around the estimated value, which gives a range in which the true value for the population is likely to fall. The standard error measures the precision with which the estimates from the sample approximate to the true population values and is used to construct the confidence interval for each survey estimate.

The 95% confidence intervals have been calculated for each estimated value presented. These are known as such, because if it were possible to repeat the survey under the same conditions a number of times, we would expect 95% of the confidence intervals calculated in this way to contain the true population value for that estimate. When assessing the results of a single survey, it is usual to assume that there is only a 5% chance that the true population value falls outside the 95% confidence interval calculated for each survey estimate. To construct the bounds of the confidence interval, 1.96 times the standard error is subtracted from, and added to, the estimated value, since under a normal distribution, 95% of values lie within 1.96 standard errors of the mean value. The confidence interval is then given by:

$$p +/- 1.96 \times se(p).$$

The 95% confidence interval for the difference between two percentages is given by:

$$(p_1-p_2) +/- 1.96 \times se(p_1-p_2).$$

If this confidence interval includes zero then the observed difference is considered to be a result of chance variation in the sample. If the interval does not include zero then it is unlikely (less than 5% probability) that the observed difference could have occurred by chance.

Standard errors for the 2006 GHS

The standard errors were calculated on weighted data using STATA[1]. Weighting for different sampling probabilities results in larger sampling errors than for an equal-probability sample without weights. However, weighting which uses population totals to control for differential non-response tends to lead to a reduction in the errors. The method used to calculate the sampling errors correctly allows for the inflation in the sampling errors caused by the first type of weighting but, in treating the second type of weighting in the same way as the first, incorrectly inflates the estimates further. Therefore the standard errors and defts presented are likely to be slight over-estimates. Weighted data were used so that the values of the percentages and means were the same as those in the substantive chapters of the report.

Tables C.1 to C.12 show the standard error, the 95% confidence interval and the deft for selected survey estimates. The tables do not cover all the topics discussed in the report but show a selection of estimates.

For the design factors of household based estimates, one was below 1, just under one fifth (17%) were below 1.1, half (50%) were less than 1.2, and about three quarters (73%) of the defts were less than 1.3. There was one case (3% of the household-based estimates) where the deft was 1.5 or greater. The higher defts were mostly for tenure and accommodation type (Table C.1) where the effects of clustering lead to a loss of precision compared with that of a simple random sample. The defts that were below 1.1 were in part for the number of persons in the household and household type, indicating that stratification has increased the precision of the sample over a simple random sample for these estimates of household variables.

For the design factors of person based estimates, 4% were below 1, about a one fifth (18%) were below 1.1, just over half (55%) were less than 1.2, and four fifths (79%) of the defts were less than 1.3. Four per cent of the defts were 1.5 or greater, including many of those for estimates of ethnicity, shown in Table C.6. As well as clustering in the same sectors, people from the same ethnic backgrounds will generally cluster within the same households, and so estimates have high sampling errors and high defts. In contrast, estimates broken down by gender will generally have lower sampling errors because there is often one man and one woman in a household; for example, the estimates of males and females in the population have defts of 0.8 (Table C.4).

Estimating standard errors for other survey measures

The standard errors of survey measures, which are not presented in the tables and for sample subgroups may be estimated by applying an appropriate value of deft to the sampling error. The choice of an appropriate value of deft will vary according to whether the basic survey measure is included in the tables. Since most deft values are relatively small (1.3 or less) the absolute effect of adjusting sampling errors to take account of the survey's complex design will be small. In most cases it will result in an increase of less than 30% over the standard error assuming a simple random sample. Whether it is considered necessary to use deft or to use the basic estimates of standard errors assuming a simple random sample is a matter of judgement and depends chiefly on the use to which the survey results are to be put.

Notes and references

1. STATA is a statistical analysis software package. For further details of the method of calculation see: Elliot D. A comparison of software for producing sampling errors on social surveys. *SSD Survey Methodology Bulletin* 1999; **44**: 27-36.

General Household Survey, 2006
Sampling Errors

Table C1 Standard errors and 95% confidence intervals for household tenure, household type and accommodation type

Base	Characteristic	% (*p*)	Unweighted sample size	Standard error of *p*	95% confidence intervals	Deft
All households						
	Household type					
	1 adult aged 16-59	14.6	9731	0.43	13.7 - 15.4	1.2
	2 adults aged 16-59	15.9	9731	0.38	15.1 - 16.6	1.0
	Youngest person aged 0-4	11.0	9731	0.35	10.3 - 11.7	1.1
	Youngest person aged 5-15	15.7	9731	0.39	14.9 - 16.4	1.1
	3 or more adults	11.5	9731	0.38	10.7 - 12.2	1.2
	2 adults, 1 or both aged 60 or over	15.6	9731	0.38	14.9 - 16.4	1.0
	1 adult aged 60 or over	15.8	9731	0.41	15.0 - 16.6	1.1
	Tenure					
	Owner occupied, owned outright	30.3	9731	0.55	29.2 - 31.4	1.2
	Owner occupied, with mortgage	39.6	9731	0.58	38.5 - 40.8	1.2
	Rented from council	11.5	9731	0.46	10.6 - 12.4	1.4
	Rented from housing association	8.1	9731	0.36	7.4 - 8.8	1.3
	Rented privately, unfurnished	7.7	9731	0.32	7.0 - 8.3	1.2
	Rented privately, furnished	2.8	9731	0.25	2.3 - 3.3	1.5
	Accommodation type					
	Detached house	22.9	9731	0.58	21.8 - 24.0	1.4
	Semi-detached house	30.3	9731	0.63	29.0 - 31.5	1.4
	Terraced house	28.0	9731	0.63	26.8 - 29.2	1.4
	Purpose-built flat or maisonette	15.3	9731	0.50	14.3 - 16.3	1.4
	Converted flat or maisonette/rooms	3.5	9731	0.29	2.9 - 4.0	1.6
	With business premises/other	0.1	9731	0.03	0.0 - 0.2	0.9

Office for National Statistics: January 2008

General Household Survey, 2006
Sampling Errors

Table C2 Standard errors and 95% confidence intervals for number of persons and cars at each household

Base	Characteristic	% (*p*)	Unweighted sample size	Standard error of *p*	95% confidence intervals	Deft
All households						
	Number of persons					
	1	30.4	9731	0.50	29.4 - 31.4	1.1
	2	34.0	9731	0.50	33.0 - 35.0	1.0
	3	15.8	9731	0.40	15.1 - 16.6	1.1
	4	13.6	9731	0.37	12.9 - 14.4	1.1
	5	4.5	9731	0.22	4.1 - 4.9	1.0
	6 or more	1.7	9731	0.16	1.4 - 2.0	1.2
	Number of cars/light vans					
	1	44.1	9731	0.53	43.0 - 45.1	1.1
	2 or more	32.7	9731	0.51	31.7 - 33.7	1.1
	none	23.3	9731	0.48	22.4 - 24.2	1.1

Table C3 Standard errors and 95% confidence intervals for households' ownership of selected consumer durables

Base	Characteristic	% (*p*)	Unweighted sample size	Standard error of *p*	95% confidence intervals	Deft
All households						
	Selected consumer durables					
	Home computer	69.1	9731	0.52	68.1 - 70.1	1.1
	Washing machine	95.3	9731	0.25	94.8 - 95.8	1.2

Office for National Statistics: January 2008

General Household Survey, 2006
Sampling Errors

Table C4 Standard errors and 95% confidence intervals for age and sex

Base	Characteristic	% (*p*)	Unweighted sample size	Standard error of *p*	95% confidence intervals		Deft
	Sex						
All persons	Male	48.9	22924	0.26	48.4 -	49.4	0.8
	Female	51.1	22924	0.26	50.6 -	51.6	0.8
	Age						
All persons	0-4	5.8	22924	0.19	5.5 -	6.2	1.2
	5-15	13.4	22924	0.29	12.8 -	13.9	1.3
	16-44	40.3	22924	0.38	39.6 -	41.1	1.2
	45-64	25.0	22924	0.35	24.3 -	25.7	1.2
	65-74	8.4	22924	0.23	7.9 -	8.8	1.3
	75 and over	7.2	22924	0.22	6.7 -	7.6	1.3
All males	0-4	6.0	11060	0.26	5.5 -	6.5	1.2
	5-15	13.8	11060	0.38	13.0 -	14.5	1.2
	16-44	41.2	11060	0.50	40.2 -	42.1	1.1
	45-64	25.2	11060	0.41	24.4 -	26.0	1.0
	65-74	8.1	11060	0.26	7.6 -	8.6	1.0
	75 and over	5.9	11060	0.23	5.4 -	6.3	1.0
All females	0-4	5.7	11864	0.23	5.2 -	6.1	1.1
	5-15	13.0	11864	0.34	12.3 -	13.6	1.1
	16-44	39.5	11864	0.45	38.6 -	40.4	1.0
	45-64	24.8	11864	0.42	24.0 -	25.6	1.1
	65-74	8.6	11864	0.28	8.1 -	9.2	1.1
	75 and over	8.4	11864	0.29	7.9 -	9.0	1.1

Office for National Statistics: January 2008

General Household Survey, 2006

Sampling Errors

C5 Standard errors and 95% confidence intervals for marital status

Base	Characteristic	% (p)	Unweighted sample size	Standard error of p	95% confidence intervals		Deft
	Marital Status						
All persons aged 16 and over	Married	51.0	22924	0.50	50.0	- 51.9	1.4
	Cohabiting	9.9	22924	0.31	9.2	- 10.5	1.4
	Single	23.5	22924	0.47	22.6	- 24.5	1.5
	Widowed	7.1	22924	0.18	6.8	- 7.5	1.0
	Divorced	6.1	22924	0.17	5.8	- 6.4	1.0
	Separated	2.1	22924	0.11	1.9	- 2.4	1.0
Men aged 16 and over	Married	52.4	8681	0.61	51.2	- 53.6	1.1
	Cohabiting	10.2	8681	0.37	9.5	- 11.0	1.1
	Single	27.2	8681	0.60	26.0	- 28.4	1.3
	Widowed	3.5	8681	0.20	3.2	- 3.9	1.0
	Divorced	4.6	8681	0.25	4.1	- 5.0	1.1
	Separated	1.6	8681	0.14	1.3	- 1.9	1.1
Women aged 16 and over	Married	49.5	9533	0.53	48.4	- 50.5	1.0
	Cohabiting	9.7	9533	0.35	9.0	- 10.3	1.2
	Single	20.6	9533	0.50	19.6	- 21.6	1.2
	Widowed	10.3	9533	0.34	9.6	- 10.9	1.1
	Divorced	7.6	9533	0.28	7.0	- 8.1	1.0
	Separated	2.2	9533	0.16	1.9	- 2.5	1.1
All persons aged 16 to 24	Married	4.1	2151	0.51	3.1	- 5.1	1.2
	Cohabiting	10.1	2151	0.85	8.4	- 11.8	1.3
	Single	85.3	2151	0.97	83.4	- 87.2	1.3
	Widowed	0.0	2151	0.00	0.0	- 0.0	0.0
	Divorced	0.1	2151	0.06	-0.1	- 0.2	1.1
	Separated	0.4	2151	0.13	0.1	- 0.6	1.0
All persons aged 25 to 34	Married	39.2	2657	1.26	36.8	- 41.7	1.3
	Cohabiting	24.5	2657	1.16	22.2	- 26.8	1.4
	Single	31.6	2657	1.17	29.3	- 33.9	1.3
	Widowed	0.1	2657	0.07	0.0	- 0.2	1.1
	Divorced	2.1	2657	0.31	1.4	- 2.7	1.1
	Separated	1.5	2657	0.23	1.0	- 1.9	1.0
All persons aged 35 to 44	Married	61.1	3404	1.07	59.0	- 63.2	1.3
	Cohabiting	12.7	3404	0.73	11.3	- 14.2	1.3
	Single	15.3	3404	0.76	13.8	- 16.8	1.2
	Widowed	0.6	3404	0.16	0.2	- 0.9	1.3
	Divorced	7.0	3404	0.47	6.1	- 7.9	1.1
	Separated	2.8	3404	0.32	2.1	- 3.4	1.1
All persons aged 45 to 54	Married	68.6	3008	1.00	66.6	- 70.5	1.2
	Cohabiting	7.3	3008	0.58	6.1	- 8.4	1.2
	Single	8.5	3008	0.58	7.4	- 9.7	1.1
	Widowed	1.1	3008	0.19	0.7	- 1.5	1.0
	Divorced	11.0	3008	0.62	9.8	- 12.3	1.1
	Separated	3.2	3008	0.35	2.5	- 3.9	1.1
All persons aged 55 to 64	Married	72.1	3060	0.96	70.2	- 74.0	1.2
	Cohabiting	4.0	3060	0.43	3.2	- 4.9	1.2
	Single	6.3	3060	0.50	5.4	- 7.3	1.1
	Widowed	4.9	3060	0.43	4.1	- 5.8	1.1
	Divorced	10.2	3060	0.65	8.9	- 11.4	1.2
	Separated	2.3	3060	0.31	1.7	- 2.9	1.2
All persons aged 65 to 74	Married	66.0	2194	1.05	64.0	- 68.1	1.0
	Cohabiting	2.2	2194	0.36	1.4	- 2.9	1.2
	Single	5.2	2194	0.42	4.4	- 6.1	0.9
	Widowed	17.4	2194	0.80	15.8	- 19.0	1.0
	Divorced	7.4	2194	0.55	6.4	- 8.5	1.0
	Separated	1.6	2194	0.24	1.1	- 2.1	0.9
All persons aged 75 and over	Married	41.3	1731	1.42	38.5	- 44.1	1.2
	Cohabiting	1.1	1731	0.34	0.4	- 1.8	1.4
	Single	5.8	1731	0.70	4.4	- 7.1	1.3
	Widowed	47.0	1731	1.39	44.3	- 49.7	1.2
	Divorced	4.1	1731	0.46	3.2	- 5.0	1.0
	Separated	0.8	1731	0.20	0.4	- 1.2	0.9

Office for National Statistics: January 2008

General Household Survey, 2006
Sampling Errors

Table C.6 Standard errors and 95% confidence intervals for ethnic origin*

Base[1]	Characteristic	% (*p*)	Unweighted sample size	Standard error of *p*	95% confidence intervals	Deft
All persons aged 16 and over	**Ethnic Origin**					
	White	91.0	18166	0.63	90.0 - 92.4	3.0
	Mixed Race	0.7	18166	0.09	0.5 - 0.9	1.4
	Asian-Indian	2.1	18166	0.33	1.4 - 2.7	3.1
	Asian- Pakistani, Bangladeshi, Other	2.1	18166	0.31	1.5 - 2.7	2.9
	Black Caribbean	1.3	18166	0.19	0.9 - 1.6	2.3
	Black African	1.4	18166	0.17	1.1 - 1.8	1.9
	Other	1.3	18166	0.15	1.0 - 1.5	1.8

* Other includes other Black groups. Information on those giving no answer has not been presented
[1] These estimates are based on 2005 data only, wheras in the report estimates are based on 2001, 2002 and 2003 data combined.
We would expect the defts to be very similar for the combined years estimates.

Table C.7 Standard errors and 95% confidence intervals for education level

Base	Characteristic	% (*p*)	Unweighted sample size	Standard error of *p*	95% confidence intervals	Deft
	Education level					
All persons aged 16 to 69	Higher education	29.5	15397	0.57	28.4 - 30.6	1.6
	Other qualifications	43.6	15397	0.53	42.6 - 44.7	1.3
	None	26.9	15397	0.50	25.9 - 27.9	1.4
All men aged 16 to 69	Higher education	30.5	6997	0.68	29.2 - 31.9	1.2
	Other qualifications	45.1	6997	0.68	43.8 - 46.4	1.1
	None	24.4	6997	0.62	23.2 - 25.6	1.2
All women aged 16 to 69	Higher education	28.6	8400	0.66	27.3 - 29.9	1.3
	Other qualifications	42.4	8400	0.62	41.2 - 43.6	1.2
	None	29.1	8400	0.58	27.9 - 30.2	1.2

Office for National Statistics: January 2008

General Household Survey, 2006

Sampling Errors

Table C8 Standard errors and 95% confidence intervals for socio-economic classification and employment status of adults

Base	Characteristic	% (p)	Unweighted sample size	Standard error of p	95% confidence intervals		Deft
	Socio-economic classification						
All persons aged 16 and over	Higher managerial and professional	12.6	16973	0.34	11.9 -	13.3	1.3
	Lower managerial and professional	22.3	16973	0.37	21.6 -	23.0	1.2
	Intermediate	12.6	16973	0.28	12.1 -	13.2	1.1
	Small employers and own account	8.2	16973	0.24	7.7 -	8.7	1.1
	Lower supervisory and technical	9.0	16973	0.25	8.5 -	9.5	1.1
	Semi-routine	17.4	16973	0.34	16.7 -	18.0	1.2
	Routine	13.4	16973	0.33	12.8 -	14.1	1.3
	Never worked and long-term unemployed	4.5	16973	0.23	4.1 -	5.0	1.4
All men aged 16 and over	Higher managerial and professional	18.5	8072	0.55	17.4 -	19.5	1.3
	Lower managerial and professional	20.7	8072	0.49	19.7 -	21.6	1.1
	Intermediate	6.0	8072	0.30	5.4 -	6.6	1.1
	Small employers and own account	11.6	8072	0.39	10.9 -	12.4	1.1
	Lower supervisory and technical	13.5	8072	0.42	12.7 -	14.3	1.1
	Semi-routine	11.9	8072	0.42	11.0 -	12.7	1.2
	Routine	14.5	8072	0.45	13.6 -	15.3	1.2
	Never worked and long-term unemployed	3.6	8072	0.26	3.1 -	4.1	1.3
All women aged 16 and over	Higher managerial and professional	7.1	8901	0.28	6.6 -	7.7	1.0
	Lower managerial and professional	23.8	8901	0.48	22.9 -	24.8	1.1
	Intermediate	18.8	8901	0.45	17.9 -	19.7	1.1
	Small employers and own account	5.0	8901	0.24	4.5 -	5.5	1.0
	Lower supervisory and technical	4.9	8901	0.26	4.4 -	5.4	1.1
	Semi-routine	22.5	8901	0.51	21.5 -	23.5	1.2
	Routine	12.5	8901	0.44	11.7 -	13.3	1.2
	Never worked and long-term unemployed	5.5	8901	0.31	4.8 -	6.1	1.3
All persons aged 16 to 44	Higher managerial and professional	13.7	7146	0.50	12.7 -	14.7	1.2
	Lower managerial and professional	23.5	7146	0.54	22.4 -	24.6	1.1
	Intermediate	12.6	7146	0.45	11.7 -	13.4	1.2
	Small employers and own account	6.5	7146	0.31	5.9 -	7.1	1.1
	Lower supervisory and technical	8.5	7146	0.36	7.7 -	9.2	1.1
	Semi-routine	16.7	7146	0.51	15.7 -	17.7	1.2
	Routine	11.7	7146	0.44	10.9 -	12.6	1.2
	Never worked and long-term unemployed	6.9	7146	0.43	6.0 -	7.7	1.4
All persons aged 45 to 64	Higher managerial and professional	13.3	5930	0.53	12.3 -	14.4	1.2
	Lower managerial and professional	23.2	5930	0.60	22.0 -	24.4	1.1
	Intermediate	11.7	5930	0.43	10.9 -	12.6	1.0
	Small employers and own account	10.4	5930	0.46	9.5 -	11.3	1.2
	Lower supervisory and technical	8.7	5930	0.40	7.9 -	9.5	1.1
	Semi-routine	17.4	5930	0.53	16.3 -	18.4	1.1
	Routine	13.2	5930	0.50	12.3 -	14.2	1.1
	Never worked and long-term unemployed	2.1	5930	0.24	1.6 -	2.6	1.3
All persons aged 65 to 74	Higher managerial and professional	9.6	2179	0.66	8.3 -	10.9	1.1
	Lower managerial and professional	17.8	2179	0.87	16.1 -	19.5	1.1
	Intermediate	13.1	2179	0.74	11.6 -	14.5	1.0
	Small employers and own account	9.8	2179	0.70	8.5 -	11.2	1.1
	Lower supervisory and technical	11.0	2179	0.77	9.5 -	12.5	1.2
	Semi-routine	18.9	2179	0.88	17.2 -	20.6	1.1
	Routine	17.1	2179	0.88	15.4 -	18.8	1.1
	Never worked and long-term unemployed	2.7	2179	0.40	1.9 -	3.5	1.2
All persons aged 75 and over	Higher managerial and professional	8.1	1718	0.61	6.9 -	9.3	0.9
	Lower managerial and professional	18.6	1718	0.93	16.8 -	20.4	1.0
	Intermediate	15.2	1718	0.92	13.4 -	17.0	1.1
	Small employers and own account	7.0	1718	0.65	5.7 -	8.3	1.1
	Lower supervisory and technical	10.8	1718	0.75	9.3 -	12.3	1.0
	Semi-routine	18.7	1718	0.97	16.8 -	20.6	1.0
	Routine	17.8	1718	1.04	15.8 -	19.8	1.1
	Never worked and long-term unemployed	3.8	1718	0.47	2.8 -	4.7	1.0
	Employment status						
All persons aged 16 & over	In employment	60.5	17941	0.46	59.6 -	61.4	1.3
	Unemployed	2.6	17941	0.15	2.3 -	2.9	1.3
	Economically inactive	36.9	17941	0.47	36.0 -	37.8	1.3
All men aged 16 and over	In employment	66.8	8511	0.56	65.7 -	67.9	1.1
	Unemployed	3.5	8511	0.23	3.0 -	3.9	1.2
	Economically inactive	29.7	8511	0.56	28.6 -	30.8	1.1

Office for National Statistics: January 2008

Table C.9 Standard errors and 95% confidence intervals for health measures

Base	Characteristic	% (p)	Unweighted sample size	Standard error of p	95% confidence intervals	Deft
	Self-reported sickness					
All persons	Longstanding illness	33.4	21346	0.45	32.5 - 34.3	1.4
	Limiting longstanding illness	32.6	10017	0.55	31.5 - 33.7	1.2
	Restricted activity in the last 14 days	34.1	11329	0.53	33.0 - 35.1	1.2
All males	Longstanding illness	11.0	1356	0.95	9.1 - 12.8	1.1
	Limiting longstanding illness	16.2	3324	0.70	14.8 - 17.5	1.1
	Restricted activity in the last 14 days	22.1	7183	0.59	20.9 - 23.2	1.2
All females	Longstanding illness	44.5	5670	0.74	43.1 - 46.0	1.1
	Limiting longstanding illness	63.0	2137	1.19	60.7 - 65.3	1.1
	Restricted activity in the last 14 days	70.0	1676	1.21	67.7 - 72.4	1.1
All persons aged 0 to 4	Longstanding illness	17.6	22666	0.32	16.9 - 18.2	1.3
	Limiting longstanding illness	15.5	10897	0.39	14.7 - 16.3	1.1
	Restricted activity in the last 14 days	19.5	11769	0.42	18.7 - 20.3	1.2
All persons aged 5 to 15	Longstanding illness	3.3	1364	0.53	2.3 - 4.4	1.1
	Limiting longstanding illness	6.5	3345	0.46	5.6 - 7.4	1.1
	Restricted activity in the last 14 days	9.7	8034	0.39	9.0 - 10.5	1.2
All persons aged 16 to 44	Longstanding illness	23.9	6006	0.61	22.7 - 25.1	1.1
	Limiting longstanding illness	37.1	2192	1.08	34.9 - 39.2	1.1
	Restricted activity in the last 14 days	48.2	1725	1.34	45.6 - 50.8	1.1
All persons aged 45 to 64	Longstanding illness	12.6	21402	0.29	12.0 - 13.2	1.3
	Limiting longstanding illness	11.0	10061	0.34	10.3 - 11.7	1.1
	Restricted activity in the last 14 days	14.0	11341	0.38	13.3 - 14.8	1.2
All persons aged 65 to 74	Longstanding illness	7.8	1356	0.80	6.2 - 9.3	1.1
	Limiting longstanding illness	7.5	3326	0.53	6.4 - 8.5	1.2
	Restricted activity in the last 14 days	10.2	7222	0.41	9.4 - 11.0	1.2
All persons aged 75+	Longstanding illness	15.4	5683	0.52	14.4 - 16.4	1.1
	Limiting longstanding illness	18.6	2138	0.95	16.8 - 20.5	1.1
	Restricted activity in the last 14 days	22.2	1677	1.20	19.9 - 24.6	1.2

General Household Survey, 2006
Sampling Errors

Table C.10 Standard errors and 95% confidence intervals for cigarette smoking

Base	Characteristic	% (p)	Unweighted sample size	Standard error of p	95% confidence intervals	Deft
	Cigarette smoking					
All persons aged 16 & over	Current cigarette smoker	22.0	16682	0.43	21.1 - 22.8	1.3
	Ex-regular cigarette smoker	23.7	16682	0.42	22.9 - 24.5	1.3
	Never regularly smoked cigarettes	54.3	16682	0.54	53.3 - 55.4	1.4
All men aged 16 and over	Current cigarette smoker	23.2	7677	0.55	22.1 - 24.3	1.1
	Ex-regular cigarette smoker	27.1	7677	0.57	26.0 - 28.2	1.1
	Never regularly smoked cigarettes	49.7	7677	0.67	48.4 - 51.0	1.2
All women aged 16 and over	Current cigarette smoker	20.9	9005	0.49	19.9 - 21.9	1.1
	Ex-regular cigarette smoker	20.7	9005	0.50	19.8 - 21.7	1.2
	Never regularly smoked cigarettes	58.4	9005	0.65	57.1 - 59.6	1.3

General Household Survey, 2006
Sampling Errors

Table C11 Standard errors and 95% confidence intervals for alcohol consumption (maximum daily amount)

Base	Characteristic	% (*p*)	Unweighted sample size	Standard error of *p*	95% confidence intervals	Deft
	Alcohol consumption (maximum daily amount)					
All men aged 16 and over	Drank nothing last week	28.9	7674	0.62	27.7 - 30.1	1.2
	Drank up to 4 units	37.8	7674	0.60	36.6 - 39.0	1.1
	Drank more than 4 and up to 8 units	15.2	7674	0.46	14.3 - 16.1	1.1
	Drank more than 8 units	18.2	7674	0.56	17.1 - 19.3	1.3
All women aged 16 and over	Drank nothing last week	44.1	9013	0.68	42.8 - 45.4	1.3
	Drank up to 3 units	36.1	9013	0.58	35.0 - 37.2	1.1
	Drank more than 3 and up to 6 units	11.8	9013	0.38	11.0 - 12.5	1.1
	Drank more than 6 units	8.0	9013	0.35	7.3 - 8.7	1.2
All aged 16 to 24	Drank nothing last week	43.9	1717	1.45	41.1 - 46.8	1.2
	Drank up to 4/3 units	19.7	1717	1.04	17.7 - 21.8	1.1
	Drank more than 4/3 and up to 8/6 units	12.6	1717	0.87	10.8 - 14.3	1.1
	Drank more than 8/6 units	23.8	1717	1.24	21.4 - 26.2	1.2
All aged 25 to 44	Drank nothing last week	33.7	5470	0.82	32.1 - 35.3	1.3
	Drank up to 4/3 units	32.5	5470	0.71	31.2 - 33.9	1.1
	Drank more than 4/3 and up to 8/6 units	16.0	5470	0.61	14.8 - 17.2	1.2
	Drank more than 8/6 units	17.8	5470	0.63	16.6 - 19.0	1.2
All aged 45 to 64	Drank nothing last week	32.3	5684	0.77	30.7 - 33.8	1.2
	Drank up to 4/3 units	42.8	5684	0.80	41.2 - 44.4	1.2
	Drank more than 4/3 and up to 8/6 units	15.3	5684	0.54	14.3 - 16.4	1.1
	Drank more than 8/6 units	9.6	5684	0.46	8.7 - 10.5	1.2
All aged 65 and over	Drank nothing last week	45.7	3816	0.96	43.8 - 47.5	1.2
	Drank up to 4/3 units	45.8	3816	0.99	43.9 - 47.8	1.2
	Drank more than 4/3 and up to 8/6 units	6.4	3816	0.41	5.6 - 7.2	1.0
	Drank more than 8/6 units	2.1	3816	0.25	1.6 - 2.6	1.1
All aged 16 and over	Drank nothing last week	37.0	16687	0.55	35.9 - 38.1	1.5
	Drank up to 4/3 units	36.9	16687	0.47	36.0 - 37.8	1.3
	Drank more than 4/3 and up to 8/6 units	13.4	16687	0.31	12.8 - 14.0	1.2
	Drank more than 8/6 units	12.8	16687	0.36	12.1 - 13.5	1.4

Office for National Statistics: January 2008

General Household Survey, 2006
Sampling Errors

Table C12 Standard errors and 95% confidence intervals for number of cohabitations

Base	Characteristic	% (p)	Unweighted sample size	Standard error of p	95% confidence intervals	Deft
	Number of cohabitations					
All women aged 16 to 59	None	84.7	4795	0.60	83.5 - 85.8	1.2
	One	10.2		0.45	9.4 - 11.1	1.0
	Two or more	5.1		0.38	4.4 - 5.8	1.2
All men aged 16 to 59	None	84.3	5838	0.56	83.2 - 85.4	1.2
	One	12.0		0.46	11.1 - 12.9	1.1
	Two or more	3.8		0.28	3.2 - 4.3	1.1
All people aged 16 to 24	None	92.2	1429	0.75	90.7 - 93.7	1.1
	One	6.8		0.70	5.4 - 8.2	1.1
	Two or more	1.0		0.26	0.5 - 1.5	1.0
All people aged 25 to 34	None	77.4	2165	0.95	75.6 - 79.3	1.1
	One	15.9		0.79	14.4 - 17.5	1.0
	Two or more	6.6		0.59	5.5 - 7.8	1.1
All people aged 35 to 44	None	77.8	2889	0.90	76.0 - 79.5	1.2
	One	15.2		0.74	13.8 - 16.7	1.1
	Two or more	7.0		0.53	6.0 - 8.0	1.1
All people aged 45 to 54	None	88.5	2669	0.67	87.2 - 89.8	1.1
	One	8.4		0.56	7.3 - 9.5	1.0
	Two or more	3.1		0.38	2.4 - 3.9	1.1
All people aged 55 to 59	None	93.6	1481	0.70	92.2 - 95.0	1.1
	One	4.9		0.61	3.7 - 6.1	1.1
	Two or more	1.5		0.35	0.8 - 2.2	1.1
All people aged 16 to 59	None	84.4	10633	0.43	83.6 - 85.3	1.2
	One	11.2		0.33	10.5 - 11.8	1.1
	Two or more	4.4		0.23	3.9 - 4.8	1.2

Office for National Statistics: January 2008

General Household Survey, 2006
Weighting and grossing

General Household Survey 2006

Weighting and grossing

Appendix D

Riaz Ali

Julia Greer

David Matthews

Liam Murray

Simon Robinson

Ghazala Sattar

© Crown copyright

Office for National Statistics

Government Buildings

Cardiff Rd

Newport

NP10 8XG

Tel: 01633 455877

Email: ghs@ons.gsi.gov.uk

All surveys accept that there will be some degree of nonresponse, although great efforts are made to keep it to a minimum[1]. The General Household Survey (GHS) compensates for **unit nonresponse** (where all survey information for a sampled household is missing) through weighting, which will be described here. The method adopted to reduce **item nonresponse** (where information for particular questions is missing as the result of conducting proxy interviews) is discussed in *Appendix B Sample design and response*.

Longitudinal surveys like the GHS experience two forms of nonresponse: non-participation to wave one and non-participation to later waves of the survey (also known as attrition). Here we will use the term nonresponse to refer to the former and attrition to refer to the latter.

The 2006 GHS is weighted using a two-step approach. The first step uses sample-based weighting to compensate for nonresponse and attrition. The second step uses population-based weighting to match the sample distribution to the population distribution in terms of age group, sex and region[2].

Weighting for nonresponse and attrition

Weighting for nonresponse and attrition can be seen as giving each respondent a weight so that they represent non-respondents who are similar to them in terms of survey characteristics. To be able to use this method consistent information about both respondents *and* non-respondents is needed to model the likelihood of response for different groups. In the case of nonresponse, little information is available on non-responding households directly from the survey[3], so an external data source is required. For attrition, we are able to match back to previous years of the survey where longitudinal households have provided information.

The sampled-based procedure for determining weighting characteristics is considered separately below for sampled households in wave one and wave two.

Sample-based weighting for nonresponse using the 2001 Census

Although we have no direct data on non-responding households to the GHS we use information from the Census to indirectly estimate the likelihood of response. Unlike the GHS, which relies upon voluntary co-operation from respondents, the Census is mandatory and therefore nonresponse is kept to an absolute minimum.

After the 2001 Census, methodological work was conducted to match Census addresses with the sampled addresses for some of the large household surveys, which included the GHS. Therefore, it was possible to match GHS respondents *and* non-respondents with corresponding information from the Census for the same address. We could then model and calculate response rates for types of household that were being under-represented in the survey. A combination of household variables available on an annual basis, such as household type, social class, region and car ownership were analysed using the software package AnswerTree (which uses chi-squared statistics to group households with similar response patterns)[4]. These chosen characteristics were used to produce the weighting classes shown in Figure D.A.

The weighting classes and weights determined from the work described above are then applied to weight the GHS data (for wave one respondents) on an annual basis.

Figure DA

Figure DA 2001 Weighting classes formed in the AnswerTree analysis

LEVEL 1 SPLIT	LEVEL 2 SPLIT	LEVEL 3 SPLIT	LEVEL 4 SPLIT	WEIGHT CLASS
Region	Number of Rooms	Number of Pensioners in the household	Sex of the Household Reference Person	
West Midlands London	More than three	All pensioners	Female	1
			Male	2
		Two or two or more persons in the household but only 1 pensioner		3
		Two or two or more persons in the household where more than one person is a pensioner No pensioners in the household		4
	Three or fewer			5
Yorkshire & Humberside East Midlands South West Wales	**Household size** More than two			6
	Two			7
	One	**Adults not employed** One		8
		Zero		9
North East North West & Merseyside East of England South East Scotland	**Number of Adults** More than two			10
	Two	**Accommodation Type** Purpose built flat Part of a converted or shared house Other		11
		Detached Semi-detached Purpose build flat		12
	One			13

Sample-based weighting for attrition using the 2005 GHS

As mentioned earlier, attrition is a form of nonresponse found on longitudinal surveys between waves. The GHS is currently in year two of its longitudinal implementation and in 2006; this meant that approximately three-quarters of sampled households had been surveyed in 2005. As these sampled household had previously participated in the survey, details of respondents and non-respondents were linked back to their corresponding wave one information. Logistic regression was used to model the likelihood of response to wave two against the characteristics of households at their wave one interview. A variety of household variables such as household type, socio-economic class, region and car ownership were tested for inclusion. Characteristics determined as significant by the logistic regression (at the five per cent significance level) were used to weight for this attrition[5]. The variables reaching significance are listed in Figure D.B below.

Figure DB

Figure DB Significant weighting variables formed in the Logistic Regression analysis

Variable	p-value
Accommodation type	0.000
Household composition	0.000
Socio-economic category of the household reference person	0.024
Educational level of the household reference person	0.000
Year of arrival into the United Kingdom of the household reference person	0.000
The household reference person is in receipt of personal income benefits	0.000

Population-based weighting (grossing)

Population-based weighting schemes address deficiencies in the data due to sample-non coverage. They can also further reduce nonresponse bias and reduce the variance (sampling error) of survey estimates.

The population-based method

The GHS sample is based on private households, which means that the population totals used in the weighting need to relate to people in private households. These totals are taken from population projections based on mid-year estimates and adjusted to exclude residents of certain institutions.

The population information and GHS data were grouped into twelve age by sex categories within six region categories to form weighting classes as shown in Figure D.B. The population-based weighting consists of adjusting the existing weights (including factors for design and nonresponse) so that the final weights ensure that weighted totals for the above demographic categories match the population totals.

This procedure, also known as calibration, was carried out using the GES SAS macro. This was implemented in such a way as to ensure that all individuals within a household were given the same final weights[6].

Figure DC

Figure DC Weighting classes used for GES analysis

Age/sex	Region
0-4	London
5-15	Scotland
16-24 male	Wales
16-24 female	Other Metropolitan
25-44 male	Other Non-metropolitan
25-44 female	South East
45-64 male	
45-64 female	
65-74 male	
65-74 female	
75+ male	
75+ female	

Presentation and interpretation of weighted data

Weighted data cannot be meaningfully compared to unweighted data from previous years without knowledge of how the weighting changes the estimates. In the GHS trend tables, weighted and unweighted data are presented for 1998 and weighted data are shown only from 2000 to 2006. Care should be taken when interpreting trend data or individual tables compared with other years as part of a time series.

It should be noted that the weighted bases used in this report are not recommended as a source for population estimates. They are the denominator for the percentages shown in tables and should not be regarded as estimates of population size[7].

Effects of weighting on data

Tables D.1 and D.2 identify the effects of weighting by comparing unweighted and weighted data for 2006. They also show the differences between the weighted and unweighted estimates for 1998 and 2000 to 2006, on a selection of household and individual level variables.

Tables D1 and D2

A comparison of the characteristics recorded on the 2001 Census forms of respondents and non-respondents in the 2001 GHS sample showed that households comprising one adult aged 16 to 59 or a couple with non-dependent children were under-represented. Households containing dependent children were over-represented in the responding sample. As would be expected, weighting has changed the value of the estimate for some variables, but the overall changes have been relatively small.

For the 2006 estimates, the most marked effect of weighting was seen in the following variables. None of the effects are large.

Increase in value of estimate.

- Household type of 1 person only from 28.1% to 30.5%.
- 1 adult households from 33.0% to 35.2%.
- Households with no access to car or light van from 21.3% to 23.3%.

Decrease in value of estimate.

- 2 adult households from 52.4% to 47.5%.
- Households containing a married couple and no children from 26.4% to 22.6%.
- Households that are owned outright from 26.4% to 22.6%

The differences between the weighted and unweighted data for 1998 and 2000 to 2006 are also shown in Tables D.1 and D.2. It can be seen that the differences produced by weighting in 2006 were similar to those in previous years for the same variables.

Notes and references

1. Appendix B describes the variation in response for the GHS since it began in 1971.

2. Barton, J. Developing a Weighting and Grossing System for the General Household Survey: *Social Survey Methodology Bulletin* (Issue 49 July 2001).

3. Some surveys collect information about the characteristics of non-responding households although this is not routinely the case on the GHS.

4. AnswerTree uses the CHAID (Chi-squared Automatic Interaction Detection) algorithm, which uses chi-squared statistics to identify optimal splits or groupings of independent variables in terms of predicting the outcome of a dependent variable, in this case response.

5. The attrition weights build on the weights calculated for when these longitudinal cases were in their first wave (2005). This means both nonresponse and attrition are covered within this weight.

6. GES, or *Generalized Estimation System*, is a SAS macro produced by Statistics Canada. The weights are formed using a form of calibration called Generalized Regression, or GREG estimation. The macro allows bounds to be set on the adjustment factors in the calibration.

7. Missing answers are excluded from the tables and in some cases this is reflected in the weighted bases, i.e. these numbers vary between tables. For this reason, the bases themselves are not recommended as a source for population estimates. Recommended data sources for population

Table D1 Weighted versus unweighted data for years 1998 to 2006 - household level

Household level variables

% of households	2006 Unweighted (a)	2006 Weighted (b)	Weighted 1998 - Unweighted 1998	Weighted 2000 - Unweighted 2000	Weighted[1] 2001 - Unweighted 2001	Weighted 2002 - Unweighted 2002	Weighted 2003 - Unweighted 2003	Weighted 2004 - Unweighted 2004	Weighted 2005 - Unweighted 2005	Weighted 2006 - Unweighted 2006 (b-a)
Household size										
1 person	11.9	13.0	1.9	2.4	1.7	2.0	2.3	2.3	2.4	1.1
2 persons	31.6	29.0	-1.3	-1.6	-1.4	-1.7	-1.0	-1.4	-1.3	-2.6
3 persons	19.1	20.3	0.2	0.2	0.2	0.1	0.0	0.0	0.0	1.2
4 persons	23.0	23.3	-0.4	-0.5	-0.3	-0.3	-0.8	-0.5	-0.7	0.3
5 persons	9.8	9.6	-0.1	-0.3	-0.1	-0.1	-0.4	-0.2	-0.4	-0.2
6 or more persons	4.6	4.8	-0.2	-0.1	-0.1	0.0	-0.1	-0.1	-0.1	0.2
Base	22924	58,041,290								
Number of adults										
1 adult	33.0	35.2	1.4	1.9	1.2	2.1	2.2	2.2	2.3	2.2
2 adults	52.4	47.5	-2.7	-2.6	-2.0	-2.5	-3.2	-3.0	-3.2	-4.9
3 adults	10.3	11.8	0.5	0.3	0.4	0.1	0.4	0.3	0.4	1.5
4 or more adults	4.3	5.4	0.7	0.5	0.3	0.5	0.6	0.5	0.5	1.1
Base	9731	24,814,647								
Number of children										
No children	72.4	73.4	1.5	1.5	1.1	0.6	2.1	1.3	1.8	1.0
1 child	12.2	13.0	0.3	0.0	0.1	0.3	0.1	0.1	0.0	0.8
2 children	11.2	10.1	-1.0	-0.8	-0.7	-0.6	-1.4	-0.9	-1.2	-1.1
3 or more children	4.2	3.6	-0.8	-0.6	-0.4	-0.4	-0.8	-0.5	-0.6	-0.6
Base	9731	24,814,647								
Household type										
1 person only	28.1	30.5	1.9	2.4	1.8	2.0	2.3	2.3	2.4	2.4
2 or more unrelated adults	2.3	2.4	0.4	0.3	0.2	0.3	0.3	0.4	0.4	0.1
Married couple, dependent children	18.6	18.1	-0.8	-0.8	-0.6	-0.6	-1.4	-1.0	-1.4	-0.5
Married couple, independent children	5.2	6.1	0.3	0.4	0.3	0.1	0.3	0.1	0.2	0.9
Married couple, no children	26.4	22.6	-2.0	-2.3	-1.7	-2.0	-1.7	-1.8	-2.0	-3.8
Lone parent, dependent children	6.8	6.8	-0.4	-0.5	-0.5	0.0	-0.1	0.0	0.0	0.0
Lone parent, independent children	2.6	3.0	0.1	0.2	0.2	0.1	0.2	0.2	0.1	0.4
2 or more families (inc. same sex cohab)	1.2	1.4	-0.1	0.0	0.0	0.0	0.0	0.0	-0.1	0.2
Cohabiting couple, with children	3.7	3.7	0.0	0.0	0.0	0.1	-0.3	-0.1	-0.3	0.0
Cohabiting couple, no children	5.3	5.4	0.4	0.4	0.3	0.1	0.2	0.0	0.3	0.1
Base	9705	24,727,335								
Tenure - harmonised										
Owns outright	33.7	30.3	-1.7	-1.9	-1.7	-1.6	-1.5	-1.1	-1.6	-3.4
Buying on mortgage	39.2	39.6	0.3	0.1	0.5	0.0	-0.1	-0.7	-0.4	0.4
Rents from Council/Local Authority	10.6	11.5	0.1	0.5	0.2	0.7	0.6	0.7	0.8	0.9
Rents from HA/Reg. Social Landlord	7.6	8.1	0.1	0.1	0.1	0.2	0.1	0.3	0.1	0.5
Rents privately - unfurnished/nk	2.2	2.8	0.4	0.5	0.4	0.4	0.6	0.4	0.5	0.6
Rents privately - furnished	6.9	7.6	0.6	0.6	0.6	0.5	0.4	0.5	0.5	0.7
Base	9731	24,814,647								
Ownership of consumer durables										
Washing machine	95.8	95.3	-0.8	-0.7	-0.6	-0.5	-0.6	-0.5	-0.4	-0.5
Telephone	99.4	99.4	-0.3	-0.2	-0.1	0.0	-0.1	0.0	-0.1	0.0
Home computer	68.9	69.1	0.4	-0.1	0.4	-0.4	-0.3	-0.7	-0.6	0.2
Base	9731	24,814,647								
Central heating	95.6	95.3	-0.3	-0.3	-0.2	-0.3	-0.1	-0.2	-0.3	-0.3
Base	9731	24,814,647								
Car or van ownership										
No car or van	21.3	23.3	0.6	1.1	0.6	1.4	1.4	1.7	1.7	2.0
One car or van	44.2	44.0	0.6	0.4	0.4	0.5	0.3	0.5	0.6	-0.2
Two cars or vans	28.0	25.9	-1.1	-1.3	-1.0	-1.7	-1.7	-2.0	-2.0	-2.1
Three or more cars or vans	6.5	6.8	-0.1	-0.2	0.0	-0.2	-0.1	-0.2	-0.2	0.3
Base	9731	24,814,647								

1 Original 2001 weighting (based on LFS 2000 population estimates).

Table D2 Weighted versus unweighted data for years 1998 to 2006 - individual level

Table D2 Weighted versus unweighted data for years 1998 to 2006 - individual level

Individual level variables

% of individuals	2006 Unweighted (a)	2006 Weighted (b)	Effect of weighting Weighted 1998 - Unweighted 1998	Weighted 2000 - Unweighted 2000	Weighted[1] 2001 - Unweighted 2001	Weighted 2002 - Unweighted 2002	Weighted 2003 - Unweighted 2003	Weighted 2004 - Unweighted 2004	Weighted 2005 - Unweighted 2005	Weighted 2006 - Unweighted 2006 (b-a)
Limiting longstanding illness										
Male	17.6	17.0	-0.2	-0.3	-0.3	-0.4	-0.1	-0.1	-0.5	-0.6
Female	20.5	20.3	-0.1	0.0	-0.1	0.1	0.3	0.3	0.3	-0.2
Total	19.1	18.8	-0.1	-0.1	-0.2	-0.2	0.1	0.2	-0.1	-0.3
Non-limiting longstanding illness										
Male	16.3	15.6	-0.4	-0.2	-0.2	-0.4	-0.2	-0.3	0	-0.7
Female	14.1	13.7	0.0	-0.1	-0.1	-0.2	-0.1	0.0	0.4	-0.4
Total	15.1	14.6	-0.2	-0.1	-0.2	-0.3	-0.2	-0.2	0.2	-0.5
No longstanding illness										
Male	66.1	67.4	0.6	0.5	0.6	0.8	0.4	0.3	0.5	1.3
Female	65.4	65.9	0.1	-0.1	0.2	0.1	-0.1	-0.2	-0.3	0.5
Total	65.7	66.6	0.4	0.2	0.4	0.4	0.1	0.0	0.1	0.9
General health										
Good										
Male	67.8	68.4	0.6	0.1	0.0	0.1	0.1	0.1	0	0.6
Female	66.0	65.8	0.2	-0.3	-0.1	-0.4	-0.4	-0.4	-0.7	-0.2
Total	66.9	67.0	0.5	-0.1	0.0	-0.1	-0.2	-0.2	-0.4	0.1
Fairly good										
Male	22.8	22.6	-0.2	0.0	0.0	0.0	-0.1	-0.2	-0.1	-0.2
Female	23.1	23.3	0.0	0.2	0.0	0.2	0.2	0.1	0.4	0.2
Total	23.0	22.9	-0.2	0.0	0.1	0.1	0.1	0.0	0.2	-0.1
Not good										
Male	9.4	9.0	-0.4	0.0	-0.1	0.0	0.0	0.1	0.1	-0.4
Female	10.9	10.9	-0.2	0.1	0.1	0.1	0.2	0.2	0.3	0.0
Total	10.2	10.0	-0.3	0.0	0.0	0.0	0.1	0.2	0.2	-0.2
Restricted activity in the last 14 days										
Male	11.2	11.0	0.0	-0.1	0.1	-0.1	-0.3	0.0	0.1	-0.2
Female	13.9	14.0	0.0	0.1	0.0	0.1	-0.1	0.1	0.3	0.1
Total	12.6	12.6	-0.1	-0.1	0.1	0.0	-0.2	0.0	0.3	0.0
Cigarette smoking by sex										
Men										
Current cigarette smokers	21.6	23.2	1.4	1.2	1.2	1.0	1.2	0.9	1.3	1.6
Ex-regular cigarette smokers	29.6	27.1	-2.0	-2.0	-1.4	-1.5	-1.2	-1.1	-1.5	-2.5
Never or (only occasionally) smoked	48.7	49.7	0.6	0.8	0.1	0.5	0.0	0.1	0.3	1.0
Women										
Current cigarette smokers	20.2	20.9	0.4	0.2	0.3	0.4	0.3	0.2	0.5	0.7
Ex-regular cigarette smokers	21.5	20.7	-0.3	-0.3	0.0	-0.2	-0.1	-0.4	-0.4	-0.8
Never or (only occasionally) smoked	58.3	58.4	-0.1	0.1	-0.3	-0.2	-0.2	0.1	-0.1	0.1
Total										
Current cigarette smokers	20.9	22.0	0.8	0.8	0.8	0.7	0.8	0.6	0.9	1.1
Ex-regular cigarette smokers	25.2	23.7	-0.9	-1.0	-0.6	-0.7	-0.7	-0.6	-0.9	-1.5
Never or (only occasionally) smoked	53.9	54.3	0.1	0.3	-0.2	0.1	-0.1	0.1	0.1	0.4
Maximum daily amount of alcohol drank last week by sex										
Men										
Drank nothing last week	27.4	28.9	0.0	0.3	0.0	0.4	0.1	0.6	0.4	1.5
Up to 4 units	33.1	37.8	-1.6	-1.2	-1.1	-1.2	-1.1	-1.3	-1.5	4.7
More than 4 units and up to 8	17.8	15.2	0.0	-0.1	0.1	-0.1	0.2	-0.1	-0.1	-2.6
More than 8 units	21.8	18.1	1.6	1.1	1.1	1.0	0.8	0.8	1.2	-3.7
Women										
Drank nothing last week	42.6	44.1	0.2	0.5	0.2	0.5	0.4	0.7	0.4	1.5
Up to 3 units	24.1	36.1	-0.7	-0.7	-0.5	-0.8	-0.7	-0.8	-0.7	12.0
More than 3 units and up to 6	18.8	11.8	0.1	0.0	0.0	-0.1	-0.1	-0.1	0	-7.0
More than 6 units	14.5	8.0	0.4	0.2	0.3	0.3	0.4	0.1	0.3	-6.5

General Household Survey, 2006
Household and Individual Questionnaires

General Household Survey 2006

Household and Individual Questionnaires

Appendix E

Riaz Ali

Julia Greer

David Matthews

Liam Murray

Simon Robinson

Ghazala Sattar

Office for National Statistics
Government Buildings
Cardiff Rd
Newport
NP10 8XG
Tel: 01633 455877
Email: ghs@ons.gsi.gov.uk

©Crown copyright

General Household Survey, 2006

Household and Individual Questionnaires

2006
GENERAL HOUSEHOLD SURVEY

HOUSEHOLD QUESTIONNAIRE

Area   Information already entered.

Address   Information already entered.

                   1..30

HHold   Information already entered.

                   1..4

StartDat   ENTER DATE INTERVIEW WITH THIS HOUSEHOLD WAS STARTED.

DateChk   IS THIS...

                   The first time you've opened this questionnaire......................................1
                   or the second or later time? 2
                   EMERGENCY CODE IF COMPUTER'S DATE IS
                   WRONG AT LATER CHECK 5

IntEdit   CODE WHETHER THIS IS THE INTERVIEW STAGE, A PROXY CONVERSION OR THE EDIT STAGE.

                   Interview 1
                   Proxy Conversion by telephone (TELEPHONE INTERVIEW UNIT ONLY)....2
                   OFFICE ONLY - EDIT 7

General Household Survey, 2006
Household and Individual Questionnaires

HOUSEHOLD INFORMATION

Information to be collected for all persons in all households

1. Name Who normally lives at this address?

RECORD A NAME / IDENTIFIER FOR EACH MEMBER OF THE HOUSEHOLD

ENTER TEXT OF AT MOST 12 CHARACTERS

2. .Curstat Code the appropriate current status for each household member for this wave.

Resident at this household 1
Under the age of 16 2
Moved, now resident locally details known can interview.................. 3
Moved, resident elsewhere in GB, details known, reallocate...............4
Moved, resident at unknown address……………………………….....5
Was resident here but has died since last call..6
Was resident here but since last call now moved to an institution
(for 6 months or more) 7
Was resident here but since last call now resident abroad
(for 6 months or more)…... 8
Ineligible. mover at GSK, new case created/ no original sampled
Members …………………………………………………………..…9

3. NewPerson Have any additional people joined the household since last wave?

Yes 1
No 2

4. NewName Record the names for each new member of the household.
(If NewPerson = 1)

5. Sex Record sex. ....
Male 1
Female 2

6. Birth What is your date of birth?

FOR DAY NOT GIVEN...........ENTER 15 FOR DAY.
FOR MONTH NOT GIVEN.....ENTER 6 FOR MONTH

Ask those who did not know, or refused to give their date of birth
(Birth = DK OR REFUSAL)

7. AgeIf What was your age last birthday?

98 or more = CODE 97

0..97

Ask if respondent is aged 16 or over
(DVAge > 15)

General Household Survey, 2006
Household and Individual Questionnaires

8. xMarSta ASK OR RECORD
CODE FIRST THAT APPLIES

Are you currently....

single, that is, never married? 1
married and living with your husband/wife? 2
a civil partner in a legally-recognised Civil Partnership 3
married and separated from your husband/wife? 4
divorced? 5
or widowed? 6
SPONTANEOUS ONLY – In a legally-recognised Civil Partnership and
separated his/her civil partner 7
SPONTANEOUS ONLY – Formerly a civil partner, the Civil Partnership
now legally dissolved 8
SPONTANEOUS ONLY – A surviving civil partner: his/her partner having
since died 9

Ask if there is more than one person in the household AND respondent is aged 16 or over AND is single, separated, divorced or widowed
(Household size > 1 & DVAge > 15 & Marstat = 1, 3, 4 or 5)

9. LiveWth ASK OR RECORD

May I just check, are you living with someone in the household as a couple?
Yes 1
No 2
SPONTANEOUS ONLY - same sex couple (but not in a formal
registered Civil Partnership) 3

Ask if there is more than one person in the household, AND the respondent is aged 16 or over
(Household size > 1 & DVAge > 15)

10. Hhldr In whose name is the accommodation owned or rented?
ASK OR RECORD

This person alone 1
This person jointly 3
NOT owner/renter 5

Ask if there is more than one person in the household, AND the accommodation is jointly owned
(Household size > 1 & Hhldr = 3)

11. HiHNum You have told me that...jointly own or rent the accommodation. Which of you/ who has the highest income (from earnings, benefits, pensions and any other sources)?

INTERVIEWER: THESE ARE THE JOINT HOUSEHOLDERS

ENTER PERSON NUMBER - IF TWO OR MORE HAVE SAME INCOME, ENTER 17

1..17

General Household Survey, 2006
Household and Individual Questionnaires

Ask if there is more than one person in the household, AND the joint householders have the same income
(Household size > 1 & HiHNum = 17)

12. JntEldA   ENTER PERSON NUMBER OF THE ELDEST JOINT HOUSEHOLDER FROM THOSE WITH THE SAME HIGHEST INCOME

ASK OR RECORD

1..17

Ask if household size is greater than one, AND the joint householders do not know, or refuse to say who has the greatest income
(Household size > 1 & HiHNum = Don't know or Refusal)

13. JntEldB   ENTER PERSON NUMBER OF THE ELDEST JOINT HOUSEHOLDER

ASK OR RECORD

1..16

Ask all households

14. DVHRPnum   PERSON NUMBER OF HRP. (Computed in Blaise)

Ask if the HRP is married or cohabiting
(HRPnum = 1..14 & MarStat = 2 or LiveWith = 1)

15. HRPPart   THE HRP IS (HRP's NAME)

ENTER THE PERSON NUMBER OF THE HRP's SPOUSE/PARTNER
NO SPOUSE/PARTNER = 17

1..17

Ask all households

16. R   I would now like to ask how the people in your household are related to each other

CODE RELATIONSHIP - ... IS ...'S ....

Spouse 1
Cohabitee..2
Son/daughter (inc. adopted)..3
Step-son/daughter 4
Foster child 5
Son- in -law/daughter - in -law...6
Parent/Guardian..7
Step-parent..8
Foster parent 9
Parent- In - law 10
Brother/sister (inc. adopted) 11
Step-brother/sister...12
Foster brother/sister..13
Brother/sister-in-law...14
Grand-child...15
Grand-parent...16
Other relative..17

Other non-relative 18
Civil Partner 20

17. CheckAdd Is this your correct address?
Yes 1
No 2

General Household Survey, 2006
Household and Individual Questionnaires

ACCOMMODATION TYPE

18. IntroAcc   The next section looks at the standard of people's housing.

All households

19. Accom   IS THE HOUSEHOLD'S ACCOMMODATION:

N.B. MUST BE SPACE USED BY HOUSEHOLD

a house or bungalow 1
a flat or maisonette 2
a room/rooms 3
other 4

Ask if respondents live in a house or bungalow
(Accom = 1)

20. HseType   IS IT (THE HOUSE/BUNGALOW)

detached 1
semi-detached 2
or terraced/end of terrace? 3

Ask if respondents live in a flat or maisonette
(Accom = 2)

21. FltTyp   IS IT (THE FLAT/MAISONETTE)

a purpose-built block 1
a converted house/some other kind of building? 2

Ask if respondents live in a room/rooms
(Accom = 3)

22. DwellNo   Is the apartment or flat….

A building is an independent structure with one or more dwellings enclosed by a roof and external walls.

Each house in a row of terraced houses counts as one building.

Flats with more than one entrance count as one building only if all flats are accessible from each entrance.

In a building with less than 10 dwellings 1
Or in a building with 10 or more dwellings? 2

Ask if respondents said their accommodation was 'something else'
(Accom = 4)

23. AccOth   IS IT (THE ACCOMMODATION)

caravan, mobile home or houseboat 1
or some other kind of accommodation? 2

General Household Survey, 2006
Household and Individual Questionnaires

Ask if respondents live in a flat/maisonette or rooms
(Accom = 2 or 3)

24. DwellNo

IS IT (THE APPARTMENT OR FLAT):

(BUILDING DEFINITION: A BUILDING IS AN INDEPENDENT STRUCTURE WITH ONE OR MORE DWELLINGS ENCLOSED BY A ROOF AND EXTERNAL WALLS. EACH HOUSE IN A ROW OF TERRACED HOUSES COUNTS AS ONE BUILDING. FLATS WITH MORE THAN ONE ENTRANCE COUNT AS ONE BUILDING ONLY IF ALL FLATS ACCESSIBLE FROM EACH ENTRANCE.)

in a building with less than 10 dwellings 1
or in a building with 10 or more dwellings? 2

Ask all households

25. AcProb

Do you have any of the following problems with your accommodation or the area you live in?

IF ASKED: BY 'AREA' I MEAN WITHIN ABOUT A 15-20 MINUTES WALK OR 5-10 MINUTES DRIVE FROM YOUR HOME.

26. Damp

[*] Leaking roof, damp walls/floors, damp foundations, or rotten floorboards or window frames?

Yes 1
No 2

IF ASKED: BY 'AREA' I MEAN WITHIN ABOUT A 15-20 MINUTES WALK OR 5-10 MINUTES DRIVE FROM YOUR HOME.

27. TooDark

[*] Too dark, or not enough light?

THIS QUESTION ASKS ABOUT PROBLEMS WITH ANY OF THE ROOMS BEING TOO DARK/NOT HAVING ENOUGH LIGHT (ON A SUNNY DAY). NOT NECESSARILY ALL OF THE ROOMS IN THE ACCOMMODATION HAVE TO BE DARK.

Yes 1
No 2

IF ASKED: BY 'AREA' I MEAN WITHIN ABOUT A 15-20 MINUTES WALK OR 5-10 MINUTES DRIVE FROM YOUR HOME.

28. Noisy

[*] Noise from neighbours or noise from the street (traffic, business, factories etc.)?

Yes 1
No 2

IF ASKED: BY 'AREA' I MEAN WITHIN ABOUT A 15-20 MINUTES WALK OR 5-10 MINUTES DRIVE FROM YOUR HOME.

29. Pollut   [*] Pollution, grime or other environmental problems in the area caused by traffic or industry?

IF ASKED: BY 'AREA' I MEAN WITHIN ABOUT A 15-20 MINUTES WALK OR 5-10 MINUTES DRIVE FROM YOUR HOME.

Yes 1
No 2

30. Crime   [*] Crime, violence or vandalism in the area?

IF ASKED: BY 'AREA' I MEAN WITHIN ABOUT A 15-20 MINUTES WALK OR 5-10 MINUTES DRIVE FROM YOUR HOME.

Yes 1
No 2

31. Rooms2   I want to ask you about all the rooms you have in your household's accommodation including any rooms you sublet to other people.

How many of the following rooms do you have in this house/flat?

A COMBINED ROOM COUNTS AS ONE ROOM

32. Bedrooms   How many bedrooms do you have?

INCLUDE BEDSITTERS, BOXROOMS OR ATTIC BEDROOMS

0..20

33. KitOver   How many kitchens over 6.5 feet wide do you have?

NARROWIST SIDE MUST BE 6.5 FEET FROM WALL TO WALL

0..20

34. KitUnder   How many kitchens under 6.5 feet wide do you have?

0..20

35. Living   How many living rooms do you have?

INCLUDE DINING ROOMS, SUNLOUNGE OR CONSERVATORY USED ALL YEAR ROUND

0..20

36. BathShow   Have you got either a bath or a shower?

Yes 1
No 2

General Household Survey, 2006
Household and Individual Questionnaires

37. Bathroom

How many bathrooms do you have with plumbed in bath/shower?

0..20

38. FlshToil

Do you have an inside flushing toilet for sole use of the household?

Yes 1
No 2

39. Utility

How many utility and other rooms do you have?

0..20

40. GHSCentH

ASK OR RECORD

Do you have any form of central heating, including electric storage heaters, in your (part of the) accommodation?

Yes 1
No 2

41. Garage

Can I just check, do you have a garage or parking space that is part of the property?

INCLUDE:
- ACCESS/USE OF GARAGE TO PROVIDE PARKING IN CONNECTION WITH THE ACCOMMODATION (FOR EXAMPLE, ACCESS TO A SHARED GARAGE WITH A PARKING SPACE ALLOCATED TO THE FLAT/APARTMENT).

42. GaragPy

Do you pay a separate fee for this garage or parking space?

Yes 1
No (included in rent or part of property) 2

43. CTBand

Could you please tell me which Council Tax band the accommodation is in; Is it in

RUNNING PROMPT

THIS MUST BE THE BAND SET BY THE COUNCIL - DO NOT ACCEPT RESPONDENT'S OWN ESTIMATE OF THE BAND OF THE PROPERTY. IF THIS HOUSEHOLD'S ACCOMMODATION IS NOT VALUED SEPARATELY (EG BECAUSE IT IS RENTED AS PART OF LARGER PREMISES) THEN USE CODE <10>

Band A 1
Band B 2
Band C 3
Band D 4
Band E 5
Band F 6
Band G 7
Band H 8
or Band I..9
Household accommodation not valued separately.............................10

General Household Survey, 2006
Household and Individual Questionnaires

CONSUMER DURABLES

Ask all households

44. IntroDur

Now I'd like to ask you about various household items you may have - this gives us an indication of how living standards are changing.

45. HasDur

Does your household have any of the following items in your (part of the) accommodation?

INCLUDE ITEMS STORED OR UNDER REPAIR.
INCLUDE ITEMS OWNED, RENTED OR ON LOAN.
IF ANY MEMBER POSSESSES AN ITEM, THE HOUSEHOLD POSSESSES IT.

46. TVcol

...Colour TV set?

Yes 1
No 2

Ask if household does not have a colour tv
(TVcol = 2)

47. TVwhy

(You said your household doesn't have a Colour TV). Is that because you...

don't want one 1
would like one but cannot afford it 2
or is there some other reason? 3

Ask all households

48. WashMash

...Washing Machine?

Yes 1
No 2

Ask if household does not have a washing machine
(WashMash = 2)

49. WashWhy

(You said your household doesn't have a washing machine). Is that because you...

don't want one 1
would like one but cannot afford it 2
or is there some other reason? 3

Ask all households

50. Telephon

Telephone?

SHARED TELEPHONES LOCATED IN PUBLIC HALLWAYS TO BE INCLUDED ONLY IF THIS HOUSEHOLD IS RESPONSIBLE FOR PAYING THE ACCOUNT.
INCLUDE MOBILE PHONES.

PROMPT AS NECESSARY

Yes, fixed telephone 1

General Household Survey, 2006
Household and Individual Questionnaires

    Yes, mobile telephone 2
    Yes, fixed and mobile telephone 3
    No 4

Ask if household does not have a telephone
(Telephon = 4)

51. TelWhy   (You said your household doesn't have a telephone). Is that because you...

    RUNNING PROMPT

    don't want one...1
    would like one but cannot afford it 2
    or is there some other reason? 3

Ask all households

52. Computer   Home computer?

    EXCLUDE: VIDEO GAMES

    Yes 1
    No 2

Ask if household does not have a home computer
(Computer = 2)

53. CompWhy   (You said your household doesn't have a computer). Is that because you...

    don't want one...1
    would like one but cannot afford it 2
    or is there some other reason? 3

Ask all households

54. UseVcl   Do you, or any members of your household, at present own or have continuous use of any motor vehicles?

    INCLUDE COMPANY CARS/VANS IF AVAILABLE FOR PRIVATE USE
    EXCLUDE COMPANY CARS/VANS IF PROVIDED ONLY FOR WORK PURPOSES

    Yes 1
    No 2

Ask if the household has use of motor vehicles(s)
(If UseVcl =1)

For each vehicle in turn

55. TypeVcl   FOR EACH VEHICLE IN TURN:
    I would now like to ask about the (Nth) vehicle. Is it…

    a car 1
    a light van..2
    a motor cycle 3
    or some other motor vehicle? 4

General Household Survey, 2006
Household and Individual Questionnaires

56. PrivVcl FOR EACH VEHICLE IN TURN:
Is the vehicle...

privately owned 1
or is it a company vehicle? 2

Ask if it is a company vehicle
(UseVcl = 1 & PrivVcl = 2)

57. WhoCar Enter the person number of whose company car it is

1..16

58. ListPr What was the manufacturer's list price of this vehicle when new, to the nearest £1,000?

IF A PRECISE FIGURE IS NOT AVAILABLE, KEY D/K.

1..99997

Ask list price is unknown
(UseVcl = 1 & PrivVcl = 2 & ListPr = DK)

59. Band SHOWCARD 1

Could you tell me in which of these bands was the list price of this vehicle when new?

Up to £10,000 1
£10,001 to £13,000 2
£13,001 to £16,000 3
£16,001 to £19,000 4
£19,001 to £22,000 5
£22,001 to £25,000 6
£25,001 to £30,000 7
£30,001 to £40,000 8
£40,001 and over 9

Ask if price band is unknown
(UseVcl = 1 & PrivVcl = 2 & ListPr = DK & Band = DK)

60. Model Could you tell me the make and model and engine size of this vehicle?

IF NO INFORMATION ON MAKE/MODEL IS AVAILABLE, TRY TO RECORD AT LEAST THE ENGINE SIZE.

TYPE IN VEHICLE INFO
NO MORE THAN 80 CHARACTERS

Ask if the household has use of motor vehicles(s)
(If UseVcl =1)

61. AnyMore   Do you, or any members of your household, at present own or have continuous use of any more motor vehicles?

INCLUDE COMPANY CARS (IF AVAILABLE FOR PRIVATE USE)

Yes 1
No 2

Ask if household does have a vehicle but not a car or light van
(UseVcl = 1 & DVNumCar = 0)

62. NoCar   This household doesn't have a car or light van. Is that because...

RUNNING PROMPT

don't want one 1
would like one but cannot afford it 2
or is there some other reason 3

Ask if household does not have continuous use of a vehicle
(UseVcl = 2)

63. CarWhy   This household doesn't have a vehicle. Is that because...

RUNNING PROMPT

don't want one 1
would like one but cannot afford it 2
or is there some other reason 3

TENURE

Ask all households

64. Ten1   SHOWCARD 2

In which of these ways do you occupy this accommodation?

MAKE SURE ANSWER APPLIES TO HRP

Own outright 1
Buying it with the help of a mortgage or loan 2
Pay part rent and part mortgage (shared ownership) 3
Rent it 4
Live here rent-free (including rent-free in relative's/friend's
property; excluding squatting) 5
Squatting 6

Ask if household rents the accommodation, or lives there rent-free
(Ten1 = 4 or 5)

65. Tied   Does the accommodation go with the job of anyone in the household?

Yes 1
No 2

General Household Survey, 2006
Household and Individual Questionnaires

66. LLord Who is your landlord?...

CODE FIRST THAT APPLIES

the local authority/council/Scottish Homes 1
a housing association, charitable trust or Local Housing Company 2
employer (organisation) of a household member 3
another organisation 4
relative/friend (before you lived here) of a household member 5
employer (individual) of a household member 6
another individual private landlord 7

67. Furn Is the accommodation provided: ...

RUNNING PROMPT

furnished 1
partly furnished (e.g. carpets and curtains only) 2
or unfurnished 3

Ask all households

68. YearBuy In which year did you (buy this accommodation / sign the first contract to rent this accommodation / move to this address)?

AN ESTIMATE IS ACCEPTABLE
ENTER YEAR IN 4 DIGIT FORMAT, E.G. 2000

1900..2006

Ask of all household members

69. NMoves How many moves have you made in the last five years, not counting moves between places outside Great Britain?

0..97

HOUSING COSTS

Ask if household rents the accommodation
(Ten1 = 4)

70. RentAmt How much rent did your household actually pay last time it was due, after deducting any Housing Benefit (rent rebate)?

ENTER TO THE NEAREST £1 (AFTER HOUSING BENEFIT).
INCLUDE COUNCIL TAX, COUNCIL WATER CHARGE, WATER RATES IF PAID AS PART OF RENT

0.00..9997.00

If RentAmt > 0

General Household Survey, 2006
Household and Individual Questionnaires

71. RentPer   How long did this cover?

one week 1
two weeks 2
three weeks 3
four weeks 4
calendar month 5
two calendar months 7
eight times a year 8
nine times a year 9
ten times a year 10
three months/13 weeks 13
six months/26 weeks 26
one year/12 months/52 weeks 52
less than one week 90
one off/lump sum 95
none of these: EXPLAIN IN A NOTE 97

72. RentHol   Do you have a rent holiday?

Yes 1
No 2

If answered yes to rent holiday
(Ten1 = 4 & RentHol = 1)

73. RentHoWk   For how many weeks a year do you have a rent holiday?

0..52

Ask if household rents the accommodation
(Ten1 = 4)

74. HBen   Some people qualify for Housing Benefit, that is, a rent rebate or allowance.

Do you or anyone else in your household receive Housing Benefit either directly or by having it paid to your landlord on your behalf?

Yes 1
No 2

Ask if receiving Housing Benefit
(Ten1 = 4 & HBen = 1)

75. AmtHB   How much Housing Benefit was allowed for the last rent?

ENTER AMOUNT TO THE NEAREST £1

0..1000

76. PerHB   How long did this cover?

one week 1
two weeks 2
three weeks 3
four weeks 4
calendar month 5
two calendar months 7
eight times a year 8

nine times a year 9
ten times a year 10
three months/13 weeks 13
six months/26 weeks 26
one year/12 months/52 weeks 52
less than one week 90
one off/lump sum..95
none of these: EXPLAIN IN A NOTE 97

Ask if household rents the accommodation
(Ten1 = 4)

77. Servic   SHOWCARD 3

Looking at this card, could you tell me whether you pay extra, that is on top of your rent, for the following items?

ENTER AT MOST 9 VALUES

Water 1
Electricity 2
Gas 3
Liquid or solid fuel (e.g. oil, coke, etc.) 4
Heating, hot water 5
Structural (building) insurance 6
Sewage removal 7
Other service charges 8
Regular maintenance and repairs 9
None of these..10

If household pays extra for services
(Servic = 1-9)

78. ServInc   Could you give me an estimate of how much you pay each month for these services (water, electricity, gas etc)?

A TOTAL FOR ALL SERVICES TOGETHER.
A ROUGH ESTIMATE IS SUFFICIENT, IN POUNDS.

0..9997

Ask if household have a mortgage/loan or part mortgage (shared ownership)
(Ten1 = 2,3)

79. MorgYr   In which year did your present mortgage begin?

1900..2006

80. MorgTyp   SHOWCARD 4

Looking at this card, which one of these options best describes your mortgage?

an endowment mortgage (where your mortgage payments cover interest only), ...1
a repayment mortgage (where your mortgage payments cover interest and part of the original loan),...2
a pension mortgage (where your mortgage payments cover interest

only),	3
or a PEP mortgage, ISA mortgage or Unit Trust mortgage,	4
or both an endowment (or any interest only) mortgage and a repayment mortgage?	5
an interest only mortgage with more than one linked investment (e.g. pension and unit trust, endowment and ISA)	6
an interest only mortgage with no linked investment (e.g. no endowment, pension, PEP or ISA)	7
or another type (not listed above)	8

81. MorgPayL   How much was your last payment on this mortgage/loan?

ENTER THIS AMOUNT TO THIS NEAREST POUND

0..9997

82. MorgPerL   How long did this cover?

one week	1
two weeks	2
three weeks	3
four weeks	4
calendar month	5
two calendar months	7
eight times a year	8
nine times a year	9
ten times a year	10
three months/13 weeks	13
six months/26 weeks	26
one year/12 months/52 weeks	52
less than one week	90
one off/lump sum	95
none of these: EXPLAIN IN A NOTE	97

Ask all households

84. AffIntro   I'd like you to think about your total housing cost.
By that I mean all the bills to do with running a house. That includes your mortgage or rent payments, bills such as gas, electricity, water and heating, house insurance, Council tax payments including sewage, water and refuse removal charges.

85. Burden   [*] To what extent are the total housing costs a financial burden or struggle for your household?

Would you say it is ...

RUNNING PROMPT

a heavy burden/ struggle,	1
a slight burden/ struggle,	2
or not a burden/ struggle at all	3

General Household Survey, 2006
Household and Individual Questionnaires

MIGRATION, CITIZENSHIP, NATIONAL IDENTITY, ETHNICITY

86. IntroCob   The next section has questions on migration and citizenship.

All persons (adult and children) individually

87. Cry1   And in which country were you/was (...) born?

        IF RESPONDENT SAYS BRITAIN, PROBE FOR COUNTRY

        COMMON CODES:

England	1
Scotland	2
Wales	3
Northern Ireland	4
Irish Republic	6
Kenya	14
Ghana	22
Nigeria	23
Jamaica	26
Bangladesh	33
India	34
Pakistan	56
Other	997

Ask if country of birth was 'other'
(Cry1 = 997)

88. CrySp   TYPE IN COUNTRY

        ENTER TEXT OF AT MOST 40 CHARACTERS

89. CryCod   PRESS <SPACE BAR> AND CHOOSE COUNTRY FROM CODING FRAME

        PRESS ENTER TO SELECT A CODE AND ENTER AGAIN TO CONTINUE

        1..144

Ask if not born in the UK
(Cry1 ≠ 1-4)

90. Arruk   In what year did you (...) first arrive in the United Kingdom?

        ENTER IN 4 DIGIT FORMAT E.G.: 2000

        1900..2004

All persons

91. CtzIntro   I've got some questions now about citizenship.

92. CkCtz   Last time you told us that you were eligible to hold a passport for [Citizen]. Is this still the case or not.

Yes, still the case..1
No, not the case..2

93. Citizen
(Wave 1 or
CkCtz = 2)

For what country or countries do you hold, or are entitled to hold, a passport?

"EU PASSPORT": CHECK IF ISSUED IN THE UK OR PROBE 'IN WHAT COUNTRY WAS THE PASSPORT ISSUED?'

CODE FIRST COUNTRY

COMMON CODES:

United Kingdom 1
Irish Republic 6
Ghana 22
Nigeria 23
Jamaica 26
India 34
South Africa 50
Pakistan 56
Poland 79
Other 997

Ask if country of citizenship is 'other'
(Citizen = 997)

94. CtzSp1   TYPE IN COUNTRY

ENTER TEXT OF AT MOST 40 CHARACTERS

95. CtzCode   PRESS <SPACE BAR> AND CHOOSE COUNTRY FROM CODING FRAME

PRESS ENTER TO SELECT A CODE AND ENTER AGAIN TO CONTINUE

CODE FIRST COUNTRY

1..144

96. OthPass   Do you hold or are you entitled to hold a passport for any other country?

Yes 1
No 2

97. Citiz2
(OthPass = 1)

For what other country or countries do you hold, or are entitled to hold, a passport?

"EU PASSPORT": CHECK IF ISSUED IN THE UK OR PROBE 'IN WHAT COUNTRY WAS THE PASSPORT ISSUED?'

COMMON CODES:

United Kingdom 1
Irish Republic 6

Ghana	22
Nigeria	23
Jamaica	26
India	34
South Africa	50
Pakistan	56
Poland	79
Other	997

Ask if second country of citizenship is 'other'
(Citiz2 = 997)

98. CtzSp2   Type in country

ENTER TEXT OF AT MOST 40 CHARACTERS

99. CtzCode2   PRESS <SPACE BAR> AND CHOOSE COUNTRY FROM CODING FRAME

PRESS ENTER TO SELECT A CODE AND ENTER AGAIN TO CONTINUE

CODE SECOND COUNTRY

1..144

All persons

100. FatCob   And in which country was your / (...'s) father born?

IF RESPONDENT SAYS BRITAIN, PROBE FOR COUNTRY

COMMON CODES:

England	1
Scotland	2
Wales	3
Northern Ireland	4
Irish Republic	6
Kenya	14
Ghana	22
Nigeria	23
Jamaica	26
Bangladesh	33
India	34
Pakistan	56
Other	997

Ask if father's country of birth was 'other'
(FatCob = 997)

101. CrySp1   TYPE IN COUNTRY

ENTER TEXT OF AT MOST 40 CHARACTERS

General Household Survey, 2006
Household and Individual Questionnaires

102. CryCd1   PRESS <SPACE BAR> TO ENTER THE CODING FRAME
PRESS ENTER TO SELECT A CODE AND ENTER AGAIN TO CONTINUE

1..144

Ask all persons

103. MotCob   And in which country was your/ (...'s) mother born?

IF RESPONDENT SAYS BRITAIN, PROBE FOR COUNTRY

COMMON CODES:

England 1
Scotland 2
Wales 3
Northern Ireland 4
Irish Republic 6
Ghana 22
Nigeria 23
Jamaica 26
Bangladesh 33
India 34
Pakistan 56
Poland 79

Other 997

Ask if mother's country of birth was 'other'
(MotCob = 997)

104. CrySp2   TYPE IN COUNTRY

ENTER TEXT OF AT MOST 40 CHARACTERS

105. CryCd2   PRESS <SPACE BAR> TO ENTER THE CODING FRAME
PRESS ENTER TO SELECT CODE AND ENTER AGAIN TO CONTINUE

1..144

All persons aged 16 and over

106. Nation   [*] What do you consider your national identity to be? Please choose your answer from this card, choose as many or as few as apply.

SHOWCARD 5(E) in England, 5(S) in Scotland, 5(W) in Wales

English 1
Scottish 2
Welsh 3
Irish 4
British 5
Other 6

If answered other
(Nation = 6)

107. NatSpec   [*] How would you describe your national identity?

                      ENTER DESCRIPTION OF NATIONAL IDENTITY

                      ENTER TEXT OF AT MOST 40 CHARACTERS

All persons

108. Ethnic   [*] To which of these ethnic groups do you consider you belong?

                      SHOWCARD 6

                      White British 1
                      Any other White background 2
                      Mixed - White and Black Caribbean 3
                      Mixed - White and Black African 4
                      Mixed - White and Asian 5
                      Any other Mixed background..6
                      Asian or Asian British - Indian..7
                      Asian or Asian British - Pakistani 8
                      Asian or Asian British - Bangladeshi 9
                      Asian or Asian British - Any other Asian background........................10
                      Black or Black British – Black Caribbean..11
                      Black or Black British – Black African..12
                      Black or Black British - Any other Black background........................13
                      Chinese 14
                      Any other...15

Ask those who describe themselves as:
  Any other White background
  Any other Mixed background
  Any other Asian background
  Any other Black background
  Any other
(Ethnic = 2, 6, 10, 13 or 15)

109. Ethdes   [*] Please can you describe your ethnic group.

                      ENTER DESCRIPTION OF ETHNIC GROUP

                      ENTER TEXT OF AT MOST 40 CHARACTERS

Answer for all households

110. SelPer   CODE PERSON NO. OF RESPONDENT WHO ANSWERED THE
                      HOUSEHOLD QUESTIONS

                      CODE ONE PERSON ONLY

                      1..17

General Household Survey, 2006
Household and Individual Questionnaires

111. Selcheck   YOU HAVE INDICATED PERSON NO. (…)
THIS IS (…).

IF THIS IS NOT CORRECT, GO BACK AND CHANGE PERSON NO. IN SELPER ABOVE

* END OF HOUSEHOLD QUESTIONNAIRE *

INDIVIDUAL QUESTIONNAIRE

Ask this section of all adults

1. Iswitch   THIS IS WHERE YOU START RECORDING ANSWERS FOR INDIVIDUALS.
DO YOU WANT TO RECORD ANSWERS FOR (name) NOW OR LATER?

Yes, now 1
Later 2
or is there no interview with this person................................... 3

Ask if answers are to be recorded now
(Iswitch = 1)

2. PersProx   INTERVIEWER: IS THE INTERVIEW ABOUT (name) BEING GIVEN:

In person 1
or by someone else 2

Ask if answers are to be recorded now, but are being answered by someone else
(Iswitch = 1 & PersProx = 2)

3. ProxyNum   ENTER PERSON NUMBER OF PERSON GIVING THE
INFORMATION

1..16

EMPLOYMENT

Ask this section of all adults

1. Wrking   Did you do any paid work in the 7 days ending Sunday the (*date*), either as an employee or as self-employed?

Yes 1
No 2

Ask if respondent is not in paid work and is a man aged 16-64, or a woman aged 16-62
(Wrking = 2 & (man aged 16-64 or woman aged 16-62))

General Household Survey, 2006
Household and Individual Questionnaires

2. SchemeET   Were you on a government scheme for employment training?

   Yes 1
   No 2

Ask if on a government scheme
(SchemeET = 1)

3. Trn   Last week were you…

   CODE FIRST THAT APPLIES

   with an employer, or on a project providing work experience or practical
   training 1
   or at a college or training centre 2

Ask if not in paid work AND not on a government scheme for employment training
(Wrking = 2 & (SchemeET = 2 or not asked SchemeET because not in the age bracket asked))

4. JbAway   Did you have a job or business that you were away from?

   Yes 1
   No 2
   Waiting to take up a new job/business already obtained 3

Ask if not in paid work AND not on a government scheme for employment training AND not away from a job
(JbAway = 2 or 3)

5. OwnBus   Did you do any unpaid work in that week for any business that you own?

   Yes 1
   No 2

Ask if the respondent did not do any unpaid work for a business that they own
(OwnBus = 2)

6. RelBus   …or that a relative owns?

   Yes 1
   No 2

Ask if not in paid work AND not on a government scheme for employment training AND not doing unpaid work
(RelBus = 2)

7. Looked   Thinking of the 4 weeks ending Sunday the (date last Sunday), were you looking for any kind of paid work or government training scheme at any time in those 4 weeks?

   Yes 1
   No 2
   SPONTANEOUS Waiting to take up a new job or business already
   Obtained 3

Ask if looking for paid work   OR waiting to take up a new job or business already obtained
(Looked = 1 or 3 OR JbAway = 3)

General Household Survey, 2006
Household and Individual Questionnaires

8. StartJ

If a job or a place on a government scheme had been available in the week ending Sunday the (date), would you have been able to start within 2 weeks?

Yes 1
No 2

Ask if looking for paid work OR waiting to take up a new job or business already obtained
(Looked = 1 or 3 OR JbAway = 3)

9. LKTime

How long have you been/were you looking for paid work/a place on a government training scheme?

Not yet started..1
Less than 1 month..2
1 month but less than 3 months 3
3 months but less than 6 months..4
6 months but less than 12 months..5
12 months or more...6

Ask if not looking for paid work, and would not be able to start work or training within 2 weeks
(Looked = 2 or StartJ = 2)

10. YInAct

What was the main reason you did not seek any work in the last 4 weeks/would not be able to start in the next 2 weeks?

Student 1
Looking after the family/home 2
Temporarily sick or injured 3
Long-term sick or disabled 4
Retired from paid work 5
None of these..6

Ask if not in paid work
(Dvilo3a = 2,3)

11. Everwk

Have you ever had a paid job, apart from casual or holiday work?

Yes 1
No 2

Ask if not in paid work, but has worked before
(Everwk = 1)

12. DtJbL

When did you leave your last PAID job?

FOR DAY NOT GIVEN................ENTER 15 FOR DAY
FOR MONTH NOT GIVEN..........ENTER 6 FOR MONTH

ENTER DATE

Ask those who are in current employment or have had a job in the past

13. IndD

CURRENT OR LAST JOB

What did the firm/organisation you worked for mainly make or do (at the place where you worked)?

DESCRIBE FULLY - PROBE MANUFACTURING or PROCESSING or DISTRIBUTING ETC. AND MAIN GOODS PRODUCED, MATERIALS USED, WHOLESALE or RETAIL ETC.

ENTER TEXT AT MOST 80 CHARACTERS

14. OccT   CURRENT OR LAST JOB

What was your (main) job (in the week ending Sunday the (n))?

ENTER TEXT AT MOST 30 CHARACTERS

15. OccD   CURRENT OR LAST JOB

What did you mainly do in your job?
RECORD SPECIAL QUALIFICATIONS/TRAINING NEEDED TO DO THE JOB

ENTER TEXT AT MOST 80 CHARACTERS

16. Stat   Were you working as an employee or were you self-employed?

Employee 1
Self-employed 2

Ask if employee
(Stat = 1)

17. Svise   In your job, did you have formal responsibility for supervising the work of other employees?

DO NOT INCLUDE PEOPLE WHO ONLY SUPERVISE:
- CHILDREN, E.G. TEACHERS, NANNIES, CHILDMINDERS
- ANIMALS
- SECURITY OF BUILDINGS, E.G. CARETAKERS, SECURITY GUARDS

Yes 1
No 2

Ask if responsible for supervising
(Svise = 1)

18. SviseDsc   Please describe the type of responsibility you have for supervising the work of other employees.

PROBE FOR WHO AND WHAT IS BEING SUPERVISED

ENTER A TEXT OF AT MOST 100 CHARACTERS

Ask if employee
(Stat = 1)

19. NEmplee   How many people worked for your employer at the place where you worked – were there......

RUNNING PROMPT

1-9	1
10	2
11-24	3
25-99	4
100-499	5
500-999	6
1000 or more?	7

Ask if 1-9 people work for employer
(NEmplee = 1)

20. NEmp1to9   Can I just check the exact number of people who worked for your employer (at the place where you worked)?

1..9

Ask if self-employed
(Stat = 2)

21. Solo   Were you working on your own or did you have employees?

on own/with partner(s) but no employees 1
with employees 2

Ask if self-employed with employees
(Solo = 2)

22. SNEmplee   How many people did you employ at the place where you worked – were there.....

RUNNING PROMPT

1-9	1
10	2
11-24	3
25-99	4
100-499	5
500-999	6
1000 or more?	7

Ask if 1-9 people work for respondent
(SNEmplee = 1)

23. SEmp1to9   Can I just check the exact number of people you employed (at the place where you worked)?

1..9

24. JbChnge   So may I just check, have you had a change of job since the date of the last interview *(date)* ?

BY 'CHANGE OF JOB' I MEAN A CHANGE IN THE NATURE OF THE ACTIVITY PERFORMED.

Yes 1
No 2

**Ask if had
change of job**

(JbChnge= 1)

25. YJbChnge   For what reason did you make your last change of job?

    To take up or seek better job 1
    End of temporary contract 2
    Obliged to stop by employer (i.e. redundancy, business closure, early
    retirement, dismissal etc) 3
    Sale or closure of family business 4
    Child care or other dependent...5
    Partner's job required move to another area or marriage 6
    Other reason 7

26. JobBYr   Which year did you begin your first regular job?

ENTER YEAR IN 4 DIGIT FORMAT E.G. 2005

THE FIRST JOB SHOULD HAVE INVOLVED WORKING FOR 15 HOURS OR MORE PER WEEK AND HAVE LASTED FOR AT LEAST SIX MONTHS (UNLESS IT ENDED BEFORE SIX MONTHS TO START ANOTHER JOB OR PERIOD OF UNEMPLOYMENT).

EXCLUDE: PART-TIME JOBS UNDERTAKEN WHILE AT SCHOOL OR UNIVERSITY.
             - HOLIDAY JOBS UNDERTAKEN DURING THE SCHOOL/UNIVERSITY HOLIDAYS (THAT IS JOBS FROM WHICH STUDENTS WILL RETURN TO THEIR STUDIES IN TERM-TIME.

27. JobBMon   Which month did you begin your first regular job?

ENTER MONTH

    January 1
    February...2
    March 3
    April 4
    May 5
    June 6
    July 7
    August 8
    September..9
    October 10
    November 11
    December...12

28. FtWk   Looking back to the time when you began your first regular job (*date*), how many years since then have you spent in paid FULL-TIME work?

ENTER TO THE NEAREST WHOLE YEAR.

FULL-TIME WORKING IS DEFINED AS MORE THAN 30 HOURS A WEEK (EXCEPT IN EDUCATION OCCUPATIONS WHERE MORE THAN 26 HOURS A WEEK IS INCLUDED AS FULL-TIME).

29. PtWk   Looking back to the time when you began your first regular job (*date*), how many years since then have you spent in paid PART-TIME work?

General Household Survey, 2006
Household and Individual Questionnaires

ENTER TO THE NEAREST WHOLE YEAR.

HOLDING TWO PART-TIME JOBS AT ONCE STILL COUNTS AS BEING IN PART-TIME WORK. IF A PART-TIME AND FULL-TIME JOB HELD CONCURRENTLY, ONLY COUNT YEARS IN FULL-TIME JOB TO AVOID DOUBLE COUNTING. PART-TIME WORKING IS DEFINED AS MORE THAN 15 HOURS A WEEK.

Ask those who are in current employment or have had a job in the past

30. FtPtWk   In your (main) job were you working…

RUNNING PROMPT

full time 1
or part time 2

Ask if employee
(Stat = 1 & (Wrking = 1 OR SchemeET = 1 OR JbAway =1))

31. EmpStY   In which year did you start working continuously for your current employer?

1900..2006

Ask if self-employed
(Stat = 2 & (Wrking = 1 OR SchemeET = 1 OR JbAway =1))

32. SEmpStY   In which year did you start working continuously as a self-employed person?

1900..2006

If less than or equal to 8 years since started working continuously for current employer/ as a self-employed person?
(EmpStY ≤ 8 less than the present date   OR   SEmpStY ≤ 8 less than the present date)

33. JobstM   and which month in (YEAR) was that?

0..12

Ask If employee
(Stat = 1 & (Wrking = 1 OR SchemeET = 1 OR JbAway =1))

34. EmpContr   SHOWCARD 7

Which of the following best describes your employment contract in your main job?

A permanent job (contract of unlimited duration) 1
Temporary job (work contract of limited duration) 2
Work without contract 3
Other working arrangement 4

Ask if in paid work, on a government scheme, or away from a job/business (not asked of proxy respondents)
(Wrking = 1 OR SchemeET = 1 OR JbAway =1)

35. EverOT Do you ever do any work which you would regard as paid or unpaid overtime?

HOURS IN MAIN JOB ONLY

Yes 1
No 2

Ask if respondent never does overtime, or if did unpaid work for own/relatives business
(EverOT = 2 OR OwnBus = 1 OR RelBus = 1)

36. Totus1 Thinking of your (main) job/business, how many hours per week do you usually work – please exclude meal breaks and overtime.

THE HOURS SHOULD BE ROUNDED TO THE NEAREST 15 MINUTES, WITH PART HOURS AS DECIMALS, FOR EXAMPLE 36 HOURS 30 MINUTES WOULD BE RECORDED AS 36.5, 40 HOURS 45 MINUTES WOULD BE RECORDED AS 40.75.
97 OR MORE = 97

0.00..97.00

Ask if respondent does overtime
(EverOT = 1)

37. Usuhr Thinking of your main job/business, how many hours per week do you usually work, please exclude mealbreaks and overtime?

THE HOURS SHOULD BE ROUNDED TO THE NEAREST 15 MINUTES, WITH PART HOURS AS DECIMALS, FOR EXAMPLE 36 HOURS 30 MINUTES WOULD BE RECORDED AS 36.5, 40 HOURS 45 MINUTES WOULD BE RECORDED AS 40.75.
97 OR MORE = 97

0.00..97.00

38. PotHr How many hours paid overtime do you usually work per week?

THE HOURS SHOULD BE ROUNDED TO THE NEAREST 15 MINUTES, WITH PART HOURS AS DECIMALS, FOR EXAMPLE 36 HOURS 30 MINUTES WOULD BE RECORDED AS 36.5, 40 HOURS 45 MINUTES WOULD BE RECORDED AS 40.75.
97 OR MORE = 97

0.00..97.00

39. UotHr How many hours unpaid overtime do you usually work per week?

THE HOURS SHOULD BE ROUNDED TO THE NEAREST 15 MINUTES, WITH PART HOURS AS DECIMALS, FOR EXAMPLE 36 HOURS 30 MINUTES WOULD BE RECORDED AS 36.5, 40 HOURS 45 MINUTES WOULD BE RECORDED AS 40.75.
97 OR MORE = 97

0.00..97.00

40. AgreeHrs   Your total usual hours come to …. Is that about right or not?

97 OR MORE = 97

Yes 1
No 2

Ask if did unpaid work for own/relatives business
(OwnBus = 1 OR RelBus = 1)

41. UnPaidHr   Thinking of the business that you did unpaid work for, how many hours per week unpaid work do you usually do for that business?

1..97

Ask all working (not asked of proxy respondents)
(Wrking = 1 OR JbAway = 1 OR SchemeET = 1 OR OwnBus = 1 OR RelBus = 1)

42. MoreJbs   Do you normally have more than one paid job in addition to the one you have just told me about?

EXCLUDE WORK AS BABY-SITTERS OR MAIL ORDER AGENTS

Yes 1
No 2

Ask if respondent has more than one paid job
(MoreJbs = 1)

43. SecondJb   What is your second job?

ENTER JOB TITLE

ENTER A TEXT OF AT MOST 80 CHARACTERS

44. EmpSE2nd   Were you working as an employee or were you self-employed?

Employee 1
Self-employed 2

45. OthJbs   Do you have any other paid jobs?

Yes 1
No 2

Ask if respondent has a third paid job
(OthJbs = 1)

46. ThirdJb   What is your third job?

ENTER JOB TITLE

ENTER A TEXT OF AT MOST 80 CHARACTERS

47. EmpSE3rd   Were you working as an employee or were you self-employed?

Employee 1
Self-employed 2

Ask if respondent has more than one job
(MoreJbs = 1)

48. TotHrOth How many hours per week do you usually work in your second (and third) jobs - please exclude meals breaks?

97 OR MORE = 97

0.00..97.00

Ask if respondent works but, in total, less than 30 hours a week
((Wrking = 1 OR JbAway = 1 OR SchemeET = 1 OR OwnBus = 1 OR RelBus = 1)
& Totus1 + Usuhr + PotHr + Uothr + UnpaidHr + TotHrOth < 30)

49. LThan30 The total number of hours you work per week on average is less than 30 hours.
What is your main reason for working less than 30 hours a week?

Undergoing education or training 1
Personal illness or disability 2
Want to work more hours but cannot find a job or work for more hours
....3
Do not want to work more hours 4
Considers number of hours in job(s) as full-time 5
Housework, looking after children or other dependant 6
Other reason 7

Ask all respondents, except proxy

50. EcStatus SHOWCARD 8

[*] I'm going to ask you about what you've been doing over the past 12 months, but first, can I just check, which of these categories best describes you at present?

Working full-time...1
Working part-time 2
Unemployed 3
Student (incl. pupil at school, those in training) 4
Looking after family home 5
Long-term sick or disabled 6
Retired from paid work 7
Not in paid work for some other reason 8

51. SameSit SHOWCARD 8

Can I just check, has your situation changed in the last 12 months (that is since (date))?

Yes 1
No 2

If situation has changed
(SameSit = 1)

52. Mths12	SHOWCARD 8

What were you doing 12 months ago, that is in (month) last year?

Working full-time..1
Working part-time 2
Unemployed 3
Student (incl. pupil at school, those in training) 4
Looking after family home 5
Long-term sick or disabled 6
Retired from paid work 7
Not in paid work for some other reason 8

53. Mths11	SHOWCARD 8

What were you doing 12 months ago, that is in (month)?

Working full-time..1
Working part-time 2
Unemployed 3
Student (incl. pupil at school, those in training) 4
Looking after family home 5
Long-term sick or disabled 6
Retired from paid work 7
Not in paid work for some other reason 8

54. Mths10	SHOWCARD 8

What were you doing 10 months ago, that is in (month)?

Working full-time..1
Working part-time 2
Unemployed 3
Student (incl. pupil at school, those in training) 4
Looking after family home 5
Long-term sick or disabled 6
Retired from paid work 7
Not in paid work for some other reason 8

55. Mths9	SHOWCARD 8

What were you doing 9 months ago, that is in (month)?

Working full time...1
Working part-time 2
Unemployed 3
Student (incl. pupil at school, those in training) 4
Looking after family home 5
Long-term sick or disabled 6
Retired from paid work 7
Not in paid work for some other reason 8

56. Mths8	SHOWCARD 8

What were you doing 8 months ago, that is in (month)?

Working full-time..1
Working part-time 2
Unemployed 3

Student (incl. pupil at school, those in training) 4
Looking after family home 5
Long-term sick or disabled 6
Retired from paid work 7
Not in paid work for some other reason 8

57. Mths7   SHOWCARD 8

What were you doing 7 months ago, that is in (month)?

Working full-time..1
Working part-time 2
Unemployed 3
Student (incl. pupil at school, those in training) 4
Looking after family home 5
Long-term sick or disabled 6
Retired from paid work 7
Not in paid work for some other reason 8

58. Mths6   SHOWCARD 8

What were you doing 6 months ago, that is in (month)?

Working full-time..1
Working part-time 2
Unemployed 3
Student (incl. pupil at school, those in training) 4
Looking after family home 5
Long-term sick or disabled 6
Retired from paid work 7
Not in paid work for some other reason 8

59. Mths5   SHOWCARD 8

What were you doing 5 months ago, that is in (month)?

Working full-time..1
Working part-time 2
Unemployed 3
Student (incl. pupil at school, those in training) 4
Looking after family home 5
Long-term sick or disabled 6
Retired from paid work 7
Not in paid work for some other reason 8

60. Mths4   SHOWCARD 8

What were you doing 4 months ago, that is in (month)?

Working full-time..1
Working part-time 2
Unemployed 3
Student (incl. pupil at school, those in training) 4
Looking after family home 5
Long-term sick or disabled 6
Retired from paid work 7
Not in paid work for some other reason 8

61. Mths3   SHOWCARD 8

What were you doing 3 months ago, that is in (month)?

Working full-time 1
Working part-time 2
Unemployed 3
Student (incl. pupil at school, those in training) 4
Looking after family home 5
Long-term sick or disabled 6
Retired from paid work 7
Not in paid work for some other reason 8

62. Mths2   SHOWCARD 8

What were you doing 2 months ago, that is in (month)?

Working full-time 1
Working part-time 2
Unemployed 3
Student (incl. pupil at school, those in training) 4
Looking after family home 5
Long-term sick or disabled 6
Retired from paid work 7
Not in paid work for some other reason 8

63. Mths1   SHOWCARD 8

What were you doing 1 month ago, that is in (month)?

Working full-time 1
Working part-time 2
Unemployed 3
Student (incl. pupil at school, those in training) 4
Looking after family home 5
Long-term sick or disabled 6
Retired from paid work 7
Not in paid work for some other reason 8

PENSIONS

The whole section on pensions (apart from the last question) is only asked of those in paid work, (including those temporarily away from job or on a government scheme), but excluding unpaid family workers.
((Wrking = 1   OR   JbAway = 1   OR   SchemeET = 1)   &   (OwnBus = 2   &   RelBus = 2))

The routing instructions above each question apply only to those who meet the above criteria.

If employee or on a government scheme
(Stat = 1   OR   SchemeET = 1)

General Household Survey, 2006

Household and Individual Questionnaires

1. PenSchm (Thinking now of your present job,) some people (will) receive a pension from their employer when they retire, as well as the state pension.
Does your present employer run an occupational pension scheme or superannuation scheme for any employees?

INCLUDE CONTRIBUTORY AND NON-CONTRIBUTORY SCHEMES
EXCLUDE EMPLOYER SPONSORED GROUP PERSONAL PENSION AND STAKEHOLDER PENSIONS

Yes 1
No 2

Ask if employer runs an occupational pension scheme
(PenSchm = 1)

2. Eligible Are you eligible to belong to your employer's occupational pension scheme?

Yes 1
No 2

Ask if eligible for employer's pension scheme
(Eligible = 1)

3. EmPenShm Do you belong to your employer's occupational pension scheme?

Yes 1
No 2

Ask if did not know or refused to say whether the employer offered an occupational pension scheme, or whether they were eligible, or whether they belonged to one
(PenSchm OR Eligible OR EmPenShm = DK / refusal)

4. PSchPoss So do you think it's possible that you belong to an occupational pension scheme run by your employer, or do you definitely not belong to one?

Possibly belongs 1
Definitely not... 2

Ask if employee OR (under pensionable age and not self-employed) - this is to select those who may have answered don't know, or refused to answer Stat
(Stat = 1 OR (under pensionable age & Stat ≠ 2))

5. PersPnt1 INTERVIEWER - INTRODUCE IF NECESSARY.
Now I would like to ask you about other types of pension arrangements (rather than employers' occupational pension schemes).

6. PersPe SHOWCARD 9

Do you at present have any of the pension arrangements shown on this card?

CODE ALL THAT APPLY

INTERVIEWER: CODE 4 ALSO INCLUDES ARRANGEMENTS WHERE SPOUSE OR SOMEONE OTHER THAN EMPLOYER HAS SET UP THE STAKEHOLDER PENSION.

A personal or private pension or retirement annuity 1
A Group Personal Pension (that is a collection of personal pensions

arranged by an employer for a group of employees) 2
A Stakeholder Pension arranged through your employer
(who may or may not contribute) 3
A Stakeholder Pension you arranged yourself 4
None of these 5
Don't know 6

Ask if has a personal/private pension and if employee (not asked of proxies)
(Stat = 1 AND PersPen = 1)

7. PPECont   Does your employer contribute to your personal/private pension?

                Yes 1
                No 2

Ask if has a personal/private pension and if employee OR (under pensionable age and not self-employed) (not asked of proxies)
(Stat = 1 OR (under pensionable age AND Stat ≠ 2) & PersPen = 1)

8. PPPCont   Do you currently make any contributions to your personal/private pension?

                Yes 1
                No 2

Ask if respondent currently makes contributions to Group Personal Pension (not asked of proxies)
(PPPCont = 1)

9. PenAmnt   In total, how much do you currently contribute each month to your personal/private pension(s)?

                IF HAVE MORE THAN ONE POLICY, SUM UP ALL POLICIES

                0.00..99997.00

10. PenPer   What period does this cover?

                one week 1
                two weeks 2
                three weeks 3
                four weeks 4
                calendar month 5
                two calendar months 6
                eight times a year 7
                nine times a year 8
                ten times a year 9
                three months/13 weeks 10
                six months/26 weeks 11
                one year/12 months/52 weeks 12
                less than one week 13
                one off/lump sum 14
                none of these: EXPLAIN IN A NOTE 15

11. PPGov   Since the date of the last interview (DATE) has any money been paid into your personal/private pension by HM Revenue and Customs (formerly the Inland Revenue)?

                Yes 1
                No 2

Don't know 3

Ask if has a Group Personal Pension and if employee (not asked of proxies)
(Stat = 1 AND PersPen = 2)

12. GPECont   Does your employer contribute to your Group Personal Pension?

                         Yes 1
                         No 2

Ask if has a Group Personal Pension and if employee OR (under pensionable age and not self-employed) (not asked of proxies)
(Stat = 1 OR (under pensionable age AND Stat ≠ 2) & PersPen = 2)

13. GPPCont   Do you currently make any contributions to your Group Personal Pension?

                         Yes 1
                         No 2

14. GPGov   Since the date of the last interview (DATE) has any money been paid into your Group Personal Pension by the HM Revenue and Customs (formerly the Inland Revenue)?

                         Yes 1
                         No 2
                         Don't know 3

Ask if has a Stakeholder Pension arranged through employer and if employee (not asked of proxies)
(Stat = 1 AND PersPen = 3)

15. SEECont   Does your employer contribute to your Stakeholder Pension arranged through your employer?

                         Yes 1
                         No 2

Ask if has a Stakeholder Pension arranged through employer and if employee OR (under pensionable age and not self-employed) (not asked of proxies)
(Stat = 1 OR (under pensionable age AND Stat ≠ 2) AND PersPen = 3)

16. SEPCont   Do you currently make any contributions to your Stakeholder Pension arranged through your employer?

                         Yes 1
                         No 2

17. SEGov   Since the date of the last interview (DATE) has any money been paid into Stakeholder Pension arranged through your employer by HM Revenue and Customs (formerly the Inland Revenue)?

                         Yes 1
                         No 2
                         Don't know 3

Ask if has a Stakeholder Pension arranged by self and if employee (not asked of proxies)
(Stat = 1 AND PersPen = 4)

18. SPECont   Does your employer contribute to your Stakeholder Pension you arranged yourself?

Yes 1
No 2

Ask if has a Stakeholder Pension arranged by self and if employee OR (under pensionable age and not self-employed) (not asked of proxies)
(Stat = 1 OR (under pensionable age AND Stat ≠ 2) AND PersPen = 4)

19. SPPCont   Do you currently make any contributions to your Stakeholder Pension you arranged yourself?

Yes 1
No 2

20. SPGov   Since the date of the last interview (DATE) has any money been paid into your Stakeholder Pension you arranged yourself by HM Revenue & Customs (formerly the Inland Revenue)?

Yes 1
No 2
Don't know 3

Ask if self-employed
(Stat = 2)

21. PersPnt2   INTERVIEWER - INTRODUCE IF NECESSARY.
Now I would like to ask you about personal pension schemes.

22. SePrsPen relief   Self-employed people may arrange pensions for themselves and get tax

on their contributions. These schemes include personal pensions, stakeholder pensions and 'self-employed pensions' (sometimes called 'Section 226 Retirement Annuities').

Do you at present contribute to one of these schemes?

Yes 1
No 2

Ask if contributes to one of the schemes and is not a proxy respondent
(SePrsPen =1 AND Persprox=1)

23. SePrsS   Which types of scheme are you contributing to – personal pension, stakeholder pension, or some other scheme?

CODE ALL THAT APPLY

Personal pension..1
Stakeholder pension...2
Other 3

Ask if respondent is contributing to a personal pension
(SePrsS = 1)

General Household Survey, 2006

Household and Individual Questionnaires

24. SEPPPAmt   In total how much do you contribute each month to your PERSONAL pension(s)?

IF HAVE MORE THAN ONE POLICY, SUM UP ALL POLICIES

0.01..99999.00

25. SEPPPer   What period does this cover?

one week 1
two weeks 2
three weeks 3
four weeks 4
calendar month 5
two calendar months 7
eight times a year 8
nine times a year 9
ten times a year 10
three months/13 weeks 13
six months/26 weeks 26
one year/12 months/52 weeks 52
less than one week 90
one off/lump sum 95
none of these: EXPLAIN IN A NOTE 97

Ask if does not, or does not know if they contribute to one of the above schemes and is not a proxy respondent
(SePrsPen = 2 or DK AND Persprox=1)

26. SeEvPers   Have you ever contributed to one of these schemes?

Yes 1
No 2

This question is asked of anyone under pensionable age who is not currently in paid work
(Under pensionable age & Wrking ≠ 1 & JbAway ≠ 1 & SchemeET ≠ 1 & OwnBus ≠ 1 & RelBus ≠1)

27. NewShp   Since April 2001, anyone can take out a stakeholder or other personal pension.

Do you at present have one of these types of pension for yourself?

PROMPT AS NECESSARY

Yes, stakeholder 1
Yes, other personal pension 2
No 3

Ask if yes at NewShp and is not a proxy respondent
(NewShp=1 OR 2 AND Persprox =1)

28. NewShpc   Have you or anyone else made any contributions to (this/either) pension in the last 12 months?

Yes 1
No 2

General Household Survey, 2006
Household and Individual Questionnaires

Ask if yes at NewShpc
(NewShpc = 1)

29. TotLst12 How much have you or anyone else contributed to your personal pension over the last 12 months?

IF HAVE MORE THAN ONE POLICY, SUM UP ALL POLICIES

0.01..99999.00

30. PpenPer What period does this cover?

one week 1
two weeks 2
three weeks 3
four weeks 4
calendar month 5
two calendar months 7
eight times a year 8
nine times a year 9
ten times a year 10
three months/13 weeks 13
six months/26 weeks 26
one year/12 months/52 weeks 52
less than one week 90
one off/lump sum 95
none of these: EXPLAIN IN A NOTE 97

EDUCATION

Ask this section of adults 16 and above (it is not asked of proxies)

1. QualCh I would now like to ask you about education and work-related training. Do you have any qualifications...

INDIVIDUAL PROMPT – CODE ALL THAT APPLY

from school, college or university, ... 1
connected with work 2
or from government schemes 3
No qualifications (Spontaneous only) 4
Don't know (Spontaneous only) 5

Ask if respondent has a qualification, or answers don't know
(QualCh = 1, 2, 3 or 5)

2. Quals SHOWCARD 10

Which qualifications do (you think) you have, starting with the highest qualifications?

CODE ALL THAT APPLY - PROMPT AS NECESSARY

Degree level qualification including foundation degrees, graduate
   membership of a professional institute or PGCE or higher 1

General Household Survey, 2006
Household and Individual Questionnaires

Diploma in higher education	2
HNC/HND	3
ONC/OND	4
BTEC/ BEC/ TEC / EdExcel	5
SCOTVEC/ SCOTEC/ SCOTBEC (Scotland)	6
Teaching qualification (excluding PGCE)	7
Nursing or other medical qualification not yet mentioned	8
Other higher education qualification below degree level	9
A level / Vocational A-level or equivalent	10
Highers (Scotland)	11
NVQ/SVQ	12
GNVQ/GSVQ	13
AS level /Vocational AS level or equivalent	14
Advanced Highers or Certificate of Sixth Year Studies (CSYS) (Scotland)	15
Access to HE	16
O level or equivalent	17
Intermediate 2 NQs (Scotland)	18
Intermediate 1 NQs (Scotland)	19
Standard Grade or O Grade (Scotland)	20
GCSE / Vocational GCSE	21
CSE	22
National Qualifications (including SGA) (Scotland)	23
RSA/ OCR	24
City and Guilds	25
YT Certificate/YTP	26
Key Skills/ Basic Skills	27
Entry Level Qualifications	28
Any other professional/vocational qualifications/ foreign qualifications	29
Don't know	30

Ask if has NVQ/SVQ
(Quals = 12 AND does NOT have a higher qualification)

3. NVQlev   What is your highest level of full NVQ/SVQ?

Level 1	1
Level 2	2
Level 3	3
Level 4	4
Level 5	5
Don't know	6

Ask if highest qualification is a degree level qualification
(Quals = 1 AND does NOT have a higher qualification)

4. Degree   Is your degree…

CODE FIRST THAT APPLIES

a higher degree (including PGCE)	1
a first degree	2
a foundation degree	3
other (e.g. graduate member of a professional institute or chartered accountant)	4
Don't know	5

Ask if has a higher degree

General Household Survey, 2006
Household and Individual Questionnaires

(Degree = 1)

5. HighO　　　　　　ASK OR RECORD

　　　　　　　　　　Was your higher degree…

　　　　　　　　　　CODE FIRST THAT APPLIES

　　　　　　　　　　a Doctorate 1
　　　　　　　　　　a Masters...2
　　　　　　　　　　a Postgraduate Certificate in Education..3
　　　　　　　　　　or some other postgraduate degree or professional qualification 4
　　　　　　　　　　Don't know 5

Ask if highest qualification is a teaching qualification excluding PGCE
(Quals = 7　AND　does NOT have a higher qualification)

6. Teach　　　　　　Was your teaching qualification for…

　　　　　　　　　　CODE ALL THAT APPLY

　　　　　　　　　　Further education..1
　　　　　　　　　　Key Stage 4 2
　　　　　　　　　　Key Stage 3 3
　　　　　　　　　　Key Stage 2 4
　　　　　　　　　　Key Stage 1 5
　　　　　　　　　　Foundation Stage 6
　　　　　　　　　　Don't know..7

Ask if highest qualification is RSA/OCR
(Quals = 24　AND　does NOT have a higher qualification)

7. RSA　　　　　　　Is your highest RSA…

　　　　　　　　　　CODE FIRST THAT APPLIES

　　　　　　　　　　a higher diploma..1
　　　　　　　　　　an advanced diploma or advanced certificate..2
　　　　　　　　　　a diploma 3
　　　　　　　　　　or some other RSA (including Stage I,II & III).......................................4
　　　　　　　　　　Don't Know..5

Ask if highest qualification is SCOTVEC
(Quals = 6　AND　does NOT have a higher qualification)

8. SCTVEC　　　　　Is your highest SCOTVEC qualification…

　　　　　　　　　　CODE FIRST THAT APPLIES

　　　　　　　　　　At a higher level (level 4)...1
　　　　　　　　　　At full National Certificate (level 3)..2
　　　　　　　　　　At a first diploma or general diploma (level 2) 3
　　　　　　　　　　At a first certificate or general certificate (below level 2).....................4
　　　　　　　　　　Modules towards a National Certificate 5
　　　　　　　　　　Don't know..6

Ask if highest qualification is BTEC, BEC, TEC or EdExcel
(Quals = 5　AND　does NOT have a higher qualification)

9. BTEC Is your highest BTEC qualification…

CODE FIRST THAT APPLIES

at higher level (level 4) 1
at National Certificate or National Diploma level (level 3) 2
a first diploma or general diploma (level 2) 3
a first certificate or general certificate (below level 2) 4
Don't know 5

Ask if highest qualification is GNVQ\GSVQ
(Quals = 13 AND does NOT have a higher qualification)

10. GNVQ Is your highest GNVQ/GSVQ at...

CODE FIRST THAT APPLIES
A FULL QUALIFICATION = 6 UNITS
A PART QUALIFICATION = 3 UNITS

Advanced 1
Full Intermediate 2
Part One Intermediate 3
Full Foundation 4
Part One Foundation 5
Don't know 6

Ask if highest qualification is City and Guilds
(Quals = 25 AND does NOT have a higher qualification)

11. CandG Is your highest City and Guilds qualification....

CODE FIRST THAT APPLIES

advanced craft/part 3 1
craft/part 2 2
foundation/part 1 3
Don't know 4

Ask if highest qualification is A levels / Vocational A-Levels
(Quals = 10 AND does NOT have a higher qualification)

12. NumAL Do you have…

one A level or equivalent 1
or more than one 2
Don't know 3

Ask if highest qualification is Scottish highers
(Quals = 11 AND does NOT have a higher qualification)

13. NumSCE Do you have…

1 or 2 Highers 1
3 or more Highers 2
Don't know 3

Ask if highest qualification is AS levels / Vocational AS levels
(Quals = 14 AND does NOT have a higher qualification)

14. NumAS   Do you have…

one AS level 1
2 or 3 AS levels 2
or 4 or more passes at this level 3
Don't know 4

Ask if highest qualification is GCSE or Standard Grade or O Grade or Intermediate 1 or Intermediate 2)
(Quals = 18, 19, 20 OR 21 AND does NOT have a higher qualification)

15. GCSE   Do you have any (GCSEs at grades A-C), (Standard Grades at 1-3 / O Grades at A-C), (Intermediate 2 at A-C and/or Intermediate 1 at A)?

Yes 1
No 2
Don't know 3

Ask if highest qualification is CSE
(Quals = 22 AND does NOT have a higher qualification)

16. CSE   Do you have any CSEs at grade 1?

Yes 1
No 2
Don't know 3

Ask if has O levels, Standard Grades 1-3, O Grades A-C, GCSEs at grade A-C, CSEs at grade 1, Intermediate 2 grade A-C or Intermediate 1 grade A
(Quals = 17 or GCSE = 1 or CSE = 1)

17. NumOL   ASK OR RECORD

You mentioned that you have passes at (GCSE at Grade A-C) (CSE Grade 1) (O level or equivalent) (Standard Grades at 1-3/O Grade at A-C) (Intermediate 2 at A-C) (Intermediate 1 at A). Do you have…

fewer than 5 passes, ...1
or 5 or more passes at this level 2
Don't know 3

18. EngMath   Do you have any (GCSEs at grades A-C), (CSE Grade 1), (Standard Grades at 1-3/ O Grades at A-C), (Intermediate 2 at A-C) or (Intermediate 1 at A) in English or Mathematics?

English 1
Maths 2
Both 3
Neither 4

Ask if respondent has any qualifications
(QualCh = 1, 2 or 3)

General Household Survey, 2006
Household and Individual Questionnaires

19. AgeHQual   How old were you when you achieved your highest qualification?

1..97

Ask all respondents

20. QulNow   Are you currently working towards or studying towards any qualifications?

Yes 1
No 2

Ask if respondent is currently working towards qualification
(QulNow = 1)

21. QulWht   SHOWCARD 10

What qualification are you studying for?

CODE ALL THAT APPLY - PROMPT AS NECESSARY

Degree level qualification including foundation degrees, graduate
  membership of a professional institute or PGCE or higher 1
Diploma in higher education 2
HNC/HND 3
ONC/OND 4
BTEC/ BEC/ TEC / EdExcel 5
SCOTVEC/ SCOTEC/ SCOTBEC (Scotland) 6
Teaching qualification (excluding PGCE) 7
Nursing or other medical qualification not yet mentioned 8
Other higher education qualification below degree level 9
A level / Vocational A-level or equivalent 10
Highers (Scotland) 11
NVQ/SVQ 12
GNVQ/GSVQ 13
AS level /Vocational AS level or equivalent 14
Advanced Highers or
Certificate of Sixth Year Studies (CSYS) (Scotland) 15
Access to HE 16
O level or equivalent 17
Intermediate 2 NQs (Scotland) 18
Intermediate 1 NQs (Scotland) 19
Standard Grade or O Grade (Scotland) 20
GCSE / Vocational GCSE 21
CSE 22
National Qualifications (including SGA) (Scotland) 23
RSA/ OCR 24
City and Guilds 25
YT Certificate/YTP 26
Key Skills/ Basic Skills 27
Entry Level Qualifications 28
Any other professional/vocational qualifications/
  foreign qualifications 29
Don't know 30

Ask if studying for a degree
(QulWht = 1)

General Household Survey, 2006
Household and Individual Questionnaires

22. DegNow   Are you studying for…

a higher degree (including PGCE) 1
a first degree 2
a foundation degree 3
other (e.g. graduate member of a professional institute or chartered
 accountant) 4
Don't know 5

Ask if studying for a higher degree
(DegNow = 1)

23. HghNow   Are you studying for…

a Doctorate 1
a Masters..2
a Postgraduate Certificate in Education...3
some other postgraduate degree or professional qualification...............4
Don't know 5

Ask if studying for a BTEC/BEC/TEC
(QulWht = 5)

24. **TECNow**   **What level BTEC/BEC/TEC are you studying for?**

at higher level (level 4)..1
at National Certificate or National Diploma level (level 3)...................2
a first diploma or general diploma (level 2) 3
a first certificate or general certificate (below level 2)........................4
Don't know 5

Ask if studying for a SCOTVEC/SCOTEC/SCOTBEC
(QulWht = 6)

25. SCNow   Are you studying for a SCOTVEC/SCOTEC/SCOTBEC…

at higher level (level 4)..1
at full National Certificate (level 3)..2
at a first diploma or general diploma (level 2) 3
at a first certificate or general certificate (below level 2)...................4
modules towards a National Certificate...5
Don't know 6

Ask if studying for an RSA
(QulWht = 24)

26. RSANow   Are you studying for a RSA at…

higher diploma level 1
advanced diploma or advanced certificate level................................2
diploma level 3
or some other RSA (including Stage I,II & III)....................................4
Don't know 5

Ask if studying for an NVQ/SVQ
(QulWht = 12)

27. NVQLe2   What is the highest level of NVQ/SVQ you are working towards?

                  Level 1 1
                  Level 2 2
                  Level 3 3
                  Level 4 4
                  Level 5 5
                  Don't know..6

Ask if answered other or don't know at quals

(Quals = 29, 30)

28. Appren   Are you doing, or have you completed, a recognised trade apprenticeship?

                  INCLUDE ADVANCED AND FOUNDATION MODERN
                  APPRENTICESHIPS (AMA/FMA) AND 'TRADE'
                  APPRENTICESHIPS

                  Yes, (completed)...1
                  Yes, (still doing) 2
                  Yes, has completed one apprenticeship and is now doing a further one 3
                  No 4

Ask all adults 16 and over (not asked of proxy respondents)

29. Enroll   Are you at present (at school or at a FE/ sixth form college) enrolled on any
                  full-time or part-time education course excluding leisure classes? (Include
                  correspondence courses and open learning as well as other forms of full-time
                  or part-time education course.)

                  Yes 1
                  No 2

Ask if enrolled on an education course
(Enroll = 1)

30. Attend   And are you …

                  RUNNING PROMPT

                  Still attending..1
                  Waiting for term to (re)start...2
                  Or have you stopped going 3

Ask if respondent is still attending school or college, or waiting for term to [re]start
(Attend = 1 or 2)

31. Course

Are you (at school or at a FE/ sixth form college), on a full or part-time course, a medical or nursing course, a sandwich course, or some other kind of course?

CODE FIRST THAT APPLIES

(School/full-time)/(CODE NOT APPLICABLE – AGED 20+)	1
(School/part-time)/(CODE NOT APPLICABLE – AGED 20+)	2
Sandwich course	3
Studying at a university or college including FE/ sixth form college FULL-TIME	4
Training for a qualification in nursing, physiotherapy, or a similar medical subject	5
On a part-time course at university or college INCLUDING day release and block release	6
On an Open College Course	7
On an Open University Course	8
Any other correspondence course	9
Any other self/open learning course	10

Ask if respondent is still attending school or college, or waiting for term to [re]start, AND is NOT under the age of 20
(Attend = 1 or 2 & Age > 19)

32. Course20

Are you (at school or at a FE/ sixth form college), on a full or part-time course, a medical or nursing course, a sandwich course, or some other kind of course?

CODE FIRST THAT APPLIES

Sandwich course	3
Studying at a university or college including FE/ sixth form college FULL-TIME	4
Training for a qualification in nursing, physiotherapy, or a similar medical subject	5
On a part-time course at university or college INCLUDING day release and block release	6
On an Open College Course	7
On an Open University Course	8
Any other correspondence course	9
Any other self/open learning course	10

Ask all adults 16 and over (not asked of proxy respondents)

33. EdAge

How old were you when you finished your continuous full-time education?

CODE AS 97 IF NO EDUCATION;
CODE AS 96 IF STILL IN EDUCATION

1..97

ADULT HEALTH

Ask this section of all adults (1(GenHlth) to 14 (ReasDen) and NHSDir and NHSDuse are not asked of proxy respondents)

Ask all (except proxy respondents)

General Household Survey, 2006
Household and Individual Questionnaires

1. GenHlth whole

[*] Over the last twelve months would you say your health has on the been good, fairly good, or not good?

Good 1
Fairly Good...2
Not Good 3

2. GenHlth2

[*] How is your health in general? Would you say it is…

RUNNING PROMPT

very good..1
good 2
fair 3
bad 4
very bad? 5

3. LSIll

[*] Do you have any long-standing illness, disability or infirmity? By long-standing, I mean anything that has troubled you over a period of time or that is likely to affect you over a period of time?

Yes 1
No 2

Ask if has a long-standing illness
(LSIll = 1)

4. LMatter

[*] What is the matter with you?

THIS IS TO ENSURE THAT THE RESPONDENT MENTIONS ALL THEIR LONGSTANDING ILLNESSES. YOU DO NOT HAVE TO RECORD VERBATIM – A SUMMARY WILL DO.

ENTER TEXT OF AT MOST 100 CHARACTERS

5. LMatNum

HOW MANY LONGSTANDING ILLNESSES OR INFIRMITIES DOES RESPONDENT HAVE?

ENTER NUMBER OF LONGSTANDING COMPLAINTS MENTIONED IF MORE THAN 6 - TAKE THE SIX THAT THE RESPONDENT CONSIDERS THE MOST IMPORTANT

1..6

For each illness

6. LMat

WHAT IS THE MATTER WITH RESPONDENT?

ENTER THE (FIRST/SECOND/etc.) CONDITION/SYMPTOM RESPONDENT MENTIONED

ENTER TEXT OF AT MOST 55 CHARACTERS

7. ICD

CODE FOR (FIRST/SECOND/etc.)COMPLAINT AT LMAT

ENTER SPACE BAR TO SEE CODES

General Household Survey, 2006
Household and Individual Questionnaires

IF CODE NOT FOUND, CHANGE ILLNESS DESCRIPTION AT BOTTOM OF LOOKUP WINDOW TO 'NONE' AND SELECT CODE FOR 'NONE OF THESE'

PRESS ENTER TO SELECT CODE AND ENTER AGAIN TO CONTINUE

ENTER TEXT OF AT MOST 2 CHARACTERS

Ask if has a long-standing illness
(Illness = 1)

8. LimitAct   Does this illness or disability (Do any of these illnesses or disabilities) limit your activities in any way?

        Yes 1
        No 2

Ask if activities are limited
(LimitAct = 1)

9. LimitL6   [*] Would you say your activities are limited or strongly limited?

        Limited 1
        Strongly limited 2

Ask all (except proxy respondents)

10. CutDown   [*] Now I'd like you to think about the 2 weeks ending yesterday. During those 2 weeks, did you have to cut down on any of the things you usually do (about the house/at work or in your free time) because of illness or injury?

        Yes 1
        No 2

Ask if had to cut down on normal activities because of illness or injury
(CutDown = 1)

11. NDysCutD   How many days was this in all during these 2 weeks, including Saturdays and Sundays?

        1..14

12. Cmatter   [*] What was the matter with you?

        ENTER TEXT OF AT MOST 40 CHARACTERS

Ask all (except proxy respondents)

13. MedRec   [*] Was there any time since the date of the last interview (*date*) when, in your opinion, you personally needed a medical examination or treatment for a health problem but you did not receive it?

        Yes, there was at least one occasion but did not receive 1
        No, there was no occasion 2

General Household Survey, 2006
Household and Individual Questionnaires

Ask if answered yes to MedRec
(MedRec = 1)

14. ReasMed — What was the main reason for not receiving the examination or treatment (the most recent time)?

DO NOT PROMPT
IF RESPONDENT SAYS CANNOT AFFORD PRESCRIPTION, USE CODE 1

Could not afford to (too expensive) 1
Waiting list 2
Could not take time because of work, care for children or for others ...3
Too far to travel/no means of transportation 4
Fear of doctor/hospitals/examination/ treatment 5
Wanted to wait and see if problem got better on its own 6
Didn't know any good doctor or specialist 7
Other reasons 8

Ask all (except proxy respondents)

15. DenRec — Was there any time since the date of the last interview (*date*) when, in your opinion, you personally needed a dental examination or treatment but you did not receive it?

Yes, there was at least one occasion but did not receive 1
No, there was no occasion..2

Ask if answered yes to DenRec
(DenRec = 1)

16. ReasDen — What was the main reason for not receiving the dental examination or treatment (the most recent time)?

SPONTANEOUS

Could not afford to (too expensive) 1
Waiting list 2
Could not take time because of work, care for children or for others ...3
Too far to travel/no means of transportation 4
Fear of dentists/hospitals/examination/ treatment 5
Wanted to wait and see if problem got better on its own 6
Didn't know any good dentist 7
Can't find NHS dentist willing to take me on as a patient 8
Other reasons 9

If respondent can't find NHS dentist
(ReasDen=8)

17. ReasNHS — May I just check, could you afford to go to a private dentist instead?

Yes 1
No 2

Ask all

18. DocTalk — During the 2 weeks ending yesterday, apart from any visit to a hospital, did you talk to a doctor for any reason at all, either in person or by telephone?

General Household Survey, 2006
Household and Individual Questionnaires

EXCLUDE: CONSULTATIONS MADE ON BEHALF OF CHILDREN UNDER 16 AND PERSONS OUTSIDE THE HOUSEHOLD.

Yes 1
No 2

Ask if contact with doctor during the last 2 weeks
(DocTalk = 1)

19. NChats   How many times did you talk to a doctor in these 2 weeks?

1..9

For each consultation

20. WhsBhlf   On whose behalf was this consultation made?

Respondent 1
Other member of household 16 or over 2

Ask if consultation was on the behalf of another member of the household
(WhsBhlf = 2)

21. ForPerNo   CODE WHO CONSULTATION WAS MADE FOR

(PERSON NUMBER)

For each consultation

22. NHS   Was this consultation...

Under the National Health Service 1
or paid for privately 2

23. GP   Was the doctor...

RUNNING PROMPT

A GP (ie a family doctor) 1
or some other kind of doctor 2

24. DocWhere   Did you talk to the doctor...

RUNNING PROMPT

By telephone 1
at your home 2
in the doctor's surgery 3
at a health centre 4
or elsewhere? 5

25. Presc   Did the doctor give (send) you a prescription?

Yes 1
No 2

General Household Survey, 2006
Household and Individual Questionnaires

Ask all adults

26. SeeChn   During the last 2 weeks ending yesterday, did you see a practice nurse at the GP surgery on your own behalf?

       Yes 1
       No 2

Ask if the respondent saw a nurse
(SeeNurse = 1)

27. NNurse   How many times did you see a practice nurse at the GP surgery in these 2 weeks?

       RECORD NUMBER OF TIMES

       1..9

Ask all adults

28. OutPatnt   During the months of (LAST 3 COMPLETE CALENDAR MONTHS) did you attend as a patient at the casualty or outpatient department of a hospital (apart from straightforward ante- or post-natal visits)?

       INCLUDE: VISITS TO PRIVATE HOSPITALS AND PRIVATE CLINICS, MINOR INJURIES UNITS AND WALK-IN CENTRES

       EXCLUDE: DOCTORS SEEN ABROAD UNLESS FORCES DOCTORS
                  - DAY PATIENTS (THEY ARE COVERED BY DAYPATNT)

       Yes 1
       No 2

Ask if respondent attended outpatients
(OutPatnt = 1)

29. NTimes1   How many times did you attend in (EARLIEST MONTH IN REFERENCE PERIOD)?

       0..97

30. NTimes2   How many times did you attend in (SECOND MONTH IN REFERENCE PERIOD)?

       0..97

31. NTimes3   How many times did you attend in (THIRD MONTH IN REFERENCE PERIOD)?

       0..97

32. Casualty   Was this visit (were any of these visits) to the Casualty department or was it (were they all) to some other part of the hospital?

       At least one visit to Casualty 1
       No Casualty visits 2

General Household Survey, 2006
Household and Individual Questionnaires

Ask if respondent visited casualty
(Casualty = 1)

33. NCasVis   (May I just check) How many times did you go to Casualty altogether?

                          1..31

Ask if respondent attended outpatients
(OutPatnt = 1)

34. PrVists   Was your outpatient visit (were any of your outpatient visits) during (REFERENCE PERIOD) made under the NHS, or was it (were any of them) paid for privately?

                          All under NHS...1
                          At least one paid for privately 2

Ask if some private visits
(PrVists = 2)

35. NPrVists   ASK OR RECORD

                          (May I just check), How many of the visits were paid for privately?

                          1..31

Ask all adults

36. DayPatnt   Since the date of the last interview (*date*), have you been in hospital for treatment as a day patient, i.e. admitted to a hospital bed or day ward, but not required to remain overnight?

                          Yes 1
                          No 2

Ask if has been a day patient AND is a women aged between 16-49
(DayPatnt = 1 & Sex = 2 & DVAge = 16-49)

37. MatDPat   May I just check, was that/were any of those day patient admissions for you to have a baby?

                          Yes 1
                          No 2

Ask if respondent was a day patient because she was having a baby
(MatDPat = Yes)

38. NumMatDP   How many separate days have you had as a day patient for having a baby since (DATE ONE YEAR AGO)?

                          97 DAYS OR MORE - CODE 97

                          1..97

39. PrMatDP   Was this day-patient stay (were any of these day-patient stays) for having a baby under the NHS, or was it (were any of them) paid for privately?

                          All under NHS...1
                          At least one paid for privately 2

General Household Survey, 2006
Household and Individual Questionnaires

Ask if day patient stay for having a baby was paid for privately AND respondent was in hospital for more than one day
(PrMatDP = 2 & NumMatDP > 1)

40. NPrMatDP   ASK OR RECORD

How many of the visits were paid for privately?

1..31

Ask if the respondent was a day patient
(DayPatnt = 1)

41. NHSPDays   (Apart from those maternity stays) how many separate days in hospital have you had as a day patient since (DATE ONE YEAR AGO)?

97 DAYS OR MORE - CODE 97

0..97

Ask if had one or more days in hospital
(NHSPDays > 0)

42. PrDptnt   Was this day-patient treatment (were any of these day-patient treatments) under the NHS, or was it (were any of them) paid for privately?

All under NHS 1
At least one paid for privately 2

Ask if day patient stay was paid for privately AND they were in hospital for more than one day
(PrDptnt = 2 & NHSPDays > 1)

43. NPrDpTnt   ASK OR RECORD

How many of the visits were paid for privately?

1..31

Ask all adults

44. InPatnt   Since the date of the last interview (*date*), have you been in hospital as an inpatient, overnight or longer?

Yes 1
No 2

Ask if respondent has been an inpatient AND she is a women aged 16-49
(InPatnt = 1 & Sex = 2 & DVAge = 16-49)

45. MatInPat   May I just check, was that/were any of those inpatient missions for you to have a baby?

Yes 1
No 2

Ask if inpatient admission was to have a baby
(MatInPat = 1)

46. NMtStay — How many separate stays in hospital as an inpatient in order to have a baby have you had since (DATE 1 YEAR AGO)?

1..6

Ask for each maternity stay

47. MtNights — How many nights altogether were you in hospital on your (no.) stay to have a baby?

1..97

48. MatNHSTr — Were you treated under the NHS or were you a private patient on that occasion?

NHS 1
Private Patient..2

If private patient
(MatNHSTr = 2)

49. MtPrvSty — Were you treated in an NHS hospital or in a private one?

NHS hospital 1
Private hospital 2

Ask if respondent has been an inpatient
(InPatnt = 1)

50. NStays — (Apart from those maternity stays) how many separate stays in hospital as an inpatient have you had since (DATE 1 YEAR AGO)?

0..6

Ask for each stay

51. Nights — How many nights altogether were you in hospital on your… (first/second/...sixth) stay?

1..97

52. NHSTreat — Were you treated under the NHS or were you a private patient on that occasion?

NHS 1
Private Patient..2

Ask if a private patient
(NHSTreat = 2)

53. PrvStay — Were you treated in an NHS hospital or in a private one?

NHS hospital 1
Private hospital 2

General Household Survey, 2006
Household and Individual Questionnaires

Ask all adults (except proxy respondents)

54. NHSDir Have you ever heard of NHS Direct? (In Scotland - NHS24)

Yes 1
No 2

Ask if respondent has heard of NHS Direct/ NHS24
(NHSDir = 1)

55. NHSDuse During the last year, that is, since (DATE ONE YEAR AGO), have you used NHS Direct? (In Scotland - NHS24)

Yes 1
No 2

CHILD HEALTH

Ask each child under 16 in household (not asked of proxy respondents)

1. AskHlth THE NEXT SECTION IS ABOUT CHILD HEALTH.
WE ONLY NEED TO COLLECT THIS INFORMATION ONCE FOR EACH CHILD IN THE HOUSEHOLD.
WHO WILL ANSWER THE CHILD HEALTH SECTION FOR (CHILD'S NAME)?

INTERVIEWER ENTER PERSON NUMBER.

1..14

2. AskNowCH INTERVIEWER: DO YOU WANT TO ASK THIS SECTION FOR (CHILD'S NAME) NOW OR LATER?

IF YOU HAVE ALREADY ASKED THIS SECTION FOR (CHILD'S NAME), DO NOT CHANGE FROM CODE 1.

Yes, now/Already asked..1
Later 2

If the section is to be asked later
(AskNowCH = 2)

3. CStill REMINDER
THE FOLLOWING ADULTS STILL NEED TO ANSWER THE CHILD HEALTH SECTION ON BEHALF OF SOME OF THE CHILDREN.

If the section is to be asked now
(AskNowCH = 1)

For each child

General Household Survey, 2006
Household and Individual Questionnaires

4. Genhlth   [*] Since the date of the last interview (*date*) would you say (NAME's) alth has on the whole been good, fairly good, or not good?

Good 1
Fairly Good..2
Not Good 3

5. Genhlth2   [*] How is (NAME's) health in general? Would you say it was…

Very good 1
Good 2
Fair 3
Bad 4
Very bad 5

6. Illness   Does (NAME) have any long-standing illness, disability or infirmity? By long-standing, I mean anything that has troubled them over a period of time or that is likely to affect them over a period of time?

Yes 1
No 2

Ask if child has a longstanding illness, disability or infirmity
(Illness =1)

7. LMatter   [*] What is the matter with (NAME)?

THIS IS TO ENSURE THAT THE RESPONDENT MENTIONS ALL LONGSTANDING ILLNESSES. YOU DO NOT HAVE TO RECORD VERBATIM HERE - A SUMMARY WILL DO.

ENTER TEXT OF AT MOST 40 CHARACTERS

8. LMatNum   HOW MANY LONGSTANDING ILLNESSES OR INFIRMITIES DOES (NAME) HAVE?

ENTER NUMBER OF LONGSTANDING COMPLAINTS MENTIONED
IF MORE THAN 6 - TAKE THE SIX THAT THE RESPONDENT CONSIDERS THE MOST IMPORTANT.

1..6

For each illness mentioned at LMatNum

9. LMatCH   WHAT IS THE MATTER WITH (NAME)?

ENTER THE (FIRST/SECOND/etc.) CONDITION/SYMPTOM RESPONDENT MENTIONED

ENTER TEXT OF AT MOST 40 CHARACTERS

10. ICDCH   CODE FOR EACH COMPLAINT AT LMatCH

ENTER SPACE BAR TO SEE CODES

IF CODE NOT FOUND, CHANGE ILLNESS DESCRIPTION AT BOTTOM OF LOOKUP WINDOW TO 'NONE' AND SELECT CODE

FOR 'NONE OF THESE'. PRESS ENTER TO SELECT CODE AND ENTER AGAIN TO CONTINUE.

If child has a longstanding illness, disability or infirmity
(Illness =1)

11. LimitAct   [*] Does this illness or disability (Do any of these illnesses or disabilities) limit (NAME)'s activities in any way?

   Yes 1
   No 2

Ask if child's activities are limited by an illness/disability
(LimitAct = 1)

12. LimitL6C   Would you say their activities are limited or strongly limited?

   Limited 1
   Strongly limited 2

For each child

13. CutDown   [*] Now I'd like you to think about the 2 weeks ending yesterday. During those 2 weeks, did (NAME) have to cut down on any of the things he/she usually does (at school or in his/her free time) because of ..... (answer at LMatter or some other) illness or injury?

   Yes 1
   No 2

Ask if child has had to cut down
(CutDown = 1)

14. NDysCutD   How many days did (NAME) have to cut down in all during these 2 weeks, including Saturdays and Sundays?

   1..14

15. Matter   [*] What was the matter with (NAME)?

   ENTER TEXT OF AT MOST 80 CHARACTERS

For each child

16. DocTalk   During the 2 weeks ending yesterday, apart from visits to a hospital, did (NAME) talk to a doctor for any reason at all, or did you or any other member of the household talk to a doctor on his/her behalf?

   Include being seen by a doctor at a school clinic, but exclude visits to a child welfare clinic run by a local authority.

   INCLUDE TELEPHONE CONSULTATIONS AND CONSULTATIONS MADE ON BEHALF OF CHILDREN.

   Yes 1
   No 2

General Household Survey, 2006
Household and Individual Questionnaires

If child consulted a doctor
(DocTalk = 1)

17. NChats — How many times did (NAME) talk to the doctor (or you or any other member of the household consult the doctor on NAME's behalf) in those 2 weeks?

1..9

For each consultation

18. NHS — Was this consultation...

RUNNING PROMPT

Under the National Health Service 1
or paid for privately 2

19. GP — Was the doctor...

RUNNING PROMPT

A GP (i.e. a family doctor) 1
or some other kind of doctor 2

20. DocWhere — Did you or any other member of the household (or NAME) talk to the doctor...

RUNNING PROMPT

By telephone 1
at your home 2
in the doctor's surgery 3
at a health centre 4
or elsewhere 5

21. Presc — Did the doctor give (send) (NAME) a prescription?

Yes 1
No 2

For each child

22. Seenurse — During the last 2 weeks ending yesterday, did (NAME)...

INDIVIDUAL PROMPT - CODE ALL THAT APPLY

see a practice nurse at the GP surgery 1
see a health visitor at the GP surgery 2
go to child health clinic 3
go to child welfare clinic 4
or did they not go to any of these 5

Ask if child saw a practice nurse
(Seenurse = 1)

23. Nnurse   How many times did (NAME) see a practice nurse at the GP surgery in these 2 weeks?

RECORD NUMBER OF TIMES

1..9

For each child

24. OutPatnt   During the months of (LAST 3 COMPLETE CALENDAR MONTHS), did (NAME) attend as a patient the casualty or outpatient department of a hospital?

INCLUDE MINOR INJURIES UNITS AND WALK-IN CENTRES

Yes 1
No 2

Ask if child has been an outpatient
(OutPatnt = 1)

25. NTimes1   How many times did (NAME) attend in (EARLIEST MONTH IN REFERENCE PERIOD)?

0..97

26. NTimes2   How many times did (NAME) attend in (SECOND MONTH IN REFERENCE PERIOD)?

0..97

27. NTimes3   How many times did (NAME) attend in (THIRD MONTH IN REFERENCE PERIOD)?

0..97

28. Casualty   Was the visit (were any of the visits) to the Casualty department or was it (were they) to some other part of the hospital?

At least one visit to Casualty 1
No Casualty visits 2

Ask if child went to casualty
(Casualty = 1)

29. NCasVis   (May I just check) How many times did (NAME) go to Casualty altogether?

1..31

For each child

General Household Survey, 2006
Household and Individual Questionnaires

30. DayPatnt   Since the date of the last interview, that is since (DATE 1 YEAR AGO) has (NAME) been in hospital for treatment as a day patient, ie admitted to a hospital bed or day ward, but not required to remain in hospital overnight?

Yes 1
No 2

Ask if child has been a day patient
(DayPatnt = 1)

31. NHSPDays   How many separate days in hospital has (NAME) had as a day patient since (DATE 1 YEAR AGO)?

1..97

For each child

32. InPatnt   During the last year, that is, since (DATE 1 YEAR AGO) has (NAME) been in hospital as an inpatient, overnight or longer?

EXCLUDE: Births unless baby stayed in hospital after mother had left.

Yes 1
No 2

Ask if child has been an inpatient
(InPatnt = 1)

33. NStays   How many separate stays in hospital as an inpatient has (NAME) had since (DATE 1 YEAR AGO)?

IF 6 OR MORE, CODE 6

1..6

For each stay

34. Nights   How many nights altogether was (NAME) in hospital during stay number (...)?

1..97

CHILDCARE

Ask this section for each child aged 0 - 12 years inclusive

1. AskCare   The next section is about your childcare needs.
We are interested in where your child is when neither you (nor your partner) are present, for example, at school, in a creche, or at some other daycare.
We are also interested in a typical term time week, that is, a week of 7 days, and which is outside school (and parents') holidays.

Who will answer the childcare section for (NAME)?

General Household Survey, 2006
Household and Individual Questionnaires

ENTER PERSON NUMBER

IF NO TYPICAL WEEK - PICK THE MOST RECENT WEEK WITHOUT HOLIDAYS.
WE ONLY NEED TO COLLECT THIS INFORMATION ONCE FOR EACH CHILD AGED 12 YEARS OR LESS IN THE HOUSEHOLD.

1..14

2. ChAtt At any time during a typical term time week did (NAME) attend any of the following?

Play group or pre-school 1
Day-care centre or workplace creche 2
Nursery school 3
School (infant to secondary) 4
Breakfast/After school club 5
Children's centres/integrated centres/combined centres 6
Boarding school 7
None of these 8

2. WkKind How many hours during a typical term time week did (NAME) spend in the playgroup, pre-school or nursery school?

INCLUDE: 'NORMAL' SCHOOL HOURS (I.E. EDUCATIONAL HOURS) AND MEAL-TIMES
EXCLUDE: BEFORE/AFTER SCHOOL CLUBS AND TRAVEL TIME ON SCHOOL TRANSPORT
0..99

3. WkSchl How many hours during that typical week did (NAME) spend in school?

INCLUDE: 'NORMAL' SCHOOL HOURS (I.E. EDUCATIONAL HOURS) AND MEAL-TIMES
EXCLUDE: BEFORE/AFTER SCHOOL CLUBS AND TRAVEL TIME ON SCHOOL TRANSPORT

0..99

4. WkBA How many hours during a typical term time week did (NAME) spend in a Breakfast or After school club or at an organised children's centre, integrated centre or combined centre?

INCLUDE:
 - BREAKFAST/AFTERSCHOOL CLUBS
 - OUTSIDE SCHOOL HOURS
 - AT AN ORGANISED CENTRE
EXCLUDE:
 - SPORTING AND CULTURAL ACTIVITIES OUTSIDE SCHOOL HOURS (E.G. MUSIC LESSONS, SPORTS CLUB). NB THESE ARE NOT USED AS A CHILDCARE SERVICE BUT RATHER FOR CHILD LEISURE.

0..99

5. WkDcare How many hours during a typical term time week did (NAME) spend in a day-care centre, crèche, family day care (even if for just a few hours)?

- INCLUDES DURING SCHOOL HOURS

- AT AN ORGANISED CENTRE
- IT IS ORGANISED, IN THAT THE CARER WILL BE EMPLOYED BY AN ORGANISATION
- TYPICALLY IN A CENTRE, THOUGH FAMILY DAY CARE USES APPROVED CARERS IN THEIR OWN HOMES

INCLUDE SPECIAL DAYCARE FOR CHILDREN WITH SPECIAL NEEDS
EXCLUDE SPORTING AND CULTURAL ACTIVITIES

0..99

6. ChPeo And during that typical term time week did any of the people listed on this card normally look after (NAME), excluding care for social occasions? (Other than resident parent(s)/guardian(s) and staff contact whilst at places previously mentioned)

NANNY REFERS TO AN EMPLOYED NANNY (DOMESTIC HELP TO LOOK AFTER CHILDREN)

CODE ALL THAT APPLY

Child's grandparents 1
Child's non-resident parent/an ex-spouse/an ex-partner 2
Child's brother or sister 3
Other relatives 4
Au Pair/Nanny (includes live-in and day nannies) 5
Friends or neighbours 6
Childminder 7
Other non-relatives 8
None of the above 9

6. WkPcare Thinking of these people, how many hours of PAID CARE did they provide for (NAME) during that typical term time week?

- CARER IS PAID DIRECTLY BY PARENT
- AT HOME OR CHILDMINDER'S HOME

0..99

7. WkUPcare ….and how many hours of UNPAID CARE did they provide for (NAME) during that typical term time week?

0..99

SMOKING

Ask this section of all adults, except proxy respondents

1. SmkIntro The next section consists of a series of questions about SMOKING (Not asked of proxy respondents)

Ask all 16 and 17 year olds
(DVAge = 16-17)

General Household Survey, 2006
Household and Individual Questionnaires

2. SelfCom1   RESPONDENT IS AGED 16 OR 17 - OFFER SELF-COMPLETION FORM AND ENTER CODE.

Respondent accepted self-completion 1
Respondent refused self-completion 2
Data now to be keyed by interviewer 3

Ask if aged 18 or over (except proxy respondents)
(DVAge ≥ 18)

3. SmokEver   Have you ever smoked a cigarette, a cigar, or a pipe?

Yes 1
No 2

Ask if respondent has ever smoked
(SmokEver = 1)

4. CigNow   Do you smoke cigarettes at all nowadays?

Yes 1
No 2

Ask if respondent smokes cigarettes now
(CigNow = 1)

5. QtyWkEnd   About how many cigarettes A DAY do you usually smoke at weekends?

IF LESS THAN 1, ENTER 0.

0..97

6. QtyWkDay   About how many cigarettes A DAY do you usually smoke on weekdays?

IF LESS THAN 1, ENTER 0.

0..97

7. CigType   Do you mainly smoke.....

RUNNING PROMPT

filter-tipped cigarettes..1
or plain or untipped cigarettes 2
or hand-rolled cigarettes? 3

Ask if cigarette types include plain or filter cigarettes
(CigType = 1 or 2)

8. CigIDesc   Which brand of cigarette do you usually smoke?

GIVE 1) FULL BRAND NAME 2) SIZE, eg King, luxury, regular.
IF NO REGULAR BRAND THEN TYPE 'no reg' HERE.
IF RESPONDENT SMOKES TWO BRANDS EQUALLY TYPE 'two' IN LETTERS HERE.

IF POSSIBLE PLEASE CHECK THE CIGARETTE PACKET

ENTER TEXT OF AT MOST 60 CHARACTERS

General Household Survey, 2006
Household and Individual Questionnaires

9. CigCODE

Code for brand at CigIDesc

ENTER SPACE BAR TO SEE CODES

PLEASE DO NOT SELECT FIRST EXAMPLE OF NAMED BRAND, BUT CHECK YOU HAVE CHOSEN THE CORRECT ONE.

IF BRAND NOT FOUND, CHANGE CIGARETTE BRAND DESCRIPTION AT BOTTOM OF LOOKUP WINDOW TO 'nf' AND SELECT CODE FOR 'BRAND NOT FOUND'. PRESS ENTER TO SELECT CODE AND ENTER AGAIN TO CONTINUE.

10. CigPack

INTERVIEWER – CODE WHETHER THE RESPONDENT'S CIGARETTE PACKET WAS CHECKED

Cigarette packet checked by respondent/ interviewer 1
Cigarette packet not checked 2

Ask if respondent smokes cigarettes now
(CigNow = 1)

11. NoSmoke

[*] How easy or difficult would you find it to go without smoking for a whole day? Would you find it...

RUNNING PROMPT

Very easy 1
Fairly easy 2
Fairly difficult or 3
Very difficult 4

12. GiveUp

[*] Would you like to give up smoking altogether?

Yes 1
No 2
Don't know 3

13. FirstCig

How soon after waking do you USUALLY smoke your first cigarette of the day?

PROMPT AS NECESSARY

Less than 5 minutes 1
5-14 minutes 2
15-29 minutes 3
30 minutes but less than 1 hour 4
1 hour but less than 2 hours 5
2 hours or more 6

Ask if respondent does not smoke cigarettes now but has smoked a cigarette or cigar or pipe
(SmokEver = 1 & CigNow = 2)

14. CigEver

Have you ever smoked cigarettes regularly?

Yes 1
No 2

Ask if respondent has ever smoked cigarettes regularly
(CigEver = 1)

15. CigUsed   About how many cigarettes did you smoke IN A DAY when you smoked them regularly?

IF LESS THAN 1, ENTER 0.

0..97

16. CigStop   How long ago did you stop smoking cigarettes regularly?

PROMPT AS NECESSARY

Less than 6 months ago 1
6 months but less than a year ago 2
1 year but less than 2 years ago 3
2 years but less than 5 years ago... 4
5 years but less than 10 years ago... 5
10 years or more ago 6

Ask of all respondents who have ever smoked cigarettes
(CigNow = 1 or   CigEver = 1)

17. CigAge   How old were you when you started to smoke cigarettes regularly?

SPONTANEOUS: NEVER SMOKED CIGARETTES REGULARLY - CODE 0

0..97

Ask respondents who have ever smoked
(SmokEver = 1)

18. CigarReg   Do you smoke at least one cigar of any kind per month nowadays?

Yes   1
No   2

Ask if respondent smokes at least one cigar per month
(CigarReg = 1)

19. CigarsWk   About how many cigars do you usually smoke in a week?

IF LESS THAN 1, ENTER 0.

0..97

Ask if respondent does not smoke at least one cigar per month
(CigarReg = 2)

20. CigarEvr   Have you ever regularly smoked at least one cigar of any kind per month?

Yes   1
No   2

Ask men who have ever smoked

(SmokEver = 1   AND   Sex = 1)

21. PipeNow   Do you smoke a pipe at all nowadays?

                      Yes 1
                      No 2

Ask if respondent doesn't currently smoke a pipe
(PipeNow = 2)

22. PipEver   Have you ever smoked a pipe regularly?

                      Yes 1
                      No 2

Ask if CigNow = Yes or CigarReg = Yes or PipeNow = Yes

23. GiveUpC   Which of the following statements best describes you?

                      CODE FIRST THAT APPLIES

                      I intend to give up smoking within the next month 1
                      I intend to give up smoking within the next 6 months 2
                      I intend to give up smoking within the next year 3
                      I intend to give up smoking but not in the next year 4
                      I intend to give up smoking, but I'm not sure when 5
                      I don't intend to give up smoking 6

DRINKING

Ask this section of all adults except proxy respondents

Ask all 16 and 17 year olds
(DVAge = 16-17)

1. Selfcom2   I'm now going to ask you a few questions about what you drink - that is if you do drink.

                      IF RESPONDENT PREFERS TO SELF-COMLPETE, OFFER SELF-COMPLETION FORM AND ENTER CODE.

                      Interviewer asked section 1
                      Respondent accepted self-completion form............................2
                      Self-completion now keyed by interviewer............................3

Ask all (except proxy respondents)
(DVAge ≥ 18   or   Selfcom2 = 1)

2. DrinkNow   Do you ever drink alcohol nowadays, including drinks you brew or make at home?

                      Yes 1
                      No 2

Ask if does not drink nowadays
(DrinkNow = 2)

General Household Survey, 2006
Household and Individual Questionnaires

3. DrinkAny   Could I just check, does that mean you never have an alcoholic drink nowadays, or do you have an alcoholic drink very occasionally, perhaps for medicinal purposes or on special occasions like Christmas or New Year?

   Very occasionally 1
   Never 2

Ask if never drinks
(DrinkAny = 2)

4. TeeTotal   Have you always been a non-drinker, or did you stop drinking for some reason?

   Always a non-drinker 1
   Used to drink but stopped..2

Ask if respondent has always been a non-drinker
(TeeTotal = 1)

5. NonDrink   [*] What would you say is the MAIN reason you have always been a non-drinker?

   Religious reasons..1
   Don't like it 2
   Parent's advice/influence 3
   Health reasons 4
   Can't afford it...5
   Other 6

Ask if respondent used to drink but stopped
(TeeTotal = 2)

6. StopDrin   [*] What would you say was the MAIN reason you stopped drinking?

   Religious reasons..1
   Don't like it 2
   Parent's advice/influence 3
   Health reasons 4
   Can't afford it...5
   Other 6

Ask if respondent drinks at all nowadays
(Drinknow = 1 or   DrinkAny = 1)

7. DrinkAmt   [*] I'm going to read out a few descriptions about the amounts of alcohol people drink, and I'd like you to say which one fits you best. Would you say you:

   hardly drink at all..1
   drink a little..2
   drink a moderate amount 3
   drink quite a lot..4
   or drink heavily 5

DRINKING OVER LAST 12 MONTHS (Questions 8 –27)

Ask if respondent drinks at all nowadays
(Drinknow = 1 or   DrinkAny = 1)

General Household Survey, 2006

Household and Individual Questionnaires

8. Intro   I'd like to ask you whether you have drunk different types of alcoholic drink in the last 12 months.

I'd like to hear about ALL types of alcoholic drinks you have had.
If you are not sure whether a drink you have had goes into a category, please let me know. I do not need to know about non-alcoholic or low alcohol drinks.

THE HELP KEYS GIVE YOU MORE INFORMATION ABOUT WHAT SHOULD BE INCLUDED AT THE DIFFERENT DRINKS CATEGORIES.

9. NBeer   SHOWCARD 11

I'd like to ask you first about NORMAL STRENGTH beer or cider which has less than 6% alcohol. How often have you had a drink of NORMAL STRENGTH BEER, LAGER, STOUT, CIDER or SHANDY (excluding cans and bottles of shandy) during the last 12 months?

NORMAL = LESS THAN 6% ALCOHOL BY VOLUME.
IF RESPONDENT DOES NOT KNOW WHETHER BEER ETC DRUNK IS STRONG OR NORMAL, INCLUDE HERE AS NORMAL.
USE HELP SCREEN FOR OTHER DRINKS TO BE INCLUDED HERE.

Almost every day 1
5 or 6 days a week 2
3 or 4 days a week 3
once or twice a week 4
once or twice a month 5
once every couple of months 6
once or twice a year 7
not at all in last 12 months 8

Ask if respondent drank normal strength beer or cider at all in last 12 months
(NBeer = 1 - 7)

10. NBeerM   How much NORMAL STRENGTH BEER, LAGER, STOUT, CIDER or SHANDY (excluding cans and bottles of shandy) have you usually drunk on any one day during the last 12 months?

CODE MEASURES THAT YOU ARE GOING TO USE.
PROBE IF NECESSARY.
CODE ALL THAT APPLY.

Half pints 1
Small cans 2
Large cans 3
Bottles 4

For each type of measure of normal strength beer

11. NBeerQ   ASK OR RECORD

How many (half pints/ small cans/ large cans/ bottles) of NORMAL STRENGTH BEER, LAGER, STOUT, CIDER or SHANDY (excluding cans and bottles of shandy) have you usually drunk on any one day during the last 12 months?

General Household Survey, 2006
Household and Individual Questionnaires

1..97

Ask if respondent drinks normal strength beers from bottles
(NBeerM = 4)

12. NBrlDesc   Which make of NORMAL STRENGTH BEER, LAGER, STOUT OR CIDER do you usually drink from bottles?

IF RESPONDENT DOES NOT KNOW WHAT MAKE, OR RESPONDENT DRINKS DIFFERENT MAKES OF NORMAL STRENGTH BEER, LAGER, STOUT OR CIDER, PROBE: 'WHAT MAKE HAVE YOU DRUNK MOST FREQUENTLY OR MOST RECENTLY?'

IF RESPONDENT DRINKS FRENCH BOTTLED BEER BUT DOES NOT KNOW THE BRAND NAME TYPE 'FRENCH', ENTER CODING FRAME AND CODE 'FRENCH BEER-BRAND NOT KNOWN'.

IF RESPONDENT DOES NOT KNOW THE BRAND NAME BUT THEY INDICATE THAT THE BOTTLE IS LARGE THEN TYPE 'LARGE', ENTER CODING FRAME AND CODE 'LARGE BOTTLE-BRAND NOT KNOWN'.

ENTER TEXT OF NO MORE THAN 21 CHARACTERS

13. NBrCode   CODE FOR BRAND AT NBRIDESC

ENTER SPACE BAR TO SEE CODES

IF BRAND NOT FOUND, CHANGE BRAND DESCRIPTION AT BOTTOM OF LOOK-UP WINDOW TO 'NF' AND SELECT CODE FOR 'BRAND NOT FOUND'

PRESS ENTER TO SELECT CODE AND ENTER AGAIN TO CONTINUE

Ask if respondent drinks at all nowadays
(Drinknow = 1 or DrinkAny = 1)

14. SBeer   SHOWCARD 11

Now I'd like to ask you about STRONG BEER OR CIDER which has 6% or more alcohol (eg Tennants Extra, Special Brew, Diamond White). How often have you had a drink of strong BEER, LAGER, STOUT or CIDER during the last 12 months?

STRONG=6% AND OVER ALCOHOL BY VOLUME.
IF RESPONDENT DOES NOT KNOW WHETHER BEER ETC DRUNK IS STRONG OR NORMAL, INCLUDE AS NORMAL STRENGTH AT NBEER ABOVE.
USE HELP SCREEN FOR OTHER DRINKS TO BE INCLUDED HERE.

Almost every day..1
5 or 6 days a week 2
3 or 4 days a week 3
once or twice a week 4
once or twice a month 5
once every couple of months 6
once or twice a year 7

not at all in last 12 months 8

Ask if respondent drank strong beer or cider at all in last 12 months
(SBeer = 1 - 7)

15. SBeerM How much STRONG BEER, LAGER, STOUT or CIDER have you usually drunk on any one day during the last 12 months?

CODE MEASURES THAT YOU ARE GOING TO USE.
PROBE IF NECESSARY.
CODE ALL THAT APPLY.

Half pints 1
Small cans 2
Large cans 3
Bottles 4

For each type of measure of strong beer

16. SBeerQ ASK OR RECORD

How many (half pints/ small cans/ large cans/ bottles) of STRONG BEER, LAGER, STOUT or CIDER have you usually drunk on any one day during the last 12 months?

1..97

Ask if respondent drinks strong beers from bottles
(SBeerM = 4)

17. SBrlDesc Which make of STRONG BEER, LAGER, STOUT OR CIDER do you usually drink from bottles?

IF RESPONDENT DOES NOT KNOW WHAT MAKE, OR RESPONDENT DRINKS DIFFERENT MAKES OF STRONG BEER, LAGER, STOUT OR CIDER, PROBE: 'WHAT MAKE HAVE YOU DRUNK MOST FREQUENTLY OR MOST RECENTLY?'

IF RESPONDENT DRINKS FRENCH BOTTLED BEER BUT DOES NOT KNOW THE BRAND NAME TYPE 'FRENCH', ENTER CODING FRAME AND CODE 'FRENCH BEER-BRAND NOT KNOWN'.

IF RESPONDENT DOES NOT KNOW THE BRAND NAME BUT THEY INDICATE THAT THE BOTTLE IS LARGE THEN TYPE 'LARGE', ENTER CODING FRAME AND CODE 'LARGE BOTTLE-BRAND NOT KNOWN'.

ENTER TEXT OF NO MORE THAN 21 CHARACTERS

18. SBrCode CODE FOR BRAND AT SBRIDESC

ENTER SPACE BAR TO SEE CODES

IF BRAND NOT FOUND, CHANGE BRAND DESCRIPTION AT BOTTOM OF LOOK-UP WINDOW TO 'NF' AND SELECT CODE FOR 'BRAND NOT FOUND'

PRESS ENTER TO SELECT CODE AND ENTER AGAIN TO CONTINUE

General Household Survey, 2006

Household and Individual Questionnaires

Ask if respondent drinks at all nowadays
(Drinknow = 1   or   DrinkAny = 1)

19. Spirits   How often have you had a drink of SPIRITS OR LIQUEURS, such as gin, whisky, brandy, rum, vodka, advocaat or cocktails during the last 12 months?

USE HELP SCREEN FOR OTHER DRINKS TO BE INCLUDED HERE

Almost every day 1
5 or 6 days a week 2
3 or 4 days a week 3
once or twice a week 4
once or twice a month 5
once every couple of months 6
once or twice a year 7
not at all in last 12 months 8

Ask if respondent drank spirits at all in the last 12 months
(Spirits = 1 – 7)

20. SpritsQ   How much SPIRITS OR LIQUEURS, such as gin, whisky, brandy, rum, vodka, advocaat or cocktails have you usually drunk on any one day during the last 12 months?

CODE THE NUMBER OF SINGLES - COUNT DOUBLES AS TWO SINGLES.

1..97

Ask if respondent drinks at all nowadays
(Drinknow = 1   or   DrinkAny = 1)

21. Sherry   How often have you had a drink of SHERRY OR MARTINI including port, vermouth, Cinzano and Dubonnet, during the last 12 months?

USE HELP SCREEN FOR OTHER DRINKS TO BE INCLUDED HERE

Almost every day 1
5 or 6 days a week 2
3 or 4 days a week 3
once or twice a week 4
once or twice a month 5
once every couple of months 6
once or twice a year 7
not at all in last 12 months 8

Ask if respondent drank sherry at all in the last 12 months
(Sherry = 1 –7)

22. SherryQ   How much SHERRY OR MARTINI, including port, vermouth, Cinzano and Dubonnet have you usually drunk on any one day during the last 12 months?

CODE THE NUMBER OF GLASSES

1..97

General Household Survey, 2006
Household and Individual Questionnaires

Ask if respondent drinks at all nowadays
(Drinknow = 1 or DrinkAny = 1)

23. Wine   How often have you had a drink of WINE, including Babycham and champagne, during the last 12 months?

USE HELP SCREEN FOR OTHER DRINKS TO BE INCLUDED HERE

Almost every day 1
5 or 6 days a week 2
3 or 4 days a week 3
once or twice a week 4
once or twice a month 5
once every couple of months 6
once or twice a year 7
not at all in last 12 months 8

Ask if respondent drank wine at all in the last 12 months
(Wine = 1 – 7)

24. WineQ   How much WINE, including Babycham and champagne, have you usually drunk on any one day during the last 12 months?

CODE THE NUMBER OF GLASSES.
1 BOTTLE = 6 GLASSES, 1 LITRE = 8 GLASSES

1..97

Ask if respondent drinks at all nowadays
(Drinknow = 1 or DrinkAny = 1)

25. Pops   How often have you had a drink of ALCOPOPS (ie. alcoholic lemonade, alcoholic colas or other alcoholic fruit-or-herb-flavoured drinks for eg Smirnoff Ice, Bacardi Breezer, WKD, Metz etc), during the last 12 months?

USE HELP SCREEN FOR OTHER DRINKS TO BE INCLUDED HERE

Almost every day 1
5 or 6 days a week 2
3 or 4 days a week 3
once or twice a week 4
once or twice a month 5
once every couple of months 6
once or twice a year 7
not at all in last 12 months 8

Ask if respondent drank alcopops at all in the last 12 months
(Pops = 1 – 7)

26. PopsQ   How much ALCOPOPS (ie. alcoholic lemonade, alcoholic colas or other alcoholic fruit-or-herb-flavoured drinks) have you usually drunk on any one day during the last 12 months?

CODE THE NUMBER OF BOTTLES

1..97

Ask if respondent drinks at all nowadays
(Drinknow = 1 or DrinkAny = 1)

27. DrnkOft   Thinking now about all kinds of drinks, how often have you had an alcoholic drink of any kind during the last 12 months?

    Almost every day 1
    5 or 6 days a week 2
    3 or 4 days a week 3
    once or twice a week 4
    once or twice a month 5
    once every couple of months 6
    once or twice a year 7
    not at all in last 12 months 8

* End of 12 month drinking section *

Ask if respondent drinks at all nowadays
(Drinknow = 1 or Drinkany = 1)

28. DrinkL7   I'd like to ask you whether you have drunk different types of alcoholic drink in the last 7 days.

I'd like to hear about ALL types of alcoholic drinks you have had.
I do not need to know about non-alcoholic or low alcohol drinks.

Did you have an alcoholic drink in the seven days ending yesterday?

    Yes   1
    No   2

Ask if respondent has had an alcoholic drink in the last week
(DrinkL7 = 1)

29. DrnkDay   On how many days out of the last seven did you have an alcoholic drink?

    1..7

Ask if respondent had an alcoholic drink on two or more days last week
(DrnkDay = 2-7)

30. DrnkSame   Did you drink more on some days than others, or did you drink about the same on each of those days?

    Drank more on one/some day(s) than other(s) 1
    Same each day 2

Ask if respondent had an alcoholic drink last week
(DrinkL7 = 1)

31. WhichDay   Which day (last week) did you last have an alcoholic drink/have the most to drink?

    Sunday 1
    Monday 2
    Tuesday 3
    Wednesday 4
    Thursday 5

Friday 6
Saturday 7

Ask if respondent has had an alcoholic drink in the last week
(DrinkL7 = 1)

32. DrnkTyp   SHOWCARD 12

Thinking about last (DAY AT WHICHDAY) what types of drink did you have that day?

If you are not sure whether a drink you have had goes into a category, please let me know.

CODE ALL THAT APPLY

Normal strength beer/lager/cider/shandy 1
Strong beer/lager/cider 2
Spirits or liqueurs 3
Sherry or martini 4
Wine 5
Alcopops 6

Ask if respondent drank 'normal strength beer/lager/cider/shandy' on that day
(DrnkTyp = 1)

33. NBrL7   Still thinking about last (DAY AT WHICHDAY), how much NORMAL STRENGTH BEER, LAGER, STOUT, CIDER or SHANDY (excluding cans and bottles of shandy) did you drink that day?

CODE MEASURES THAT YOU ARE GOING TO USE,
CODE ALL THAT APPLY.
PROBE IF NECESSARY.

Half pints 1
Small cans 2
Large cans 3
Bottles 4

For each measure mentioned at NBrL7

34. NBrL7Q   ASK OR RECORD

How many (Answer AT NBrL7) of NORMAL STRENGTH BEER, LAGER, STOUT, CIDER OR SHANDY (EXCLUDING CANS AND BOTTLES OF SHANDY) did you drink that day?

1..97

Ask if respondent described measures in 'Bottles'
(NBrL7 = 4)

35. NB7lDesc   ASK OR RECORD

Which make of NORMAL STRENGTH BEER, LAGER, STOUT or CIDER did you drink from bottles on that day?

IF RESPONDENT DRANK DIFFERENT MAKES CODE WHICH THEY DRANK MOST

General Household Survey, 2006
Household and Individual Questionnaires

IF RESPONDENT DRINKS FRENCH BOTTLED BEER, BUT DOES NOT KNOW THE BRAND NAME TYPE 'FRENCH', ENTER CODING FRAME AND CODE 'FRENCH BEER – BRAND NOT KNOWN'

IF RESPONDENT DOES NOT KNOW THE BRAND NAME BUT THEY INDICATE THAT THE BOTTLE IS LARGE THEN TYPE 'LARGE, ENTER CODING FRAME AND CODE 'LARGE BOTTLE – BRAND NOT KNOWN'

ENTER TEXT OF AT MOST 21 CHARACTERS

36. NB7CODE   Code for brand at NB7IDesc

ENTER SPACE BAR TO SEE CODES

IF BRAND NOT FOUND, CHANGE BRAND DESCRIPTION AT BOTTOM OF LOOKUP WINDOW TO 'nf' AND SELECT CODE FOR 'BRAND NOT FOUND'

Ask if respondent drank 'strong beer/lager/cider' on that day
(DrnkTyp = 2)

37. SBrL7   Still thinking about last (DAY AT WHICHDAY), how much STRONG BEER, LAGER, STOUT OR CIDER did you drink that day?

CODE MEASURES THAT YOU ARE GOING TO USE
CODE ALL THAT APPLY. PROBE IF NECESSARY.

Half pints 1
Small cans..2
Large cans..3
Bottles 4

For each measure mentioned at SBrL7

38. SBrL7Q   ASK OR RECORD

How many (Answer AT SBrL7) of STRONG BEER, LAGER, STOUT or CIDER did you drink on that day?

1..97

Ask if respondent described measures in 'Bottles'
(SBrL7 = 4)

39. SB7IDesc   ASK OR RECORD

Which make of STRONG BEER, LAGER, STOUT or CIDER did you drink from bottles on that day?

IF RESPONDENT DRANK DIFFERENT MAKES CODE WHICH THEY DRANK MOST

IF RESPONDENT DRINKS FRENCH BOTTLED BEER BUT DOES NOT KNOW THE BRAND NAME TYPE 'FRENCH', ENTER CODING FRAME AND CODE 'FRENCH BEER – BRAND NOT KNOWN'

IF RESPONDENT DOES NOT KNOW THE BRAND NAME BUT THEY

INDICATE THAT THE BOTTLE IS LARGE THEN TYPE 'LARGE', ENTER CODING FRAME AND CODE 'LARGE BOTTLE – BRAND NOT KNOWN'

ENTER TEXT OF AT MOST 21 CHARACTERS

40. SB7CODE   Code for brand at SB7IDesc

ENTER SPACE BAR TO SEE CODES

IF BRAND NOT FOUND, CHANGE BRAND DESCRIPTION AT BOTTOM OF LOOKUP WINDOW TO 'nf' AND SELECT CODE FOR 'BRAND NOT FOUND'

Ask if respondent drank spirits or liqueurs on that day
(DrnkTyp = 3)

41. SpirL7   Still thinking about last (DAY AT WHICHDAY), how much spirits or liqueurs, such as whisky, brandy, rum, vodka, advocaat or cocktails did you drink on that day?

CODE THE NUMBER OF SINGLES - COUNT DOUBLES AS TWO SINGLES

1..97

Ask if respondent drank sherry or martini on that day
(DrnkTyp = 4)

42. ShryL7   Still thinking about last (DAY AT WHICHDAY), how much sherry or martini, including port, vermouth, Cinzano and Dubonnet did you drink on that day?

CODE THE NUMBER OF GLASSES

1..97

Ask if respondent drank wine on that day
(DrnkTyp = 5)

43. WineL7   Still thinking about last (DAY AT WHICHDAY), how much wine, including Babycham and champagne, did you drink on that day?

CODE THE NUMBER OF GLASSES
1 BOTTLE = 6 GLASSES. 1 LITRE = 8 GLASSES.

1..97

Ask if respondent drank alcopops on that day
(DrnkTyp = 6)

44. PopsL7   Still thinking about last (DAY AT WHICHDAY), how much alcopops (ie. Smirnoff Ice, Bacardi Breezer, WKD, Metz) did you drink on that day?

CODE THE NUMBER OF BOTTLES

1..97

Ask if respondent drinks at all nowadays
(Drinknow = 1 or Drinkany = 1)

45. DrAmount [*] Compared to five years ago, would you say that, on the whole, you drink...

RUNNING PROMPT

more 1
about the same 2
or less nowadays 3

FAMILY INFORMATION

Ask this section of all aged 16-59 (except proxy respondents)

1. FamIntro THE NEXT SECTION CONSISTS OF A SERIES OF QUESTIONS ABOUT FAMILY INFORMATION
(Not asked of proxy respondents)

To all married couples
(MarStat = 2 or 3)

2. ChkFIA INTERVIEWER CODE

Respondent is married and spouse IS a household member 1
Respondent is married but their spouse is NOT a household member ... 2

Ask if married, but partner NOT a household member
(ChkFIA = 2)

3. HusbAway INTRODUCE AS NECESSARY

May I check, is your husband/ wife absent because he/ she usually works away from home, the marriage has broken down or for some other reason?

Usually works away (include Armed Forces, Merchant Navy) 1
Marriage broken down/ separated 2
Other reason 3

To all

4. SelfCom3 The next set of questions is about family information, which you may wish to complete on your own.

EXPLAIN THAT THESE QUESTIONS COVER ANY CURRENT AND PREVIOUS MARRIAGES AND PERIODS OF LIVING TOGETHER AND FOR WOMEN THERE ARE QUESTIONS ABOUT ANY CHILDREN THEY HAVE HAD.

-IF YOU HAVE CHOSEN LAPTOP SELF-COMPLETION PLEASE EXPLAIN THAT INSTRUCTIONS WILL APPEAR ON THE SCREEN AND THEN WORK THROUGH THE FIRST 5 QUESTIONS WITH RESPONDENT. REMEMBER TO PRESS <F2> TO SAVE WORK BEFORE HANDING OVER LAPTOP.

General Household Survey, 2006
Household and Individual Questionnaires

-IF YOU HAVE CHOSEN PAPER SELF-COMPLETION OFFER THE (COLOURED) FORM TO RESPONDENT.

Respondent accepted self-completion by laptop NOW	1
Will complete later	2
Respondent accepted self-completion by paper	3
Section read and entered by interviewer	4
Interviewer now entering data from paper questionnaire	5
Respondent refused whole section	6

Ask if Respondent accepts self-completion by laptop (CASI)
(SelfCom3 = 1)

[TestQ1 to TestQ5 are **only** test questions for the respondent so that they can practice answering different types of questions on the laptop.]

5. TestQ1   The next section consists of a series of questions for you to go through with the interviewer

PRESS <1> TO CONTINUE

6. TestQ2   How old are you?

Please enter the number of years

00..99

7. TestQ3   Can you tell me the year in which you were born?

Enter year in a 4 digit format e.g. 2000

1900..2006

8. TestQ4   Can you tell me the month in which you were born?

Enter the month

January	1
February	2
March	3
April	4
May	5
June	6
July	7
August	8
September	9
October	10
November	11
December	12

9. TestQ5   Can you tell me your date of birth?

Enter in full e.g. 01/02/1976

………………………………

10. TESTEND   The next section consists of questions on family information and are for you to complete alone. If you have any problems please ask the interviewer for help

General Household Survey, 2006
Household and Individual Questionnaires

PRESS <1> TO CONTINUE

Ask people who have been married
(Marstat = 2, 3, 4 or 5)

11. AreWed   Earlier you said you were married/ separated/ divorced/ widowed. Thinking of this/ your last marriage, are/ were you legally married or are/ were you simply living together as a couple?

Legally married 1
Living together as a couple..2

Ask if respondent was legally married
(WhereWed = 1)

12. HowWed   Did you get married with a religious ceremony of some kind, or at a register office or approved premises?

Religious ceremony of some kind...1
Civil marriage in register office or approved premises 2
Both religious ceremony and register office/ approved premises...........3

Ask if respondent has been legally married
(WhereWed = 1)

13. NumMar   How many times have you been legally married?

PLEASE INCLUDE PRESENT MARRIAGE

1..7

Ask all cohabiting couples, including same sex couples (exc. couples now separated)
(Livewith = 1 or 3   or   WhereWed = 2)

14. ClYr   THIS QUESTION REFERS TO YOUR CURRENT RELATIONSHIP WHERE YOU ARE LIVING WITH SOMEONE AS A COUPLE BUT ARE NOT LEGALLY MARRIED

ENTER YEAR IN 4 DIGIT FORMAT E.G. 2000

Which year did you and your partner start living together as a couple?

1900..2006

15. CLMon   THIS QUESTION REFERS TO YOUR CURRENT RELATIONSHIP WHERE YOU ARE LIVING WITH SOMEONE AS A COUPLE BUT ARE NOT LEGALLY MARRIED

ENTER MONTH

Which month did you and your partner start living together as a couple?

1..12

Ask cohabiting couples, including same sex couples, but not separated, divorced or widowed respondents
(Livewith = 1 or 3 OR Wherewed = 2 AND is NOT separated, divorced or widowed)

General Household Survey, 2006

Household and Individual Questionnaires

16. ClMar   Have you ever been legally married?

                               Yes 1
                               No 2

Ask if respondent has been legally married
(ClMar = 1)

17. ClNumMar   How many times have you been legally married altogether?

                               1..7

Ask of all who are, or have been, legally married
(NumMar ≥ 1 or ClNumMar ≥ 1)

18. Intro   THE NEXT SCREEN CONSISTS OF A TABLE OF MARRIAGES FOR (NAME). PLEASE ENTER DETAILS OF MARRIAGES STARTING WITH THE EARLIEST AND ENDING WITH THE CURRENT OR MOST RECENT.

For each marriage

19. YrMar   In which year were you married?

                               ENTER YEAR IN 4 DIGIT FORMAT E.G. 2000

                               1900..2006

20. MonMar   .....which month in that year (*year*) were you married?

                               ENTER MONTH

                               1..12

21. LvTgthr   Before getting married did you and your husband/wife live together as a couple?

                               Yes 1
                               No 2

Ask if lived as a couple before getting married
(LvTgthr = 1)

22. YrLvTg   .....in which year did you start living together?

                               ENTER YEAR IN 4 DIGIT FORM E.G.2000

                               1900..2006

23. MonLvTg   ......which month in that year (*year*) did you start living together?

                               ENTER MONTH

                               1..12

Ask all who are or have been legally married
(NumMar ≥ 1 or ClNumMar ≥ 1)

For last marriage entered

24. Current   Thinking about your present/ most recent/ first marriage, is this marriage current, or has it ended through death, divorce or separation?

    Current 1
    Ended through death, divorce or separation 2

Ask if marriage ended
(Current = 2   or   marriage number less than total marriages)

25. HowEnded   ASK OR RECORD

    Did your marriage end in ...

    RUNNING PROMPT

    death 1
    divorce 2
    or separation. 3

Ask if marriage ended in death
(HowEnded = 1)

26. YrDie   …..in which year did your husband/wife die?

    ENTER YEAR IN 4 DIGIT FORMAT E.G. 2000

    1900-2006

27. MonDie   ……in which month of that year (*year*) did your husband/wife die?

    ENTER MONTH

    1..12

Ask if marriage ended in divorce or separation
(HowEnded = 2 or 3)

28. YrSep   ….in which year did you stop living together as a couple?

    ENTER YEAR IN 4 DIGIT FORMAT E.G. 2000

    1900-2006

29. MonSep   …..in which month of that year (*year*) did you stop living together?

    ENTER MONTH

    1..12

Ask if marriage ended in divorce
(HowEnded =2)

30. YrDiv   …..in which year was your decree absolute granted?

    ENTER YEAR IN 4 DIGIT FORMAT E.G. 2000

1900-2005

31. MonDiv …..in which month of that year (*year*) was your decree absolute granted?

ENTER MONTH

1..12

Ask if respondent is aged 16-59
(DVAge = 16-59)

32. Cohab Have you had any (other) relationships in which you lived together with someone as a couple but did not get married?

Yes. 1
No 2

Ask if respondent is aged 16-59, and has had previous cohabiting relationships
(DVAge = 16-59 & Cohab = 1)

33. Numcohab How many relationships have you had altogether in which you lived together with someone as a couple but did not get married?
(Please exclude your present relationship)

1..7

34. Intro Now I would like to ask you some questions about (the first three of these relationships/ this relationship/ these relationships).

(RECORD DETAILS OF THE FIRST THREE RELATIONSHIPS, STARTING WITH THE FIRST)

Ask each question for the first, second and third relationship (questions 35 – 47)

35. TimeCoy1 Thinking about the first/second/third relationships where you lived with someone as a couple but did not get married……how long did you live together in terms of years and months?

PLEASE ENTER NUMBER OF YEARS ONLY

0..99

36. Timecom1 NOW ENTER NUMBER OF MONTHS

0..11

37. WhencoY1 Can you tell me the year in which you started or stopped living together as a couple with your partner?

ENTER THE YEAR

1950..2005

38. WhencoM1 Can you tell me the month in which you started or stopped living together as a couple with your partner?

INTERVIEWER ENTER THE MONTH

1..12

39. Starten1   INTERVIEWER: IS THIS WHEN THE RESPONDENT AND HIS/HER PARTNER STARTED LIVING TOGETHER OR FINISHED LIVING TOGETHER AS A COUPLE?

Started living together 1
Finished living together 2

40. Othdate1   If that was the date you started/stopped living together, then you stopped/started living together in …(month) …(year)
Does that seem about right?

Yes 1
No 2

Ask if computed start/end date not correct
(Othdate1 = 2)

41. RghtdtY1   What is the correct date?

ENTER THE YEAR

1950..2005

42. RghtdtM1   (What is the correct date?)

INTERVIEWER ENTER THE MONTH

1..12

Ask if respondent is aged 16-59, and has had previous cohabiting relationships
(DVAge = 16-59   &   Cohab = 1)

Ask each question for the first, second and third relationship

43. EndCoh1   You said you stopped living together in …(month) …(year).
May I just check, was this when you stopped living together in the same accommodation, when the relationship ended or both?

Stopped living in the same accommodation only 1
End of the relationship only 2
Both happened at the same time 3
Partner died..4

Ask if date given is when they stopped living together
(EndCoh1 = 1)

44. EndrelY1   In which year did the relationship end?

ENTER THE YEAR

1950..2005

45. EndrelM1   In which month did the relationship end?

ENTER THE MONTH

1..12

General Household Survey, 2006
Household and Individual Questionnaires

Ask if the date given is when relationship ended
(EndCoh1 = 2)

46. EndlivY1

In which year did you stop living in the same accommodation?

ENTER THE YEAR

1950..2005

47. EndlivM1

In which month did you stop living in the same accommodation?

INTERVIEWER ENTER THE MONTH

1..12

Ask respondents aged 16-59
(DVAge = 16-59)

If respondent is female (sex=2) *also include* *

48. Children

ASK OR RECORD

Do you have any children in the household?

(INCLUDES ADULT CHILDREN)
(*AND/OR STEP OR FOSTER CHILDREN.)

(Respondent instruction)
(IF APPROPRIATE PLEASE INCLUDE ANY CHILDREN FROM YOUR PARTNER'S PREVIOUS RELATIONSHIP)

Yes 1
No 2

Ask women who have a child in the household
(Sex = 2 & Children =1)

49. StpChldF

(The next questions are about the family.)
Do you have any step, foster, or adopted children of any age living with you? (For stepchildren, please include any children of any age from your husband/partner's previous marriage or relationship.)

Yes 1
No 2

Ask men who have a child in the household
(Sex = 1 & Children =1)

50. StpChldM

Do you have any stepchildren of any age living with you? (Please include any children of any age from your wife/partner's previous marriage or relationship.)

Yes 1
No 2

Ask women with a step, foster or adopted child
(StpChldF = 1)

General Household Survey, 2006
Household and Individual Questionnaires

51. NumStepF   How many step children live with you? (Please include children of any age from your husband/partner's previous marriage or relationship.)

0..7

Ask men with a stepchild living with them
(StpChldM=1)

52. NumStepM   How many step children live with you? (Please include children of any age from your husband/partner's previous marriage or relationship.)

1..7

Ask women with a step, foster or adopted child living with them
(StpChldF = 1)

53. NumFost   How many foster children have you living with you altogether?

0..7

54. NumAdop   How many adopted children have you living with you altogether?

0..7

Ask women with a step, foster or adopted child, or a man with a stepchild living with them
(StpChldF = 1 or StpChldM=1)

55. StepInt   THE NEXT SCREEN CONSISTS OF A TABLE FOR THE STEP-CHILDREN (AND ADOPTED AND FOSTER- CHILDREN) OF (NAME) PLEASE ENTER DETAILS FOR EACH CHILD.

Ask for each step/foster/adopted child

56. ChildNo   From the list below, please copy the number of the first/second… step/foster/adopted child.

INCLUDES ADULT CHILDREN

1..20

57. ChldType   Thinking about *name*. Is *name* your step-child, your adopted-child or are they your foster-child?

Step 1
Foster 2
Adopted 3

58. ChLivYr   Please state the year (CHILDS NAME) started living with you.

ENTER YEAR (IN 4 DIGIT FORMAT, E.G. 2000)

1900..2006

59. ChLivMon   Please state the month (CHILDS NAME) started living with you.

ENTER MONTH

1..12

General Household Survey, 2006
Household and Individual Questionnaires

Ask all women
(Sex = 2)

60. Baby   ASK OR RECORD
                     EXCLUDE: ANY STILLBORN.
                     INCLUDE: ALL CHILDREN RESPONDENT HAS GIVEN BIRTH TO.

                     Have you ever given birth to a baby - even one who only lived for a short time?

                     Yes 1
                     No 2

Ask women who have had a baby
(Baby = 1)

61. NumBaby   EXCLUDE: ANY STILLBORN

                     How many children have you given birth to, including any who live somewhere else and any who have died since birth?

                     1..20

62. BirthInt   In the next set of questions, we want you to record some details about all the children you have ever given birth to. This includes children:
                     -   who live here;
                     -   children who live somewhere else, such as adult children; and
                     -   children who have died since birth.
                     Please enter details for each child, in the order in which they were born, starting with the one you gave birth to first.

For each child

63. BirthDte   Please enter the Date of birth

                     PLEASE ENTER IN DATE OF BIRTH ORDER - ELDEST FIRST, YOUNGEST LAST.

                     AS A GUIDE, THE D.O.B. OF EACH HOUSEHOLD MEMBER IS LISTED BELOW

64. BirthSex   Sex of child

                     Male 1
                     Female 2

65. ChldLive   Is child living with you?

                     Yes 1
                     No, lives elsewhere..2
                     No, deceased...3

Ask all women aged 16-49
(Sex = 2 & DVAge = 16-49)

66. Pregnant	Are you pregnant now?	

Yes 1
No/unsure 2

67. MoreChld SHOWCARD 13

[*] Do you think that you will have any (more) children (after the one you are expecting)? Could you choose your answers from this card.

Yes 1
Probably yes 2
Probably not 3
No 4
Don't know..5

Ask if respondent answered don't know above
(MoreChld = DK)

68. ProbMore [*] On the whole do you think...

You will probably have any/more children 1
Or you will probably not have any/more children?................2

Ask if respondent is likely to have more children
(MoreChld = 1 or 2 or ProbMore = 1)

69. TotChld [*] How many children do you think you will have born to you in total, including those you have had already who are still alive/ (and) the one you are expecting?

1..14

70. NextAge [*] How old do you think you will be when you have your first/next baby (after the one you are expecting)?

1..97

FINANCIAL SITUATION

This section to be asked of household, not individuals
Ask HRP

1. Repay The next section has questions on your HOUSEHOLD's financial situation.

Do you or anyone in your household have to repay any credit card, hire purchase or other loans (that is, excluding mortgage repayments or other loans connected with the accommodation)?

Yes 1
No 2

Ask if anyone in the household has any loans to repay
(Repay = 1)

2. BurdRepy   [*] To what extent is the repayment of such loans and the interest a financial burden or struggle for your household?

Would you say it is...

RUNNING PROMPT

a heavy burden/ struggle, ...1
a slight burden/ struggle, ...2
or not a burden/ struggle at all? 3

Ask all households

3. Afford   SHOWCARD 14

Looking at this card, can I just check whether your household could afford the following?

CODE ALL THAT APPLY

To pay for a week's annual holiday away from home? 1
To eat meat, chicken or fish (or vegetarian equivalent) every second day?
....2
To pay an unexpected, but necessary, expense of £500? 3
To keep your home adequately warm? 4
Afford none of these (SPONTANEOUS ONLY) 5

4. FinArr   SHOWCARD 15

Looking at this card, has your household been in arrears at any time since *date*, that is, unable to pay any of the following ON TIME?

CODE ALL THAT APPLY

rent for accommodation? 1
mortgage payments? 2
utility bills, such as for electricity, water or gas? 3
hire purchase instalments or other loan payments? 4
In arrears on none of these (SPONTANEOUS ONLY) 5

5. EndsMeet   [*] A household may have different sources of income and more than one household member may contribute to it.

Thinking of your household's total monthly or weekly income, is your household able to make ends meet, that is pay your usual expenses.....

IF NOT MAKING ENDS MEET, CODE AS 'WITH GREAT DIFFICULTY' (CODE 1)

RUNNING PROMPT

with great difficulty, ...1
with difficulty, ...2
with some difficulty, ...3
fairly easily, ...4
easily, ...5
or very easily? 6

6. LowestIn [*] Thinking of the household's basic needs, what is the very minimum amount of money the household needs each month to pay its usual expenses?

Please answer in relation to the present circumstances of your household, and what you consider as usual expenses.

0.00..999999.99

7. HowLong For how long did this cover?

one week 1
two weeks 2
three weeks 3
four weeks 4
calendar month 5
two calendar months 7
eight times a year 8
nine times a year 9
ten times a year 10
three months/13 weeks 13
six months/26 weeks 26
one year/12 months/52 weeks 52
less than one week 90
one off/lump sum 95
none of these: EXPLAIN IN A NOTE 97

INCOME

Ask all adults (except proxy respondents)

1. Intro SHOWCARD 16

Looking at this card, are you at present receiving any of these benefits in your own right: that is, where you are the named recipient?

PRESS <1> TO CONTINUE (AND ENTER ANSWERS AT THE NEXT QUESTION), OR
PRESS <7> IF RESPONDENT REFUSES BENEFITS QUESTIONS

Continue with benefits questions 1
Refused whole income section 7

Ask if continuing with benefits questions
(Intro = 1)

2. Ben1Q SHOWCARD 17

Looking at this card, are you at present receiving any state benefits in your own right: that is, where you are the named recipient?

CODE ALL THAT APPLY

Child Benefit 1

Guardian's Allowance	2
Carer's Allowance	3
Retirement pension (National Insurance), or Old Person's pension	4
Widow's pension, Bereavement Allowance or Widowed Parents (formerly Widowed Mother's) Allowance	5
War Disablement Pension or War Widow's/Widower's Pension (and any related allowances)	6
Severe Disablement Allowance	7
None of these	8

3. DisBen

SHOWCARD 17

And looking at this card, are you at present receiving any of the state benefits shown on this card - either in your own right, or on behalf of someone else in the household?

CODE ALL THAT APPLY

Care component of Disability Living Allowance	1
Mobility component of Disability Living Allowance	2
Attendance Allowance	3
None of these	4

Ask if receiving Attendance Allowance (DisBen = 3)

4. Attall

Is this paid as part of your retirement pension or do you receive a separate payment?

Paid as part of pension	1
Separate payment	2

Ask all except proxy respondents

5. Ben2Q

SHOWCARD 18

CODE ALL THAT APPLY

Now looking at this card, are you at present receiving any of these benefits in your own right - that is, where you are the named recipient?

Job Seekers' Allowance (JSA)	1
Pension Credit	2
Income Support	3
Incapacity Benefit	4
Maternity Allowance	5
Industrial Injury Disablement Benefit	6
None of these	7

Ask all except proxy respondents

6. TxCred

SHOWCARD 19

received

Now looking at this card, are you at present receiving any of these Tax Credits, in your own right? Please include any lump sum payments

in the last six months?

General Household Survey, 2006
Household and Individual Questionnaires

	Working Tax Credit (excluding any childcare tax credit) 4
	Child Tax Credit (including any childcare tax credit) 5
	None of these..6

7. Inclus   SHOWCARD 20

Did your last wage/salary include any of the items on this card?

CODE ALL THAT APPLY.

Statutory Sick Pay 1
Statutory Maternity Pay...2
Statutory Paternity Pay..3
Statutory Adoption Pay 4
Income Tax Refund 5
Mileage Allowance or fixed allowance for motoring............................6
Motoring Expenses Refund 7
Tax credit..8
None of these...9

Ask women under 55 years
(Sex = 2 & DVAge <55 years)

7. MatAll   SHOWCARD 20

Are you currently getting either of the things shown on this card, in your own right?

Maternity Allowance 1
Statutory Maternity Pay from an employer or former employer 2
Neither of these 3

Ask all (except proxy respondents)

8. Ben12m   SHOWCARD 21

In the last 12 months, have you received any of the things shown on this card, in your own right?

A grant from the Social Fund for funeral expenses...............................1
A grant from Social Fund for maternity expenses/Sure Start
     Maternity Grant 2
A Social Fund loan or Community Care grant 3
None of these...4

Ask all aged 60 or over
(DVAge > 60)

9. Winter   In the last 12 months have you received a winter fuel payment in your own right?

Yes 1
No 2

Ask all (except proxy respondents)

10. Ben6m   SHOWCARD 22

In the last 6 months have you received any of the things on this card in your own right?

'Extended payment' of Housing Benefit/rent rebate, or Council Tax Benefit (4 week payment only)	2
Widow's payment or Bereavement payment – lump sum	3
Child Maintenance Bonus	4
Lone Parent's Benefit Run-On	5
Any National Insurance or State Benefit not mentioned earlier	6
None of these	7

Code for each benefit mentioned
(Ben1Q, DisBen, Ben2Q, Makall, TxCred, Ben12m, Ben6m)

11. BAmt   How much did you get last time?

IF COMBINED WITH ANOTHER BENEFIT AND UNABLE TO GIVE SEPARATE AMOUNT, ENTER `Don't know` <CTRL> + <K>

0.00..997.00

If don't know or refusal at the amount of benefit received
(Ben1Amt = DK or Refusal)

12. BAmtDK   INTERVIEWER: IS THIS `DON'T KNOW` BECAUSE IT'S PAID IN COMBINATION WITH ANOTHER BENEFIT, AND YOU CANNOT ESTABLISH A SEPARATE AMOUNT?

Yes (Please give full details in a Note)	1
No	2

Ask if amount of benefit received was greater than zero
(Ben1Amt > 0.00)

13. Bpd   How long did this cover?

one week	1
two weeks	2
three weeks	3
four weeks	4
calendar month	5
two calendar months	7
eight times a year	8
nine times a year	9
ten times a year	10
three months/13 weeks	13
six months/26 weeks	26
one year/12 months/52 weeks	52
less than one week	90
one off lump sum	95
none of these	97

Ask if receiving Retirement pension or old person's pension
(Ben1Q = 4)

14. BenUs   Is this the amount you usually get?

General Household Survey, 2006
Household and Individual Questionnaires

Yes 1
No 2
No such thing as a usual amount 3

Ask if not usual amount
(BenUs = 2)

15. BUAmt   How much do you usually get?

IF COMBINED WITH ANOTHER BENEFIT AND UNABLE TO GIVE SEPARATE AMOUNT, ENTER DON'T KNOW <CTRL+K>

0.00..997.00

16. BUPd   How long does this cover?

one week..1
two weeks...2
three weeks..3
four weeks..4
calendar month 5
two calendar months...7
eight times a year..8
nine times a year...9
ten times a year...10
three months/13 weeks 13
six months/26 weeks...26
one year/12 months/52 weeks...52
less than one week 90
one off lump sum...95
none of these 97

Ask if receiving War Disablement Pension or War Widow's Pension
(Ben1Q = 6)

17. WPentype   Do you receive...

RUNNING PROMPT

War Disablement Pension 1
or War Widow's Pension 2

Ask if receiving CARE component of Disability Living Allowance
(DisBen = 1)

18. WhoReCar   Whom do you receive it for?

IF CURRENT HOUSEHOLD MEMBER, ENTER PERSON NUMBER OTHERWISE ENTER 97

1..16, 97

Ask if receiving MOBILITY component of Disability Living Allowance
(DisBen = 2)

19. WhoReMob   Whom do you receive it for?

General Household Survey, 2006
Household and Individual Questionnaires

> IF CURRENT HOUSEHOLD MEMBER, ENTER PERSON NUMBER OTHERWISE ENTER 97
>
> 1..16, 97

Ask if receiving Attendance Allowance
(DisBen = 3)

20. WhoReAtt   Whom do you receive it for?

> IF CURRENT HOUSEHOLD MEMBER, ENTER PERSON NUMBER OTHERWISE ENTER 97
>
> 1..16, 97

Ask all (except proxy respondents)

21. OthSrc   SHOWCARD 23

> Taking your response from this card, pleases tell me if you are receiving any of the following type of regular payments
>
> CODE ALL THAT APPLY
>
> Occupational pensions from former employer(s) 1
> Occupational pensions from a spouse's former employer(s) 2
> Private pensions or annuities 3
> Regular redundancy payments from former employer(s) 4
> Government Training Schemes such as YT allowance 5
> None of these 6

Ask if receiving payments from occupational pensions from employer(s)
(OthSrc = 1)

23. PFEmpNet   In total how much do you receive each month from the occupational pension(s) from your former employer(s) AFTER tax is deducted? (ie NET)

> DO NOT PROBE MONTH. ACCEPT CALENDAR MONTH OR 4 WEEKLY.
>
> 0.01..99999.97

24. PFEmpGrs   In total how much do you receive each month from the occupational pension(s) from your former employer(s) BEFORE tax is deducted? (i.e. GROSS)?

> DO NOT PROBE MONTH. ACCEPT CALENDAR MONTH OR 4 WEEKLY.
>
> 0.01..99999.97

Ask if receiving payments from occupational pensions from a spouse's former employer(s)
(OthSrc = 2)

25. SpousNet   In total how much do you receive each month from the occupational pension(s) from your spouse's former employer(s) AFTER tax is deducted? (ie NET)

DO NOT PROBE MONTH. ACCEPT CALENDAR MONTH OR 4 WEEKLY.

0.01..99999.97

26. SpousGrs — In total how much do you receive each month from the occupational pension(s) from your spouse's former employer(s) BEFORE tax is deducted? (i.e. GROSS)?

DO NOT PROBE MONTH. ACCEPT CALENDAR MONTH OR 4 WEEKLY.

0.01..99999.97

Ask if receiving payments from private pension(s) or annuities
(OthSrc = 3)

27. PrivPNet — In total how much do you receive each month from your private pension(s) or annuities AFTER tax is deducted? (ie NET)

DO NOT PROBE MONTH. ACCEPT CALENDAR MONTH OR 4 WEEKLY.

0.01..99999.97

28. PrivPGrs — In total how much do you receive each month from your private pension(s) or annuities BEFORE tax is deducted? (i.e. GROSS)?

DO NOT PROBE MONTH. ACCEPT CALENDAR MONTH OR 4 WEEKLY.

0.01..99999.97

Ask if receiving regular redundancy payments from a former employer(s)
(OthSrc = 4)

29. RedunNet — In total how much do you receive each month from regular redundancy payments from your former employer(s) AFTER tax is deducted? (ie NET)

DO NOT PROBE MONTH. ACCEPT CALENDAR MONTH OR 4 WEEKLY.

0.01..99999.97

30. RedunGrs — In total how much do you receive each month from regular redundancy payments from your former employer(s) BEFORE tax is deducted? (i.e. GROSS)?

DO NOT PROBE MONTH. ACCEPT CALENDAR MONTH OR 4 WEEKLY.

0.01..99999.97

Ask if receiving payments from Government Training Schemes
(OthSrc = 5)

31. TrainNet — In total how much do you receive each month from Government Training Schemes, such as YT allowance, AFTER tax is deducted? (ie NET)

DO NOT PROBE MONTH. ACCEPT CALENDAR MONTH OR 4 WEEKLY.

0.01..99999.97

32. TrainGrs Schemes, In total how much do you receive each from Government Training such as YT allowance, BEFORE tax is deducted? (i.e. GROSS)?

DO NOT PROBE MONTH. ACCEPT CALENDAR MONTH OR 4 WEEKLY.

0.01..99999.97

Ask all (except proxy respondents)

33. ReglrP SHOWCARD 24

Taking your response from this card, please tell me if you are receiving any of the following type of regular payments

CODE ALL THAT APPLY

Educational grant..1
Regular payments from friends or relatives
outside the household 2
Maintenance, alimony or separation allowance..................3
None of these...4

Ask if receiving an educational grant
(RegIrP = 1)

35. EdGrnt In total how much do you receive from your educational grant EACH MONTH?

0.01..99999.97

Ask if receiving regular payments from friends or relatives outside household
(RegIrP = 2)

36. RegFr In total how much do you receive from friends or relatives EACH MONTH?

0.01..99999.97

Ask if receiving maintenance, alimony or separation allowance
(RegIrP = 3)

37. RegMa In total how much do you receive maintenance, alimony or separation allowances EACH MONTH?

0.01..99999.97

Ask all (except proxy respondents)

38. RegOPM SHOWCARD 25

Taking your response from this card, please tell me if you are making any of the following type of regular payments

General Household Survey, 2006
Household and Individual Questionnaires

> Regular payments to friends or relatives outside the household 2
> Maintenance, alimony or separation allowance 3
> None of these..4

Ask if making regular payments to friends or relatives outside household
(RegOPM = 1)

40. RegFro   In total how much do you give to friends or relatives EACH MONTH?

                 0.01..99999.97

Ask if making payments of maintenance, alimony or separation allowance
(RegOPM = 2)

41. RegMaO   In total how much do you pay out in terms of maintenance, alimony or separation allowances EACH MONTH?

                 0.01..99999.97

Ask all (except proxy respondents)

42. Rentpay   In the 12 months since (DATE), have you received any rent from property, for example, renting out a building, house, a flat, a room or land?

> Yes 1
> No 2

Ask if respondent is receiving rent
(Rentpay = 1)

43. RentBF   How much did you receive in the 12 months since (DATE), BEFORE deducting income tax, but after deducting all allowable expenses?

                 ALLOWABLE EXPENSES INCLUDE INTEREST PAYMENTS, INSURANCE, REPAIRS AND MAINTENANCE, RATES AND SERVICES.

                 0.01..99999.97

Ask if answered 'Don't know' to above question
(RentBF = Don't know)

44. RentBApx   SHOWCARD 25

                 Perhaps you could give an approximate range?

                 REMIND: LAST 12 MONTHS... BEFORE TAX IS DEDUCTED... BUT AFTER DEDUCTING EXPENSES

> Less than £1,000 per year 1
> £1,001 to £3,000 per year 2
> £3,001 to £5,000 per year 3
> £5,001 to £10,000 per year 4
> More than £10,000 per year 5

Ask if respondent is receiving rent
(Rentpay = 1)

45. RentAFT How much did you receive in the 12 months since (DATE), AFTER deducting income tax, but after deducting all allowable expenses?

ALLOWABLE EXPENSES INCLUDE INTEREST PAYMENTS, INSURANCE, REPAIRS AND MAINTENANCE, RATES AND SERVICES.

0.01..99999.97

Ask if answered 'Don't know' to above question
(RentAFT = Don't know)

46. RentAApx Perhaps you could give an approximate range?

REMIND: LAST 12 MONTHS... AFTER TAX IS DEDUCTED... AFTER DEDUCTING EXPENSES

Less than £1,000 per year 1
£1,001 to £3,000 per year 2
£3,001 to £5,000 per year 3
£5,001 to £10,000 per year 4
More than £10,000 per year 5

The next group of questions (Q47-Q84) are only asked of those in paid work, (including those temporarily away from job or on a government scheme), but excluding unpaid family workers.
(Wrking = 1 OR JbAway = 1 OR SchemeET = 1)

The routing instructions above each question apply only to those who meet the above criteria.

Ask if an employee
(Stat = 1)

47. PyPeriod The next questions are about earnings from your main job.

How long a period does your wage/salary usually cover?

one week...1
two weeks...2
three weeks...3
four weeks..4
calendar month..5
two calendar months..7
eight times a year...8
nine times a year..9
ten times a year..10
three months/13 weeks..13
six months/26 weeks..26
one year/12 months/52 weeks...52
less than one week...90
one off lump sum...95
none of these...97

Ask all, except those who are paid less than once a week, or in a one off sum, or answered 'none of these'
(PyPeriod <= 52)

48. TakeHome How much is your usual take home pay per (period at PyPeriod) after all deductions? (Please do not include any Working Families' Tax Credit / Disabled Person's Tax Credit payment that you received)

0.00..99999.97

Ask if paid less than once a week, or in a one off sum, or in none of these ways, or did not know how much money they usually took home
(PyPeriod = 90, 95 or 97 or TakeHome = DK)

49. TakHmEst   SHOWCARD 26

Please look at this card and estimate your usual take home pay per (period at PyPeriod) after all deductions? (Please do not include any Working Families Tax Credit / Disabled Person's Tax Credit payment that you received)

0..32

Ask if an employee
(Stat = 1)

50. GrossAm   How much are your usual gross earnings per (period at PyPeriod) before any deductions?

0.01..99999.97

Ask if respondent does not know how much their usual gross earnings are
(GrossAm = DK)

51. GrossEst   SHOWCARD 26

Please look at this card and estimate your usual gross earnings per (period at PyPeriod) before any deductions?

0..32

Ask if an employee
(Stat = 1)

52. PaySlip   INTERVIEWER - CODE WHETHER PAYSLIP WAS CONSULTED

Pay slip consulted by respondent, but not by interviewer 1
Pay slip consulted by interviewer 2
Pay slip not consulted 3

Ask if answered PyPeriod
(1 α PyPeriod α 97)

53. PayBonus   In your present job, have you ever received an occasional addition to pay in the last 12 months (that is since DATE 1 YEAR AGO) such as a Christmas bonus or a quarterly bonus?

EXCLUDE SHARES AND VOUCHERS.

Yes   1
No   2

Ask if respondent received a pay bonus
(PayBonus = 1)

54. HowBonus

Was the bonus or commission paid...

RUNNING PROMPT

after tax was deducted (net)..1
or before tax was deducted (gross)...2
or some before and some after?...3

If some or all tax was deducted, or they did not know if tax was deducted from pay bonus (HowBonus = 1 or 3 or DK)

55. NetBonus

What was the total amount you received in the last 12 months (that is since DATE 1 YEAR AGO) AFTER tax was deducted (ie net)?

0.01..99999.97

Ask if some or all tax was deducted from the pay bonus (HowBonus = 2 or 3)

56. GrsBonus

What was the total amount you received in the last 12 months (that is since DATE A YEAR AGO) before tax was deducted (ie gross)?

0.01..99999.97

Ask if self-employed (Stat = 2)

57. BusAccts

You said earlier you were self-employed in your main job.

In this job/business are annual business accounts prepared for HM Revenue & Customs for tax purposes?

INCLUDE IF PREPARED BY ACCOUNTANT

Yes 1
No 2
Not yet but will be 3

Ask if business accounts are prepared or will be (BusAccts = 1,3)

58. Se1

The questions that follow are about just your own share of the business. What is the most recent period for which accounts have been prepared for HM Revenue and Customs (formerly the Inland Revenue)?

ENTER BEGINNING OF PERIOD
IF DAY OF MONTH NOT NONE, ENTER 15

ENTER DATE

59. Se2

ENTER END OF PERIOD (FOR WHICH ACCOUNTS HAVE BEEN PREPARED)
IF DAY OF MONTH NOT NONE, ENTER 15

ENTER DATE

60. SeWeeks   May I check, how many weeks does this cover?

IF COVERS FULL 12 MONTHS ENTER 52

1..104

61. Profit1   What was your share of the profit or loss figure shown on these accounts for this period?

0..9999997

62. Profit2   Did the answer in the previous question refer to profit or loss?

Profit/Earnings............1
Loss 2

63. ProfTax   Can I just check, is that figure before the reduction of income tax?

Yes (before tax)............1
No (after tax)............2

64. PrBefore   What was (your share of) the profit before tax and lump sum National Insurance deductions?

0..9999997

Ask if business accounts are not prepared
(BusAccts = 2)

65. GrsSEMJb   You said earlier that you are self-employed in your main job.

Now I'd like to ask you some questions about your income from your job/ business; that is, after paying for any materials, equipment or goods that you use(d) in your work.

On average, what was your WEEKLY or MONTHLY (or ANNUAL) income from this job/ business over the last 12 months - BEFORE deducting Income Tax and National Insurance contributions?

IF NOTHING OR MADE A LOSS, ENTER ZERO.
IF BUSINESS PARTNERSHIP, ENTER PERSON'S SHARE OF INCOME ONLY.
IF SELF-EMPLOYED LESS THAN 12 MONTHS, REFER JUST TO PERIOD SELF-EMPLOYED.

0.00..999999.97

Ask if respondent does not know gross income from job/ business
(GrsSEMJb = DK)

66. WorkAcc   Do you have separate bank or building society accounts for your work and your private finances?

Yes 1
No 2

Ask if respondent doesn't know gross income and has a separate account
(WorkAcc = 1)

67. OwnSum   SHOWCARD 29

(CARD SHOWS:
MONEY FORM THE WORK ACCOUNT
- USED FOR PAYMENTS TO YOURSELF AND ANY OTHER PERSONAL SPENDING
- USED TO PAY DOMESTIC BILLS (INCLUDING STANDING ORDERS)
- TRANSFERRED TO A PRIVATE ACCOUNT
- USED FOR ANY OTHER NON-BUSINESS USE)

Do you draw money from your work account for any non-business purposes, such as any of the things shown on this card?

(CODE 'YES' IF ANY APPLY)

Yes 1
No 2

Ask if respondent does draw money from business account for non-business purposes
(OwnSum = 1)

68. OwnAmt   Thinking of the last 12 months, on average how much did you take each month for these non-business purposes?

1..9997

69. OwnOther   Apart from drawings from the bank/building society, do/did you receive any other income from this job/business, for personal use?

Yes 1
No 2

Ask if respondent receives other income from job/business
(OwnOther = 1)

70. OwnOtAmt   On average, how much is that each month?

0..9997

Ask if a value is given above (including zero)
(GrsSEMJb ≠ DK)

71. GrossPer   How long does this cover?

one week 1
two weeks 2
three weeks 3
four weeks 4
calendar month 5
two calendar months 7
eight times a year 8
nine times a year 9
ten times a year 10
three months/13 weeks 13
six months/26 weeks 26
one year/12 months/52 weeks 52
less than one week 90

one off lump sum	95
none of these: EXPLAIN IN A NOTE	97

Ask if gross income is above zero
(GrsSEMJb >= 1)

72. NetSEMJb   On average, what is your WEEKLY or MONTHLY (or ANNUAL) income from this job/ business over the last 12 months - AFTER deducting Income Tax and National Insurance contributions?

0.00..999999.97

73. NetPer   What period does this cover?

one week	1
two weeks	2
three weeks	3
four weeks	4
calendar month	5
two calendar months	7
eight times a year	8
nine times a year	9
ten times a year	10
three months/13 weeks	13
six months/26 weeks	26
one year/12 months/52 weeks	52
less than one week	90
one off lump sum	95
none of these: EXPLAIN IN A NOTE <CTRL> + <M>	97

Ask if gross income is zero
(GrsSEMJb = 0)

74. LosSEMJb   On average, how much have you been LOSING WEEKLY or MONTHLY (or ANNUALLY) from this job/ business over the last 12 months after deducting all business expenses?

0.00..999999.97

75. LossPer   What period does this cover?

one week	1
two weeks	2
three weeks	3
four weeks	4
calendar month	5
two calendar months	7
eight times a year	8
nine times a year	9
ten times a year	10
three months/13 weeks	13
six months/26 weeks	26
one year/12 months/52 weeks	52
less than one week	90
one off lump sum	95
none of these: EXPLAIN IN A NOTE	97

Ask if respondent has a second or third job as an employee
(EmpSE2nd = 1 or EmpSE3rd = 1)

76. PaySecJb   The next questions are about earnings from your second (and third) jobs.

77. SjNetAm   You said earlier you had another job(s), in which you were an employee.

In the last month, how much did you earn from your other/occasional job(s), AFTER DEDUCTIONS for tax and National Insurance (ie NET)?

0.01..99999.97

78. SjGrsAm   You said earlier you had another job(s), in which you were an employee.

In the last month, how much did you earn from your other/occasional job(s), BEFORE DEDUCTIONS for tax and National Insurance (ie GROSS)?

0.01..99999.97

Ask if respondent has a second or third job as self-employed
(*EmpSE2nd* = 2 or EmpSE3rd = 2)

79. SjPrfGrs   You said earlier you had another job(s), in which you were self-employed.

In the last 12 months (that is since DATE 1 YEAR AGO) how much have you earned from this work, BEFORE deducting income tax and National Insurance contributions, but...

- after deducting all business expenses, and
- before deducting money drawn for your own use?

IF MADE NO PROFIT ENTER 0.

0.00..99999.97

80. SjPrfNet   In the last 12 months (that is since DATE 1 YEAR AGO) how much have you earned from this work, AFTER deducting income tax and National Insurance contributions, but...

- after deducting all business expenses, and
- before deducting money drawn for your own use?

IF MADE NO PROFIT ENTER 0.

0.00..99999.97

Ask if respondent has a second or third job as self-employed and gross profit is zero
(Stat = 2 AND SjPrfGrs = 0)

81. SjLssGrs   In the last 12 months (that is since DATE 1 YEAR AGO) how much have you lost from this work, after deducting all business expenses?

IF MADE NO LOSS ENTER 0.

0.00..99999.97

Ask all

82. OddJob (Apart from your main job) do you earn any money from odd jobs or from work that you do from time to time?

PROMPT AS NECESSARY

INCLUDE BABYSITTING, MAIL ORDER AGENT, POOLS AGENT ETC.

Yes 1
No 2

Ask if respondent has odd/other jobs
(OddJob = 1)

83. OddJEmp In these job(s) do you work as an employee or are you self-employed?

employee..1
self-employed 2

84. OddJAmnt In the last month (that is since DATE 1 MONTH AGO) how much have you earned from your odd/ occasional jobs?

0.00..99999.97

Ask all (except proxy respondents)
– NOTE: End of the section of questions that are only asked of those in paid work

85. OthPay Apart from anything you have already mentioned, have you received any payment from interest from savings, Bank or Building Society accounts, income from shares, bonds, unit trusts or gilt-edged stock or any unincorporated business in the last 12 months (that is since DATE 1 YEAR AGO)?

PROMPT AS NECESSARY
CODE ALL THAT APPLY - EXCLUDE BENEFITS NO LONGER RECEIVED.

Interest from savings, Bank or Building Society accounts....................1
Income from shares, bonds, unit trusts or gilt-edged stock 2
Income from unincorporated business..3
Other 4
None of these...5

Ask if respondent is receiving income from some other source
(OthPay = 4)

86. OthSourc Please specify other source

ENTER TEXT OF AT MOST 100 CHARACTERS

Ask if respondent is receiving interest from savings
(OthPay = 1)

87. Investpy How much have you received in total from interest on savings, Bank or Building Society accounts in the last 12 months?

0.01..99999.97

Ask if respondent is receiving income from an unincorporated business

(OthPay = 4)

88. Unincpy   How much have you received in total from the unincorporated business in the last 12 months?

                       0.01..99999.97

Ask if respondent is receiving income from shares, bonds, unit trusts or gilt-edged stock
(OthPay = 2)

89. Sharepy   How much have you received in total from shares, bonds, unit trusts or gilt-edged stock in the last 12 months?

                       0.01..99999.97

Ask if respondent is receiving income from another source
(OthPay = 4)

90. OthRgPAm   (Apart from interest and income from shares)
How much have you received from other sources in the last 12 months?

                       0.01 .. 99999.97

Ask if proxy respondent

91. NtIncEst   SHOWCARD 26

                       I would now like to ask you about the income of (NAME).
Please could you look at this card and estimate their total NET income (that is after deduction of tax, National Insurance and any expenses) (NAME) brings into the household in a year from all sources (benefits, employment, investments etc)?

                       0..32

Ask if household contains one or more children (under the age of 16)

92. Askwlth   The next section is about child income. We only need to collect this information once for each child in the household. Who will answer the child income section for (child's name)?

                       ENTER PERSON NUMBER

                       1..16

93. AskNowCH   INTERVIEWER: DO YOU WANT TO ASK THIS SECTION FOR (CHILD'S NAME) NOW OR LATER?

                       IF YOU HAVE ALREADY ASKED THIS SECTION FOR (CHILD'S NAME), DO NOT CHANGE FROM CODE 1.

                       IF YOU WANT TO JUMP PAST THIS SECTION THEN PRESS <END>.

                       Yes, now/Already asked 1
                       Later 2

Ask if section is to be asked now
(AskNowCH =1)

General Household Survey, 2006
Household and Individual Questionnaires

94. INCSOR Has (CHILD'S NAME) received or earned an independent source of income in the last 12 months?

Please ignore anything less than £30.00 per month, or money received from other household members.

Yes 1
No 2

Ask if child has received or earned an independent source of income
(INCSOR = 1)

95. ChInc How much was this income?

0.00..99999.00

96. Chpypd What period did this cover?

one week..1
two weeks...2
three weeks..3
four weeks..4
calendar month 5
two calendar months...7
eight times a year..8
nine times a year...9
ten times a year...10
three months/13 weeks 13
six months/26 weeks...26
one year/12 months/52 weeks...52
less than one week 90
one off lump sum..95
none of these: EXPLAIN IN A NOTE..97

SOCIAL AND CULTURAL PARTICIPATION

1. CPIntro Now I'd like to ask you some questions about how often you attend cultural events. Thinking about the last twelve months, how many times did you do any of the following?

SHOWCARD 30

2. FreqCin …..go to the cinema?

No visits during the last 12 months 1
One to three times...2
Four to six times 3
Seven to twelve times 4
More than twelve times...5

3. FreqPerf …...go to live performances?

No visits during the last 12 months 1
One to three times...2
Four to six times 3

General Household Survey, 2006
Household and Individual Questionnaires

Seven to twelve times 4
More than twelve times 5

General Household Survey, 2006
Household and Individual Questionnaires

4. FreqArt ……visit historical monuments, museums, art galleries or archaeological sites?

No visits during the last 12 months 1
One to three times 2
Four to six times 3
Seven to twelve times 4
More than twelve times 5

5. FreqLSpt …..attend a live sporting event?

No visits during the last 12 months 1
One to three times 2
Four to six times 3
Seven to twelve times 4
More than twelve times 5

6. SNIntro The next few questions are about how often you personally contact your family, relatives and friends in your spare time. Not counting the people you live with, how often do you do any of the following?

SHOWCARD 31

7. FreqMtR How often do you see family and relatives?

Daily 1
Every week (not every day) 2
Several times a month (not every week) 3
Once a month 4
At least once a year (less than once a month) 5
Never 6

8. FreqMtF How often do you see friends?

Daily 1
Every week (not every day) 2
Several times a month (not every week) 3
Once a month 4
At least once a year (less than once a month) 5
Never 6

9. FreqCnR How often do you contact family and relatives on the phone, by letter, fax, email or text or use chatrooms or the internet to talk to relatives?

Daily 1
Every week (not every day) 2
Several times a month (not every week) 3
Once a month 4
At least once a year (less than once a month) 5
Never 6

10. FreqCnF How often do you contact friends on the phone, by letter, fax, email or text or use chatrooms or the internet to talk to friends?

Daily 1
Every week (not every day) 2
Several times a month (not every week) 3
Once a month 4
At least once a year (less than once a month) 5
Never 6

11. SitIntro   I am going to describe two situations where people might need help. For each one, could you tell me if you would ask any of your neighbours for help?

12. IllBed   You are ill in bed and need help at home. Would you ask any of your neighbours for help?

   Yes 1
   No 2

13. Money   You are in financial difficulty and need to borrow some money to see you through the next few days. Would you ask any of your neighbours for help?

   Yes 1
   No 2

14. Care   Now I'd like to talk about any unpaid help you may have given to people who do not live with you. In the past 12 months have you given any unpaid help, such as providing transport or running errands? For example, cooking or shopping. Please do not count any help you gave through a group, club or organisation.

   Yes 1
   No 2

If respondent has given unpaid help

SHOWCARD 31

14. VolFreq   Thinking about the unpaid help you have mentioned, how often would you say you give this kind of help?

   Daily 1
   Every week (not every day) 2
   Several times a month (not every week) 3
   Once a month 4
   At least once a year (less than once a month) 5
   Never 6

15. GrpIntro   The next question is about involvement in groups, clubs and organisations. These could be formally organised groups or just groups of people who get together to do an activity or talk about things. Please exclude just paying a subscription, giving money, and anything that was a requirement of your job.

   In the last 12 months, have you been involved with any groups such as the ones shown on this card?

   Yes 1
   No 2

If respondent has been involved with any groups
GrpIntro = 1

8. GrpInf   Which of the categories on this card best describe the groups you have taken part in?

SHOWCARD 32

Code all that apply.

Hobbies/social clubs 1
Sports/exercise groups, including taking part, coaching or going to watch 2
Local community or neighbourhood groups 3
Environmental groups 4
Political groups 5
Trade union groups 6
Religious groups..7
Charitable organisations or groups..8
Professional associations...9
Other group 10

General Household Survey 2006

Summary of main topics included in GHS questionnaires: 1971 to 2006

Appendix F

Riaz Ali	Office for National Statistics
Julia Greer	Government Buildings
David Matthews	Cardiff Rd
Liam Murray	Newport
Simon Robinson	NP10 8XG
Ghazala Sattar	Tel: 01633 455877
© Crown copyright	Email: ghs@ons.gsi.gov.uk

Office for National Statistics: January 2008

General Household Survey, 2006

Summary of main topics included in GHS questionnaires: 1971 to 2006

Topic Area	Year included
ACCIDENTS THAT RESULTED IN SEEING A GP OR GOING TO HOSPITAL	1981, 1984 1987-89
ATTENDANCE AT ACTIVITIES ON SCHOOL PREMISES IN LAST 12 MONTHS	1984
BURGLARIES AND THEFTS FROM PRIVATE HOUSEHOLDS	1972-73, 1979-80, 1985-86, 1991, 1993, 1996
BUS TRAVEL	1982
CAR OWNERSHIP	1971-96, 1998, 2000-06
Driving licences and private motoring	1980
CAREER OPPORTUNITIES	1972
CARERS	1985, 1990, 1995, 2000
Carers aged 8-17	1996
CHILD CARE	
Pre-school children (aged under 5)	1971-79, 1986
Children aged 0-11	1991
Children under 14	1998
Children aged 0-12	
ETHNICITY AND COUNTRY OF BIRTH	
Colour, assessment of persons seen*	1971-92
Country of birth:	
of adults and their parents	1971-96, 1998, 2000-06
of children	1979-96, 1998, 2000-06
Year of entry to UK:	
Adults	1971-96, 1998, 2000-06
Children	1979-96, 1998, 2000-06
Ethnic origin	1983-96, 1998, 2000-06
National identity	2001-06
CONSUMER DURABLES	
Possession of various consumer durables	1972-76, 1978-96, 1998, 2000-06
Possession of a telephone	1972-76, 1979-96, 1998, 2000-06
Possession of a mobile telephone:	1992, 2000-06
Access to the Internet	2000-04

Office for National Statistics: January 2008

General Household Survey, 2006

Summary of main topics included in GHS questionnaires: 1971 to 2006

CONTRACEPTION AND STERILISATION

Whether woman/partner has been sterilised for contraceptive reasons	1983-84, 1986-87, 1989, 1991, 1993, 1995, 1998, 2002
Whether woman/partner has had other sterilising operation	
Details of any reversal of sterilisation operations	1983-84, 1986-87
Current use of contraception/reason for not using contraception	1983, 1986, 1989, 1991, 1993, 1995, 1998, 2002
Previous usual method of contraception	1989, 1991, 1993, 1995, 1998, 2002
Use of contraception	1989, 1991, 1993, 1995, 1998, 2002
Use of emergency contraception in previous 2 years	1993, 1995, 1998, 2002
Whether woman/partner would have difficulties in having (more) children	1983-84, 1986-87, 1989, 1991, 1993, 1995, 1998, 2002
Reasons for difficulties in getting pregnant and whether consulted a doctor	

DRINKING

Rating of drinking behaviour according to quantity-frequency (QF) index based on reported alcohol consumption in the 12 months before interview	1978, 1980, 1982, 1984
Rating of drinking behaviour according to average weekly alcohol consumption (AC) rating	1986-98 (alternate yrs), 2000-02
Alcohol consumption on the heaviest drinking day in the 7 days before interview	1998, 2000-06
Personal rating of own drinking behaviour	1978-98 (alternate yrs), 2000-06
Whether think drinking/smoking can damage health	1978-90 (alternate yrs)
Whether non-drinkers have always been non-drinkers or used to drink but stopped, and reasons	1992-98 (alternate years), 2000-06
Whether drink more or less than the recommended sensible amount	1992-96 (alternate years)
Whether drink more or less now than 5 years ago	1998, 2000-06

EDUCATION

Current education

Current education status	1971-96, 1998, 2000-06
Type of educational establishment currently attended:	
- by adults aged under 50	1971-81, 1984-90
- by adults aged under 70	1991-96, 1998, 2000-06
- by children aged 5-15	1971-77
Qualification/examination aimed at	1971, 1974-76
Expected date of completion of full-time education	1971-76
Whether intend to do any paid work while still in full-time education, and if so, when	1971-76
Whether currently attending any leisure or recreation classes	1973-78, 1981, 1983, 1993-96

Past education

Age on leaving school	1972-96, 1998
Age on leaving last place of full-time education	1971-96, 1998, 2000-06
Type of educational establishment last attended full time	1971-96, 1998
Qualifications obtained	1971-96, 1998, 2000-06

Job training — 1971-84

Students in institutional accommodation — 1981-87

Office for National Statistics: January 2008

General Household Survey, 2006

Summary of main topics included in GHS questionnaires: 1971 to 2006

ELDERLY	**1979-80, 1985, 1987, 1991, 1994, 1996, 1998, 2001**
EMPLOYMENT	
Those currently working	
Main job - occupation and industry and employee/self-employed	**1971-96, 1998, 2000-06**
Subsidiary job - occupation and industry and	**1971-78, 1980-84**
- employee/self-employed	**1987-91**
Last job - occupation and industry and employee/self-employed	**1986**
Whether has a second job	**1992-96, 1998**
Whether present job was obtained through a government scheme	**1989-92**
Youth Opportunities Programme Schemes	**1982-84**
Youth Training Scheme	**1985-95**
Journey time to work	**1971-76, 1978**
Usual number of hours worked per week (excluding overtime)	**1971-96, 1998, 2005-06**
Hours of paid/unpaid overtime usually worked per week	**1973-83, 1998**
Usual number of hours worked per week (including paid/unpaid overtime)	**2000-06**
Usual number of days worked per week	**1973, 1979-84**
Number of days worked in reference week	**1977-78**
Length of time with present employer/present spell of self-employment	**1971-96, 1998, 2000-06**
Whether self-employed during the previous 12 months	**1986-91**
Number of changes of employer in 12 months before interview	**1971-76, 1979-91**
Number of new employee jobs started in 12 months before interview	**1977-78, 1983-91**
Source of hearing about job(S) started in 12 months before interview	**1971-77, 1980-84**
Whether paid by employer when sick	**1971-76, 1979-81**
Whether employer is in the public/private sector	**1983, 1985, 1987**
Trade Union and Staff Association membership	**1983**
Whether people work all or part of the time at home	**1993**
Whether does any unpaid work for members of the family	**1993-95**
Whether has ever been a company director	**1987**
Type of National Insurance contribution paid by:	
- married and widowed women	**1972-1980**
- married, widowed and separated women	**1981-1983**
Level of satisfaction with present job as a whole	**1971-83**
Level of satisfaction with specific aspects of present job	**1974-83**
Whether thinking of leaving present employer, and if so, why	**1971-76**

Office for National Statistics: January 2008

General Household Survey, 2006

Summary of main topics included in GHS questionnaires: 1971 to 2006

Whether signed on at an Unemployment Benefit Office in the reference week	**1984-90, 1994-96**
Absence from work in the reference week	**1971-72, 1974-84**
Sickness absence in the four weeks before interview	**1981-84**
Sickness absence in the 3 months before interview	**1992**
Whether registered as unemployed in the reference week	**1977-82**
Unemployment experience in 12 months before interview	**1975-77, 1983-84**
Economic activity status 12 months before interview	**1979-91, 2005-06**
Whether in employment prior to present job	**1986**
Whether on any government schemes	**1985-96**
Usual job of father	**1971-91**

Those currently unemployed

Most recent job - occupation and industry and employee/self-employed	**1971-96, 1998, 2000-06**
Whether most recent job was obtained through a government scheme	**1989-92**
Whether has ever had a paid job	**1986-96, 1998, 2000-06**
Whether has ever worked for an employer as part of a government scheme	**1989-91**
Whether registered as unemployed in the reference week	**1971-83**
Methods of seeking work in the reference week	
Whether signed on at an Unemployment Benefit Office in the reference week	**1984-90, 1994-96**
Whether looking for full or part-time work	**1983**
Whether taking part in Youth Training Scheme or Youth Opportunities Programmes	**1984**
Whether last job was organised through Youth Opportunities Programme (16-19 yrs olds)	**1982**
For those who in the reference week were looking for work	**1991-96, 1998, 2000-06**
- would they have been able to start within 2 weeks if a job had been available	
For those who in the reference week were waiting to take up a new job already obtained	
- would they have started that job in the reference week if it had been available then	**1977-82**
- when was the new job obtained and when did they expect to start it	**1979**
Whether paid unemployment benefit (and supplementary allowances) for reference week	**1971-74**
When last worked and reasons for stopping work	**1971-73, 1974-79, 1986**
Reasons for leaving last job	**1981-82, 1986**
Whether last job was full/part time	**1986**
Length of current spell of unemployment	**1974-96, 1998**
Unemployment experience in 12 months before interview	**1975-77, 1983-84**
Economic activity status 12 months before interview	**1979-91**
Number of new employee jobs started in 12 months before interview	**1977, 1982-91**
Source of hearing about all jobs started in 12 months before interview	**1982-84**

Office for National Statistics: January 2008

General Household Survey, 2006

Summary of main topics included in GHS questionnaires: 1971 to 2006

Whether on any government schemes	**1985-96, 1998, 2000-06**
Whether does any unpaid work for members of the family and if so:	
number of hours a week and where	**1993-96, 1998, 2000-06**
for whom and type of work	**1993-96**
Whether has ever been a company director	**1987**
Type of National Insurance contribution paid in preceding two completed tax years by married/widowed/separated women aged 20-59, not working in week before interview	**1982-83**
Usual job of father	**1971-92**
The economically inactive	
Major activity in the reference week	
Last job - occupation and industry and employee/self-employed	**1971-96, 1998, 2000-06**
Usual job (of retired persons)	
- occupation and industry	**1973-76, 1979-88**
- employee/self-employed	
When finished last job	**1971-73, 1977-78, 1986**
Reasons for stopping work	**1971-73, 1978-82, 1986**
Whether registered as unemployed in the reference week	**1972-83**
Whether signed on at an Unemployment Benefit Office in the reference week	**1984-90, 1994-96**
Whether paid unemployment benefit (and supplementary allowance) for reference week	**1972-74**
Whether would like a regular paid job, whether looking for work, and if a job had been available would they have been able to start within 2 weeks	**1991-96, 1998, 2000-06**
Length of time currently out of employment	**1993-96, 1998, 2000-06**
Main reason for not looking for work	**1986-87**
Whether would like regular paid job	**1986-87**
Whether has ever had a paid job	**1986-96, 1998, 2000-06**
Whether has had a paid job in last 12 months	**1987-91**
Whether has ever worked for an employer as part of a government scheme	**1989-91**
Whether has had a paid job in previous 3 years	**1986**
Whether last job was full/part time	**1986**
Unemployment experience in 12 months before interview	**1975-77, 1983-84**
Economic activity status 12 months before interview	**1980-91**
Number of new employee jobs started in 12 months before interview	**1977, 1984-91**
Source of hearing about all jobs started in 12 months before interview	**1977**
Whether on any government schemes	**1985-96, 1998, 2000-06**
Whether does any unpaid work for members of the family and if so:	
number of hours a week and where	**1993-96, 1998, 2000-06**

Office for National Statistics: January 2008

General Household Survey, 2006

Summary of main topics included in GHS questionnaires: 1971 to 2006

for whom and type of work	**1993-96**
Whether has ever been a company director	**1987**
Type of National Insurance contribution paid in the preceding two completed tax years by Married/widowed/separated women aged 20-59, not working in week before interview	**1982**
Future work intentions	**1971-76**
Usual job of father	**1971-92**
FORESTS	
Whether ever visits forests or woodland areas, facilities visitors would like to see there	**1987**
HEALTH	
Acute sickness (restricted activity in a two-week reference period)*	1971-76, 1979-96, 1998, 2000-06
Chronic health problems	1977-78
Chronic sickness (longstanding illness or disability)	
Prevalence of longstanding illness or disability*	1971-76, 1979-96, 1998, 2000-06
Cause of the illness or disability*	1971-75
When the illness or disability started*	1971
Type of illness or disability*	1988-89, 1994-96, 1998, 2000-06
Prevalence of limiting longstanding illness or disability*	1972-76, 1979-96, 1998, 2000-06
When it started to limit activities and whether housebound or bedfast because of it*	1972-76
Dental health	1983, 1985, 1987, 1989, 1991, 1993, 1995, 2003
General health in the 12 months before interview	1977-96, 1998, 2000-06
Hearing	1977-81, 1985, 1991-2, 1994-5, 1998, 2001-2,
Medicine-taking in seven days before interview	4th qtr 1972, 1973
Short-term health problems	1977-78
Sight	1977-82, 1985, 1987, 1994, 1990-94, 1998, 2001
Tinnitus (sensation of noise in the ears or head)	1981
HEALTH SERVICES	
Day patient visits	1971-96, 1998, 2000-

Office for National Statistics: January 2008

General Household Survey, 2006

Summary of main topics included in GHS questionnaires: 1971 to 2006

	06
GP consultations	1971-96, 1998, 2000-06
Consultations in the two weeks before interview:	
number of consultations*	
NHS or private*	
type of doctor*	
site of consultation*	
cause of consultation*	1971-75
whether consulted because something was the matter, or for some other reason*	1981
whether consultation about reported longstanding illness or restricted activity*	1983-84, 1986-87
whether was given a prescription*	1981-96, 1998, 2000-06
whether was referred to hospital*	1981-85, 1988-90
whether was given National Insurance medical certificate	1981-85
whether saw a practice nurse and, if so, the number of times*	2000-06
Access to GPs	1977
Whether has used NHS Direct in the last year	2004-06
Health and personal social services	1971-76, 1970-85, 1991, 1994, 1998, 2001
Inpatient spells	
Spells in hospital as an inpatient in a three-month reference period:	1971-1976
Stays in hospital as an inpatient in a 12-month reference period:	
number of stays*	1982-96, 1998, 2000-06
number of nights on each stay*	1992-96, 1998, 2000-06
NHS or private patient	1982-83, 1985-87, 1995-96, 1998, 2000-06
whether private patients were treated in an NHS/private hospital	1998, 2000-06
whether claimed for under private medical insurance	1982-83, 1987
Whether on waiting list for admission to hospital and length of time on list*	1973-76
Outpatient (OP) attendances	
Attendances at hospital OP departments in a three-month reference period:	
number of attendances*	1971-96, 1998, 2000-06
NHS or private	1973-76, 1982-83, 1985-87, 1995-96, 1998, 2000-06
nature of complaint causing attendance*	1974-76
whether claimed for under private medical insurance	1982-83, 1987, 1995
number of casualty visits*	1995-96, 1998, 2000-06
Appointments with OP departments:	1973-76

Office for National Statistics: January 2008

General Household Survey, 2006

Summary of main topics included in GHS questionnaires: 1971 to 2006

HOUSEHOLD COMPOSITION	1971-96, 1998, 2000-06
Age*, sex*, marital status of household members	
Relationship to head of household*	
Family unit(s)	
Housewife	1971-80
HOUSING (see also MIGRATION)	
Accommodation: amenities	
Length of residence at present address*	
Age of building	1971-96, 1998, 2000-04
Type of accommodation	
Number of rooms and bedrooms	
Whether have separate kitchen	
Bath/WC: sole use, shared, none	1971-90
WC: inside or outside the accommodation	
Installation/replacement of bath or WC	1971-76
Cost of improvements made to the accommodation	
Floor level of main accommodation	1973-96, 1998, 2000-04
Whether there is a lift	2001-04
Central heating and fuel use	
Accommodation problems including damp, noise, pollution and crime	2004-2006
Housing costs	
Gross value	1971-86
Type of mortgage	1972-77, 1979, 1981, 1984-86
Current mortgage payments	1972-77, 1979, 1981, 1984
Purchase price of present home, amount of mortgage or loan and date mortgage started	1985-86, 1992-93
Current rent	1972-77, 1979, 1981
Amount of any rent rebate/allowance and/or rate rebate received	
Whether in receipt of housing benefit	1985-95, 1998, 2000-06
Whether rent paid by DSS or local authority	1998, 2000-06
Council Tax band for households containing person(s) aged 65 or over	2001
Tenure	
Whether present home is owned or rented	1971-96, 1998, 2000-06
Whether in co-ownership housing association scheme	1981-95
Change of tenure on divorce or remarriage	1991-93
Change of tenure on marriage or cohabitation	1998
Housing history of local authority tenants and owner occupiers who had become owners in the previous five years	1985-86
Whether ever rented from local authority, and if so, whether bought that accommodation	1991-93

Office for National Statistics: January 2008

General Household Survey, 2006

Summary of main topics included in GHS questionnaires: 1971 to 2006

Owner occupiers:	
- in whose name the property is owned	**1978-96, 1998, 2000-06**
- whether property is owned outright or being bought with a mortgage or loan	**1971-96, 1998, 2000-06**
- how outright owners originally acquired their home	**1978-80, 1982-83, 1985-86**
- source of mortgage or loan	**1978-80, 1982-86, 1992-93**
- whether currently using present home as security for a (second) mortgage or loan of any kind, and if so, details	**1980-82, 1992-93**
- whether owner occupiers with mortgage have taken out a remortgage on present home	**1985-87, 1992-93**
- whether recent owner occupiers had previously rented this accommodation	**1981-82, 1985-86**
- whether had rented present accommodation before deciding to buy	**1992-93**
- whether previous accommodation was owned and if so, details of the sale	**1992-93**
Renters:	
- in whose name the property is rented	**1985-96, 1998, 2000-05**
- from whom the accommodation is rented	**1971-96, 1998, 2000-06**
- whether landlord lives in the same building	**1971-72, 1975-76, 1979-96, 1998, 2000-04**
- whether have considered buying present home and, if not, why not	**1980-89**
- tenure preference	**1985-88**
- whether previously owned/buying accommodation and reasons for leaving	**1995-96**
Local authority renters	**1990-91**
HOUSING SATISFACTION	**1978, 1988, 1990**
INCOME	
Income over 12 months before interview	**1971-78**
Current income	**1979-96, 1998, 2000-06**
INHERITANCE	**1995**
LEISURE / SOCIAL AND CULTURAL PARTICIPATION	
Arts and entertainments, museums, galleries, historic buildings	**1987, 2006**
Holidays away from home in four weeks before interview	**1973, 1977, 1980, 1983, 1986**
Leisure activities in the four weeks before interview	**1973, 1977, 1980, 1983,**
Social activities and hobbies in the four weeks before interview	**1973, 1977, 1980, 1983, 1986, 1987 1990, 1993, 1996, 2002**
Involvement in groups, clubs and organisations, in last 12 months	**2006**
Whether did any voluntary arts/cultural work in the 4 weeks before interview	**2002**
Personal contact with family, relatives and friends in spare time	**2006**

Office for National Statistics: January 2008

General Household Survey, 2006

Summary of main topics included in GHS questionnaires: 1971 to 2006

Asking for help from neighbours	**2006**
Unpaid help given to others outside the home, in last 12 months	**2006**
LIBRARIES	**1987**
LONG-DISTANCE TRAVEL	**1971-72**
MARRIAGE, COHABITATION AND CHILDBIRTH	
Marital history	1979-96, 1998, 2000-06
Date of present marriage	1971-78
Expected family size	1974-78
Date of birth and sex of all liveborn children and whether they live with mother	1979-96, 1998, 2000-06
Date of birth of step, foster and adopted children living in the household, and how long they have lived there	1979-87, 1989-96, 1998, 2000-06
Whether women think they will have any (more) children, how many in all, and age at which they think will have their first/next baby	1979-96, 1998, 2000-06
Current cohabitation	1979-96, 1998, 2000-06
Cohabitation before current or most recent marriage	1979, 1981-88
Cohabitation before all marriages	1989-96, 1998, 2000-06
Number of cohabiting relationships that did not lead to marriage	1998, 2000-06
MIGRATION	
Past movement	
Length of residence at previous address*	1971-77
Details of previous accommodation	1971-73, 1978-80
Number of moves in last five years*	1971-77, 1979-96, 1998, 2000-06
Potential movement – people thinking of moving*	1971-78, 1980-81, 1983
Frustrated potential movement – people who had previously thought of moving*	1974-76, 1980, 1983
MOBILITY AIDS – difficult getting about without assistance	1993, 1996, 2001
PENSIONS	
Whether covered by employer's pension scheme	1971-76, 1979, 1982-83, 1985, 1987-96, 1998, 2000-06
Whether the scheme is contributory, reasons for not belonging to the scheme	1971-76, 1979, 1982-83, 1985, 1987
Whether ever belonged to present employer's pension scheme	1985, 1987
Whether in receipt of a pension from a previous employer	1983, 1985, 1987
Whether ever belonged to a previous employer's pension scheme	
Length of time in last employer's pension scheme and in last job	1985
Whether retained any pension rights from any previous employer	1971-76, 1979, 1982-83, 1985, 1987
Whether pays Additional Voluntary Contributions into employer's pension scheme	1987
Whether has a stakeholder pension	2001-06
Whether has a group personal pension	2003-06

Office for National Statistics: January 2008

General Household Survey, 2006

Summary of main topics included in GHS questionnaires: 1971 to 2006

Whether currently belongs to a personal pension scheme and whether employer contributes	**1991-96, 1998, 2000-06**
Whether has ever contributed towards a personal pension	**1987-96, 1998, 2000-06**
Date the personal pension was taken out	
Whether belonged to an employer's pension scheme during the 6 months prior to taking out a personal pension	**1989-90**
Whether makes any other income tax deductible pension contributions	**1993-96, 1998, 2000-06**
- whether free standing additional voluntary contributions	**2000-06**
PRIVATE MEDICAL INSURANCE	**1982-83, 1986-87, 1995**
SHARE OWNERSHIP	**1987-88, 1992-96, 1998**
SMOKING	
Cigarette smoking	
Prevalence of cigarette smoking	**1972-76, 1978-98 (alternate years), 2000-06**
Current cigarette smokers:	
number of cigarettes smoked per day	**1972-76, 1978-98 (alternate years), 2000-06**
type of cigarette smoked mainly	
usual brand of cigarette smoked	**1984-98 (alternate years), 2000-06**
age started smoking cigarettes regularly	**1988-98 (alternate years), 2000-06**
whether would find it difficult to not smoke for a day	
whether would like to give up smoking altogether	**1992-98 (alternate years), 2000-06**
when is the first cigarette of the day smoked	
whether or not intends to give up smoking in the future	**2003-06**
Regular cigarette smokers:	
age started smoking cigarettes regularly	**1972-73**
Occasional cigarette smokers:	
whether ever smoked cigarettes regularly	
age started smoking cigarettes regularly	**1972-73**
how long ago stopped smoking cigarettes regularly	
Current non-smokers:	
whether ever smoked cigarettes regularly	**1972-76, 1978-98 (alternate years), 2000-06**
age started smoking cigarettes regularly	
number smoked per day when smoking regularly	**1972-73, 1980-98 (alternate years), 2000-06**
how long ago stopped smoking cigarettes regularly	
Cigar smoking	**1972-76, 1978-98 (alternate years), 2000-06**

Office for National Statistics: January 2008

General Household Survey, 2006

Summary of main topics included in GHS questionnaires: 1971 to 2006

Pipe smoking	**1972, 1978, 1986-98 (alternate years), 2000-06**
SOCIAL CAPITAL	
Opinion of local services, amenities, organisations, safety in the area, local problems	**2000, 2004**
SOCIAL MOBILITY	**2005**
Opinion of living standards across the generations	
- mother and father's main job, year of birth and qualifications	
- household financial problems	
SPORT	**1987, 1990, 1993, 1996, 2002**
VOLUNTARY WORK	**1981, 1987, 1992, 2002**

* including children

Office for National Statistics: January 2008

General Household Survey 2006

List of tables: 2006

Appendix G

Riaz Ali

Julia Greer

David Matthews

Liam Murray

Simon Robinson

Ghazala Sattar

© Crown copyright

Office for National Statistics

Government Buildings

Cardiff Rd

Newport

NP10 8XG

Tel: 01633 455877

Email: ghs@ons.gsi.gov.uk

General Household Survey 2006
List of Tables

1 Smoking

1.1 Prevalence of cigarette smoking by sex and age: 1974 to 2006
1.2 Ex-regular cigarette smokers by sex and age: 1974 to 2006
1.3 Percentage who have never smoked cigarettes regularly by sex and age: 1974 to 2006
1.4 Cigarette-smoking status by sex and marital status
1.5 Cigarette-smoking status by age and marital status
1.6 Prevalence of cigarette smoking by sex and whether household reference person is in a non-manual or manual socio-economic group: England 1992 to 2006
1.7 Prevalence of cigarette smoking by sex and socio-economic classification of the household reference person: England: 2001 to 2006
1.8 Prevalence of cigarette smoking by sex and socio-economic classification based on the current or last job of the household reference person
1.9 Prevalence of cigarette smoking by sex and socio-economic classification based on own current or last job, whether economically active or inactive, and, for economically inactive persons, age
1.10 Prevalence of cigarette smoking by sex and country of Great Britain:1978 to 2006
1.11 Prevalence of cigarette smoking by sex, country and region of England: 1998 to 2006
1.12 Cigarette-smoking status by sex, country and region: 20044 to 2006 combined
1.13 Cigarette-smoking status by sex: 1974 to 2006
1.14 Cigarette-smoking status by sex and age
1.15 Average daily cigarette consumption per smoker by sex and age: 1974 to 2006
1.16 Average daily cigarette consumption per smoker by sex, and socio- economic classification based on the current or last job of the household reference person
1.17 Type of cigarette smoked by sex: 1974 to 2006
1.18 Type of cigarette smoked by sex and age
1.19 Grouped tar yield per cigarette: 1986 to 2006
1.20 Tar yield per cigarette: 1986 to 2006
1.21 Tar yields by sex and age of smoker
1.22 Tar yields by sex and socio-economic classification based on the current or last job of the household reference person
1.23 Prevalence of smoking by sex and type of product smoked: 1974 to 2006
1.24 Prevalence of smoking among men by sex age and type of product smoked
1.25 Age started smoking regularly by sex: 1992 to 2006
1.26 Age started smoking regularly by sex and socio-economic classification based on the current or last job of the household reference person
1.27 Age started smoking regularly by sex, whether current smoker and if so, cigarettes smoked a day
1.28 Proportion of smokers who would like to give up smoking altogether, by sex and number of cigarettes smoked per day: 1992 to 2006
1.29 Proportion of smokers who would find it difficult to go without smoking for a day, by sex and number of cigarettes smoked per day: 1992 to 2006
1.30 Proportion of smokers who have their first cigarette within five minutes of waking, by sex and number of cigarettes smoked per day: 1992 to 2006
1.31 Proportion of smokers who would like to give up smoking altogether, by sex, socio-economic classification of household reference person, and number of cigarettes smoked a day
1.32 Proportion of smokers who would find it difficult to go without smoking for a day, by sex, socio-economic classification of household reference person, and number of cigarettes smoked a day
1.33 Proportion of smokers who have their first cigarette within five minutes of waking, by sex, socio-economic classification of household reference person, and number of cigarettes smoked a day

2 Drinking

2.1 Average weekly consumption (units), by sex and age: 1992-2006
2.2 Weekly alcohol consumption level: percentage exceeding specified amounts by sex and age: 1988-2006
2.3 Drinking last week by sex and age: 1998 to 2006
2.4 Maximum drunk on any one day last week by sex and age: 1998 to 2006
2.5 Average weekly alcohol consumption (units), by sex and age
2.6 Average weekly alcohol consumption (units), by sex and socio-economic class based on the current or last job of the household reference person
2.7 Average weekly alcohol consumption (units), by sex and usual gross weekly household income (£)
2.8 Average weekly alcohol consumption (units), by sex and economic activity status
2.9 Average weekly alcohol consumption (units), by sex and usual gross weekly earnings (£)
2.10 Average weekly alcohol consumption (units), by sex and Government Office Region
2.11 Whether drank last week and number of drinking days by sex and age

2.12 Maximum drunk on any one day last week, by sex and age
2.13 Drinking last week, by sex, and socio-economic classification based on the current or last job of the household reference person
2.14 Maximum number of units drunk on at least one day last week, by sex and socio-economic classification based on the current or last job of the household reference person
2.15 Drinking last week, by sex and usual gross weekly household income
2.16 Maximum drunk on any one day last week by sex and usual gross weekly household income
2.17 Drinking last week, by sex and economic activity status
2.18 Maximum drunk on any one day last week, by sex and economic activity status
2.19 Drinking last week, by sex and usual gross weekly earnings
2.20 Maximum drunk on any one day last week, by sex and usual gross weekly earnings
2.21 Drinking last week, by sex and Government Office Region
2.22 Maximum drunk on any one day last week, by sex and Government Office Region

3 Households, families and people

3.1   Trends in household size: 1971 to 2006
3.2   Trends in household type: 1971 to 2006
3.3   Percentage living alone by age: 1973 to 2006
3.4   Percentage of men and women living alone by age
3.5   Type of household: 1979 to 2006
3.6   Family type and marital status of lone parents: 1971 to 2006
3.7   Families with dependent children: 1972 to 2006
3.8   Average (mean) number of dependent children by family type: 1971 to 2006
3.9   Age of youngest dependent child by family type
3.10 Stepfamilies with dependent children by family type
3.11 Usual gross weekly household income of families with dependent children by family type
3.12 The distribution of the population by sex and age: 1971 to 2006
3.13 Percentage of males and females by age
3.14 Socio-economic classification based on own current or last job by sex and age
3.15 Ethnic group of GHS respondents: 2001 to 2006
3.16 GHS respondents: age by ethnic group
3.17 GHS respondents: sex by ethnic group
3.18 GHS respondents: Ethnic group by Government Office Region
3.19 GHS respondents: average household size by ethnic group of household reference person
3.20 GHS respondents: percentage born in the UK by age and ethnic group
3.21 Weighted bases for Tables 3.3 and 3.8

4 Housing and consumer durables

4.1   Tenure: 1971 to 2006
4.2   Type of accommodation: 1971 to 2006
4.3   Type of accommodation occupied by households renting from a council compared with other households: 1981 to 2006
4.4   (a) Type of accommodation by tenure
      (b) Tenure by type of accommodation
4.5   (a) Household type by tenure
      (b) Tenure by household type
4.6   Housing profile by family type: lone parent families compared with other families
4.7   Type of accommodation by household type
4.8   Usual gross weekly income by tenure
4.9   (a) Age of household reference person by tenure
      (b) Tenure by age of household reference person
4.10 Tenure by sex and marital status of household reference person
4.11 Housing tenure by ethnic group of household reference person
4.12 (a) Socio-economic classification and economic activity status of household reference person by tenure
      (b) Tenure by socio-economic classification and economic activity status of household reference person
4.13 (a) Length of residence of household reference person by tenure
      (b) Tenure by length of residence of household reference person
4.14 Persons per room: 1971 to 2006
4.15 Persons per room and mean household size by tenure
4.16 Closeness of fit relative to the bedroom standard by tenure

4.17	Cars or vans: 1972 to 2006
4.18	Availability of a car or van by socio-economic classification of household reference person
4.19	Consumer durables, central heating and cars: 1972 to 2006
4.20	Consumer durables, central heating and cars by socio-economic classification of household reference person
4.21	Consumer durables, central heating and cars by usual gross weekly household income
4.22	Consumer durables, central heating and cars by household type
4.23	Consumer durables, central heating and cars by family type: lone parent families compared with other families

5 Marriage and cohabitation

5.1	Sex by marital status
5.2	(a) Age by sex and marital status
	(b) Marital status by sex and age
5.3	Percentage currently cohabiting by sex and age
5.4	Percentage currently cohabiting by legal marital status and age
5.5	Cohabitees: age by legal marital status
5.6	Cohabitees: age by sex
5.7	Legal marital status of women aged 18-49: 1979 to 2006
5.8	Percentage of women aged 18-49 cohabiting by legal marital status: 1979 to 2006
5.9	Women aged 16-59:
	(a) Whether has dependent children in the household by marital status
	(b) Marital status by whether has dependent children in the household
5.10	Women aged 16-59: percentage cohabiting by legal marital status and whether has dependent children in the household
5.11	Cohabiting women aged 16-59: whether has dependent children in the household by legal marital status
5.12	Number of past cohabitations not ending in marriage by sex and age
5.13	Number of past cohabitations not ending in marriage by current marital status and sex
5.14	Age at first cohabitation which did not end in marriage by year cohabitation began and sex
5.15	Duration of past cohabitations which did not end in marriage by number of past cohabitations and sex

6 Occupational and personal pension schemes

6.1	Current pension scheme membership by age and sex
6.2	Membership of current employer's pension scheme by sex and whether working full time or part time
6.3	Membership of current employer's pension scheme by sex: 1983 to 2006
6.4	Current pension scheme membership by sex and socio-economic classification
6.5	Membership of current employer's pension scheme by sex and socio-economic classification
6.6	Current pension scheme membership by sex and usual gross weekly earnings
6.7	Current pension scheme membership by sex and length of time with current employer
6.8	Whether or not current employer has a pension scheme by sex and length of time with current employer
6.9	Current pension scheme membership by sex and number of employees in the establishment
6.10	Membership of current employer's pension scheme by sex and number of employees at the establishment
6.11	Membership of current employer's pension scheme by sex and industry group
6.12	Membership of personal pension scheme by sex and whether working full time or part time: self-employed persons
6.13	Membership of a personal pension scheme for self-employed men working full time: 1991 to 2006
6.14	Membership of personal pension scheme by sex and length of time in self-employment

7 General health and use of health services

7.1	Self perception of general health during the last 12 months: 1977 to 2006
7.2	Trends in self-reported sickness by sex and age, 1972 to 2006: percentage of persons who reported
	(a) longstanding illness
	(b) limiting longstanding illness
	(c) restricted activity in the 14 days before interview
7.3	Acute sickness: average number of restricted activity days per person per year, by sex and age
7.4	Chronic sickness: prevalence of reported longstanding illness by sex, age and socio-economic classification of household reference person
7.5	Chronic sickness: prevalence of reported limiting longstanding illness by sex, age and socio-economic classification of household reference person
7.6	Acute sickness
	(a) Prevalence of reported restricted activity in the 14 days before interview by sex, age and socio-economic

classification of household reference person
(b) Average number of restricted activity days per person per year, by sex, age and socio-economic classification of household reference person

7.7 Chronic sickness: prevalence of reported longstanding illness by sex, age and economic activity status
7.8 Chronic sickness: prevalence of reported limiting longstanding illness by sex, age and economic activity status
7.9 Acute sickness
(a) Prevalence of reported restricted activity in the 14 days before interview by sex, age and economic activity status
(b) Average number of restricted activity days per person per year, by sex, age and economic activity status
7.10 Self-reported sickness by sex and Government Office Region: percentage of persons who reported:
(a) longstanding illness
(b) limiting longstanding illness
(c) restricted activity in the 14 days before interview
7.11 Chronic sickness: rate per 1000 reporting longstanding condition groups, by sex
7.12 Chronic sickness: rate per 1000 reporting longstanding condition groups, by age
7.13 Chronic sickness: rate per 1000 reporting selected longstanding condition group, by age and sex
7.14 Chronic sickness: rate per 1000 reporting selected longstanding conditions, by sex and age
7.15 Chronic sickness: rate per 1000 reporting selected longstanding condition groups, by socio-economic classification of household reference person
7.16 Chronic sickness: rate per 1000 reporting selected longstanding condition groups, by sex and age and socio-economic classification of household reference person
7.17 Trends in consultations with an NHS GP in the 14 days before interview by sex and age: 1972 to 2006
7.18 Average number of NHS GP consultations per person per year by sex and age: 1972 to 2006
7.19 NHS GP consultations: trends in site of consultation: 1971 to 2006
7.20 Percentage of persons who consulted an NHS GP in the 14 days before interview by sex and site of consultation, and by age and site of consultation
7.21 NHS GP consultations
(a) Percentage of persons who consulted a doctor in the 14 days before interview by sex, age and economic activity status
(b) Average number of consultations per person per year, by sex, age and economic activity status
7.22 Percentage of persons consulting an NHS GP in the 14 days before interview who obtained a prescription from the doctor, by sex, age and socio-economic classification of household reference person
7.23 GP consultations: consultations with a doctor in the 14 days before interview by sex of person consulting and whether consultation was NHS or private
7.24 Trends in reported consultations with a practice nurse by sex and age: 2000 to 2006
(a) percentage consulting a practice nurse in the 14 days before interview
(b) average number of consultations with a practice nurse per person per year
7.25 Percentage of children using health services other than a doctor in the 14 days before interview
7.26 Trends in percentages of persons who reported attending an outpatient or casualty department in the 3 months before interview by sex and age: 1972 to 2006
7.27 Trends in day-patient treatment in the 12 months before interview by sex and age: 1992 to 2006
7.28 Trends in inpatient stays in the 12 months before interview by sex and age: 1982 to 2006
7.29 Average number of nights spent in hospital as an inpatient during the 12 months before interview, by sex and age
7.30 Inpatient stays and outpatient attendances
(a) average number of inpatient stays per 100 persons in a 12-month reference period, by sex and age
(b) average number of outpatient attendances per 100 persons per year, by sex and age

General Household Survey, 2006

Smoking and drinking among adults 2006

General Household Survey 2006

Smoking and drinking among adults, 2006

Eileen Goddard

© Crown copyright

Office for National Statistics

Government Buildings

Cardiff Rd

Newport

NP10 8XG

Tel: 01633 812630

Email: ghs@ons.gsi.gov.uk

Contents

Introduction

An overview of the General Household Survey	1
Other GHS results for 2006	2

1 Smoking

The reliability of smoking estimates	3
The effect of weighting on the smoking data	3
Government policy and targets for the reduction of smoking	4

The prevalence of cigarette smoking

Trends in the prevalence of cigarette smoking	4
Cigarette smoking and marital status	6
Cigarette smoking and socio-economic classification	6
Cigarette smoking and economic activity status	8
Regional variation in cigarette smoking	8
Cigarette consumption	9
Cigarette type	9
Tar yield	11
Cigar and pipe smoking	11
Age started smoking	12
Dependence on cigarette smoking	13
Notes and references	14

2 Drinking   47

Measuring alcohol consumption

Average weekly alcohol consumption	47
Maximum daily amount drunk last week	48

Updated method of converting volumes to units

Effect on GHS data of updated conversion factors	50
Average weekly alcohol consumption	50

Maximum drunk on any one day in the previous week	51

Trends in alcohol consumption

Trends in average weekly alcohol consumption	52
Trends in last week's drinking	54

Alcohol consumption in 2006

Weekly alcohol consumption and sex and age	55
Weekly alcohol consumption and household socio-economic class	55
Weekly alcohol consumption, income and economic activity status	56
Regional variation in average weekly alcohol consumption	56

Last week's drinking in 2006

Frequency of drinking during the last week	56
Maximum daily amount drunk last week	57
Drinking last week and socio-economic characteristics	57
Drinking last week and household income	58
Drinking last week, economic activity status and earnings from employment	58
Regional variation in drinking last week	59
Notes and references	59

Introduction

This report provides information about smoking and drinking based on data collected by the General Household Survey in 2006. It also includes tables showing data on the trends and changes in smoking and drinking measured by the GHS over several decades.

An overview of the General Household Survey

The General Household Survey (GHS) is a multi-purpose continuous survey carried out by the Office for National Statistics (ONS). It collects information on a range of topics from people living in private households in Great Britain. The survey started in 1971 and has been carried out continuously since then, except for breaks to review it in 1997/1998 and to re-develop it in 1999/2000.

The survey presents a picture of households, families and people living in Great Britain. This information is used by government departments and other organisations, such as educational establishments, businesses and charities, to contribute to policy decisions and for planning and monitoring purposes.

The interview consists of questions relating to the household, answered by the household reference person or spouse, and an individual questionnaire, asked of all resident adults aged 16 and over. Demographic and health information is also collected about children in the household. The GHS collects data on a wide range of core topics which are included on the survey every year. These are:

- demographic information about households, families and people;
- housing tenure and household accommodation;
- access to and ownership of consumer durables, including vehicles;
- migration;
- employment;
- education;
- health and use of health services;
- smoking;
- drinking;
- family information, including marriage, cohabitation and fertility;
- income.

The modular structure of the GHS allows for a number of additional topics to be included each year to a plan agreed by its sponsors. Only one such topic, on social and cultural participation, was included in the 2006 survey.

The 2006 GHS was sponsored by the Office for National Statistics, Information Centre for health and social care, Department for Work and Pensions, HM Revenue & Customs, Scottish Government and Eurostat.

Since April 1994, the GHS has been conducted on a financial year basis, with fieldwork spread evenly across the year April-March. However, in 2005 the survey period reverted to a calendar year to bring it in line with other ONS continuous surveys.

Another change in 2005 was that, in line with European requirements, the GHS adopted a longitudinal sample design, in which households remain in the sample for four years (waves) with one quarter of the sample being replaced each year. Thus approximately three quarters of the 2005 sample were re-interviewed in 2006. More details are given in Appendix B.

A major advantage of the longitudinal component of the design is that it is more efficient at detecting statistically significant estimates of change over time than the previous cross-sectional design. This is because an individual's responses to the same question at different points in time tend to be positively correlated, and this reduces the standard errors of estimates of change.

The response rate for the 2006 survey was 76 per cent, giving an achieved sample size of 9,731 households and 18,214 adults aged 16 and over, of whom 16,736 gave a full interview in person (interviews obtained by proxy from another member of the household do not include questions on smoking and drinking).

Other GHS results for 2006

Results for other GHS topics will be combined with those from other sources in *Social Trends* and other reports due to be published in 2008. Tables from all GHS topic areas are published on the National Statistics website: www.statistics.gov.uk/ghs. Technical information about the GHS in the form of appendices is also available at www.statistics.gov.uk/ghs, including:

- a glossary of definitions and terms used throughout the report and notes on how these have changed over time (Appendix A);

- information about the sample design and response (Appendix B);

- sampling errors (Appendix C);

- weighting and grossing (Appendix D);

- the household and individual questionnaires used in 2006, excluding self-completion forms and prompt cards (Appendix E);

- a list of the main topics covered by the survey since 1971 (Appendix F).

Smoking and drinking among adults 2006

1 Smoking

Questions about smoking behaviour have been asked of GHS respondents aged 16 and over in alternate years since 1974. Following the review of the GHS carried out in 1997, the smoking questions became part of the continuous survey and have been included every year from 2000 onwards. Note, however, that the tables in this report show data for every four years from 1974 to 1998.

This report updates information about trends in cigarette smoking presented in earlier GHS reports and on the National Statistics website. It also discusses variations according to personal characteristics such as sex, age, socio-economic classification and economic activity status, and comments briefly on the prevalence of cigarette smoking in different parts of Great Britain. Smoking prevalence in relation to ethnicity is not included in this report: the 2005 report[i] included this topic in some detail, based on five years combined data, to give large enough samples for analysis in minority ethnic groups. Other topics covered in 2006 include cigarette consumption, type of cigarette smoked, how old respondents were when they started smoking, and dependence on cigarettes.

The reliability of smoking estimates

As noted in earlier GHS reports, it is likely that the GHS underestimates cigarette consumption and (perhaps to a lesser extent) prevalence (the proportion of people who smoke). For example, evidence suggests that when respondents are asked how many cigarettes they smoke each day, there is a tendency to round the figure down to the nearest multiple of 10. Underestimates of consumption are likely to occur in all age groups.

Under-reporting of prevalence, however, is most likely to occur among young people. To protect their privacy, particularly when they are being interviewed in their parents' home, young people aged 16 and 17 complete the smoking and drinking sections of the questionnaire themselves, so that neither the questions nor their responses are heard by anyone else who may be present. This is probably only partially successful in encouraging honest answers[ii].

When considering trends in smoking, it is usually assumed that any under-reporting remains constant over time. However, since the prevalence of smoking has fallen, this assumption may not be entirely justified. As smoking has become less acceptable as a social habit, some people may have become less inclined to admit how much they smoke – or, indeed, to admit to smoking at all.

The effect of weighting on the smoking data

Weighting to compensate for non-response was introduced on the GHS in 2000 and was described in detail in the GHS 2000 report[iii]. The effect of weighting on the smoking data is slight, increasing the overall prevalence of cigarette smoking by one percentage point. The change occurs because weighting reduces the contribution to the overall figure of those aged 60 and over, among whom prevalence is relatively low.

Government policy and targets for the reduction of smoking

In December 1998 *Smoking Kills – a White Paper on tobacco*[iv] was released, which included targets for reducing the prevalence of cigarette smoking among adults in England to 24 per cent by 2010. In 2004, the Department of Health agreed a new Public Service Agreement (PSA) which revised the target downwards: the aim now is to reduce the prevalence of cigarette smoking among adults to 21 per cent or less by 2010[v].

Since smoking is estimated to be the cause of about one third of all cancers, reducing smoking is also one of three key commitments at the heart of the NHS *Cancer Plan*, which was published in 2000[vi]. In particular, the *Cancer Plan* focuses on the need to reduce the comparatively high rates of smoking among those in manual socio-economic groups, which result in much higher death rates from cancer among unskilled workers than among professionals. The more recent PSA targets mentioned in the previous paragraph also included reducing prevalence among routine and manual groups to 26 per cent or less.

Legislation came into force in February 2003 which banned cigarette advertising on billboards and in the press and magazines, and further restrictions on advertising at the point of sale were introduced in December 2004. A ban on smoking in enclosed public places came into force in Scotland during the spring of 2006: similar bans in England and Wales were introduced in 2007.

The GHS interview cannot accommodate extensive questions about people's views on smoking, but the Information Centre for health and social care regularly commissions the inclusion of such questions on the ONS Omnibus Survey, most recently in October/November 2006[vii].

The prevalence of cigarette smoking

Trends in the prevalence of cigarette smoking

The overall prevalence of smoking among the adult population was 22 per cent in 2006, compared with 24 per cent the previous year. The fall of two percentage points is statistically significant, and it occurred among both men and women.

This downturn follows a period of little change since the second half of the 1990s: the prevalence of cigarette smoking fell substantially in the 1970s and the early 1980s, from 45 per cent in 1974 to 35 per cent in 1982. The rate of decline then slowed, with prevalence falling by only about one percentage point every two years until 1994, after which it levelled out at about 27 per cent before resuming a slow decline in the 2000s.

It should be noted that during periods when the prevalence of smoking in the general population is changing little, upward and downward movements in survey estimates are to be expected, and this can make the detection of trends over a short period difficult.

Throughout the period during which the GHS has been monitoring cigarette smoking, prevalence has been higher among men than among women, and this continues to be the case: in 2006, 23 per cent of men and 21 per cent of women were cigarette smokers.

Figure 1.1: Prevalence of cigarette smoking: Great Britain, 1974 to 2006

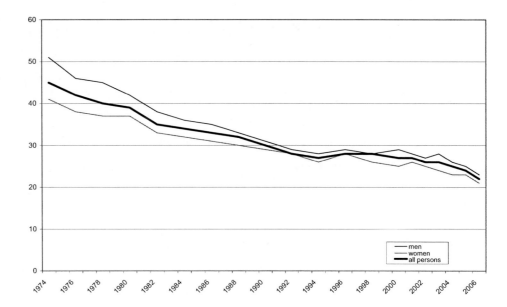

The present difference of two percentage points in prevalence between men and women is considerably less than it was in the 1970s. In 1974, for example, 51 per cent of men smoked cigarettes, compared with 41 per cent of women. The reduction in the difference results mainly from a combination of two factors.

1. First, there is a cohort effect resulting from the fact that smoking became common among men several decades before it did among women. In the 1970s there was a fall in the proportion of women aged 60 and over who had never smoked regularly.

2. Second, men are more likely than women to have given up smoking cigarettes. It should be noted, however, that this difference conceals the fact that some men who give up smoking cigarettes remain smokers (by continuing to smoke cigars and pipes). This is very rare among women who stop smoking cigarettes.

It should be noted that the proportion of respondents saying that they used to smoke regularly was the same in 2005 and 2006 (27 per cent of men and 21 per cent of women). However, the proportion saying that they had never smoked regularly did rise, suggesting that the measured fall in prevalence between 2005 and 2006 may be due to people becoming more reluctant to admit to smoking, rather than to more people giving up.

Smoking among different age groups is another key area of interest. Since the early 1990s, the prevalence of cigarette smoking has been higher among those aged 20 to 24 than among those in other age groups, but the difference relative to the next age group, those aged 25 to 34, has reduced in recent years. Up to the early twenties, more young people are starting to smoke than are giving up (as shown later, only about one in six of those who have smoked at some time in their lives took up the habit at age 20 or older).

Since the survey began, the GHS has shown considerable fluctuation in prevalence rates among those aged 16 to 19, particularly if young men and young women are considered separately. However, this is mainly because of the relatively small sample size in this age group and has occurred within a pattern of overall decline in smoking prevalence in this age group. The year on year fall in prevalence among those aged 16 to 19 from 24 per cent in 2005 to 20 per cent in 2006, although marked, is on the borderline of statistical significance, but is significantly lower than the rate of 31 per cent in 1998. Sampling fluctuations have also affected comparisons between young men and women in this age group. In recent years, prevalence has tended to be higher among young women than among young men, but this was not the case in 2006, when it was at the same level, 20 per cent, for both sexes.

At 12 per cent in 2006, prevalence continues to be lowest among men and women aged 60 and over. Although they are more likely than younger people to have ever been smokers, they are also much more likely to have given up.

Figure 1.1, Tables 1.1-1.3

Cigarette smoking and marital status

The prevalence of cigarette smoking varies considerably according to marital status. It is much lower among married people than among those in any of the three other marital status categories (single, cohabiting, and widowed, divorced or separated). This is not explained by the association between age and marital status (for example, married people and those who are widowed, divorced or separated are older, on average, than single people). Table 1.5 shows that in every age group except the youngest, married people were less likely to be smokers than were other respondents (although the difference is not statistically significant among those aged 60 and over). For example, among those aged 25 to 34, 34 per cent of those who were single and 35 per cent of those who were cohabiting were smokers, compared with only 21 per cent of those who were married.

Tables 1.4-1.5

Cigarette smoking and socio-economic classification

The National Statistics Socio-economic Classification (NS-SEC), which was introduced in 2001, does not allow categories to be collapsed into broad non-manual and manual groupings. So, since the *Cancer Plan* targets for England relate particularly to those in the manual socio-economic groups, the old socio-economic groupings have been recreated for this report in Table 1.6. Because of the new occupation coding, the classifications are not exactly the same, and comparisons with previous years should be treated with caution.

The GHS has consistently shown striking differences in the prevalence of cigarette smoking in relation to socio-economic group, with smoking being considerably more prevalent among those in manual groups than among those in non-manual groups. In the 1970s and 1980s, the prevalence of cigarette smoking fell more sharply among those in non-manual than in manual groups, so that differences between the groups became proportionately greater (table not shown). There was little further change in the relative proportions smoking cigarettes during the 1990s.

In England in 2006, 28 per cent of those in manual groups were cigarette smokers, compared with 33 per cent in 1998, confirming progress towards the targets set out in the *Cancer Plan*. These are to reduce prevalence among those in the manual group to 26 per cent in 2010. However, since the proportion of those in non-manual groups who are cigarette smokers has fallen by a similar amount (from 22 per cent in 1998 to 17 per cent in 2006) the differential between non-manual and manual has not reduced.

Figure 1.2: Prevalence of cigarette smoking by socio-economic group: England, 1992 to 2006*

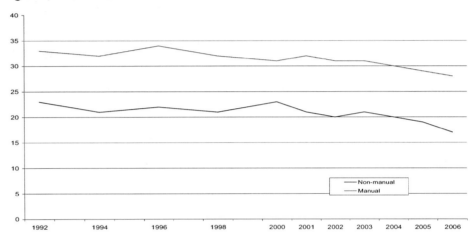

* weighted data are shown from 1998 onwards

However, caution is advisable when making comparisons over this period: the re-created socio-economic groups may have been affected by the change from head of household to household reference person as the basis for assessing socio-economic group, and by revisions to the way in which occupation is coded.

Table 1.7 shows similar trends in England since 2001 using the new socio-economic classification of the household reference person. It was noted earlier that there is a PSA target to reduce the prevalence of smoking among those in households classified as routine or manual to 26 per cent or lower by 2010. Over the period 2001 to 2006, the prevalence of cigarette smoking fell by four percentage points among those in routine and manual households, from 33 per cent to 29 per cent. Prevalence also fell by four percentage points among those in managerial and professional households (from 19 per cent in 2001 to 15 per cent in 2006), but the decrease in prevalence was somewhat greater among those in intermediate households, where it fell from 27 per cent to 21 per cent over the same period.

The prevalence of cigarette smoking in Great Britain in 2006 in relation to the eight- and three- category versions of NS-SEC is shown in Table 1.8. As was the case with the socio-economic groupings used previously, there were striking differences between the various classes. Prevalence was lowest among those in higher professional and higher managerial households (11 per cent and 14 per cent respectively) and highest, at 31-32 per cent, among those whose household reference person was in a routine or semi-routine occupation.

Figure 1.2, Tables 1.6-1.8

Cigarette smoking and economic activity status

Those who were economically active were more likely to smoke than those who were not, but this is largely explained by the lower prevalence of smoking among those aged 60 and over, who form the majority of economically inactive people.

Indeed, prevalence was highest among economically inactive people aged 16 to 59: 30 per cent of this group were smokers, compared with 24 per cent of economically active people aged 16-59 and only 12 per cent of economically inactive people aged 60 and over. Prevalence was particularly high among economically inactive people aged 16 to 59 whose last job was a routine or manual one, 46 per cent of whom were cigarette smokers.

Table 1.9

Regional variation in cigarette smoking

The data presented so far have been mainly for Great Britain, but the PSA targets and those included in the NHS *Cancer Plan* relate to England only. Table 1.10 shows that in 2006, overall prevalence in England was 22 per cent, the same as in Great Britain as a whole.

In every previous year except 2004, prevalence has been higher in Scotland than in England, although the difference has not always been large enough to be statistically significant. In 2006, 25 per cent of adults in Scotland were smokers, a significantly higher proportion than in England. In Wales, 20 per cent of adults were smokers, significantly fewer than in Scotland, but not significantly different from the proportion in England.

Care should be taken in interpreting differences between the regions of England in any one year, because sample sizes are small in some cases, making them subject to relatively high levels of sampling error. However, the inclusion of questions on smoking in every year since 2000, together with the relative stability of smoking prevalence in recent years, facilitates the combination of several years of data to enable more robust regional comparisons to be made. Table 1.12 shows data for the three years 2004 to 2006 combined, giving a sample of more than 50,000 adults in Great Britain.

This shows the same differences between the three countries of Great Britain as were described above for 2006 alone: prevalence among men and women is significantly higher in Scotland than in England and Wales. It also shows that the proportion of adults who have never smoked regularly is the same in all three countries, 53 per cent, so that the variation in prevalence is entirely due to different proportions having stopped smoking.

For men in England, the three regions of England with the highest prevalence were the North East, the North West and Yorkshire and the Humber, where 26-27 per cent of men were cigarette smokers (similar to the level in Scotland, and significantly higher than in all other regions except London). Among women, prevalence in the North East, at 28 per cent, was significantly higher than in every other region of England, and also significantly higher than in Wales and Scotland. The prevalence of cigarette smoking was lowest, at 20 per cent, among women in the West Midlands, the East of England, London and the South East.

Much of the overall regional variation in prevalence is contributed by differences in the proportions of smokers smoking 20 or more cigarettes a day. Among men, this ranges from 6 per cent in London and the South West to 12 per cent in Scotland. Among women, the range is even greater, from 4 per cent in the East of England, London and the South East to 10 per cent in the North East.

Tables 1.10-1.12

Cigarette consumption

The overall decline in smoking prevalence since the mid 1970s has been due to a fall in the proportions of both light smokers (defined as fewer than 20 cigarettes per day) and heavy smokers (20 cigarettes or more per day). The proportion of all adults smoking on average 20 or more cigarettes a day has fallen among men from 26 per cent in 1974 to 8 per cent in 2006, and from 13 per cent to 5 per cent of women over the same period.

In all age groups, respondents are much more likely to be light than heavy smokers, the difference being most pronounced among those aged under 35. For example, 17 per cent of young men and 19 per cent of young women aged 16 to 19 were light smokers in 2006, and only 2 per cent and 1 per cent respectively were heavy smokers.

The overall reported number of cigarettes smoked per male and female smoker has changed little since the early 1980s: the apparent slight fall among men smokers since the 1990s appears to be due to the introduction of weighting.

As in previous years, male smokers smoked more cigarettes a day on average than female smokers: in 2006, men smoked on average 15 cigarettes a day, compared with 13 for women. Cigarette consumption also varied by age. Among both men and women smokers, those aged 35 to 59 smoked the most – men smokers in this age group smoked on average 16 cigarettes a day and women smoked 14-15 a day.

GHS reports have consistently shown cigarette consumption levels to be higher among male and female smokers in manual socio-economic groups than among those in non-manual groups. A similar pattern is evident in relation to NS-SEC. In 2006, smokers in households where the household reference person was in a routine or manual occupation smoked an average of 15 cigarettes a day, compared with 12 a day for those in managerial or professional households.

Tables 1.13-1.16

Cigarette type

Filter cigarettes continue to be the most widely smoked type of cigarette, especially among women, but there has been a marked increase since the early 1990s in the proportion of smokers who smoke mainly hand-rolled tobacco. In 1990, 18 per cent of men smokers and 2 per cent of women smokers said they smoked mainly hand-rolled cigarettes, but by 2006 this had risen to 35 per cent and 16 per cent respectively. It should be noted that this increase in the proportion of smokers smoking mainly hand-rolled tobacco coincides with a fall in the prevalence of cigarette smoking from 30 per cent in 1990 to 22 per cent in 2006, so that the proportion of all adults who smoke hand-rolled tobacco has not increased so sharply: it has risen from about 3 per cent to about 5 per cent (no table shown).

There are likely to be two main reasons for this increase in the use of hand-rolled cigarettes:

- the rise in the real price of packaged cigarettes - hand-rolled ones are cheaper;

- the reduced tar and nicotine yield of packaged cigarettes: depending on how they are rolled and smoked, hand-rolled ones can give a higher tar and nicotine yield.

The use of hand-rolled tobacco was more common among men aged 35 and over than among younger men. Among women smokers there was less variation with age, except that only 9 per cent of women smokers aged 60 or over used hand-rolled tobacco.

Figure 1.3, Tables 1.17-1.18

Figure 1.3: Type of cigarette smoked, by sex: Great Britain, 1974 to 2006

(a) Men

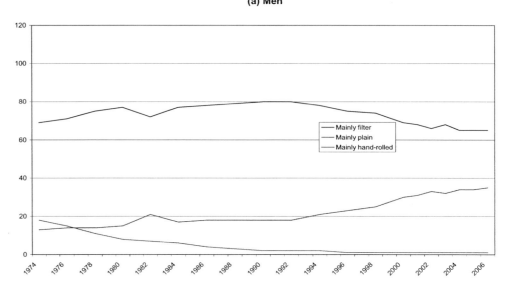

(b) Women

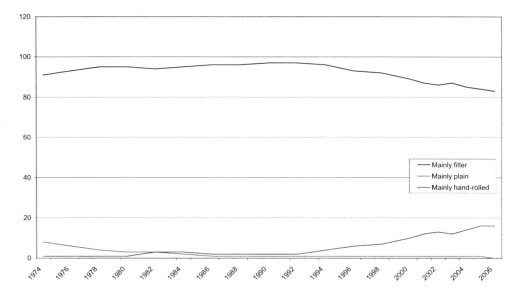

Tar yield[viii]

Table 1.19 shows the very marked reduction in the tar yield of cigarettes over the period during which the GHS has been collecting information about brand smoked. In 1986, 40 per cent of those who smoked manufactured cigarettes smoked brands yielding 15mg or more of tar per cigarette. In the following decade, the proportion smoking this type of cigarette fell to zero. Initially, this was partly due to smokers switching to lower tar brands, but the main factor has been the requirement for manufacturers to reduce substantially the tar yields of existing brands. Following legislation in 1992, they were required to reduce the tar yield to no more than 12mg per cigarette by the beginning of 1998. An EU Directive which came into force at the end of 2002 further reduced the maximum tar yield to 10 mg per cigarette from January 2004.

The effect of the recent changes in legislation can be seen in Table 1.20, in that there have been no brands with a yield of 12mg or more since 2003, even though these were the main brand of more than one third of smokers in previous years. There has been a compensating increase in the next highest category: the proportion of smokers smoking brands with a yield of 10 but less than 12mg increased from 13 per cent in 1998 to 71 per cent in 2002. since when it has remained at about the same level. Although this may seem surprising in view of the maximum legal declared yield of 10mg, the Directive relates to the tar yield as declared by the manufacturer, and this is permitted to vary by up to 15 per cent from the yield as measured for the Laboratory of the Government Chemist. Thus the yield as measured, which is what the GHS tables show, may be up to 11.5mg for a declared value of 10mg.

Among smokers aged under 60, differences between men and women in the tar yield of their usual brand were small. Among those aged 60 and over, however, women were much less likely to smoke brands in the highest tar band: 84 per cent of men smokers but only 65 per cent of women smokers in that age group did so.

There was also a difference in tar yield of cigarettes smoked according to the socio-economic class of the smoker's household reference person. Those in managerial and professional households were more likely than other smokers to smoke lower tar cigarettes: 27 per cent of smokers in managerial and professional households smoked cigarettes with a tar yield less than 8mg, compared with only 13 per cent of smokers in routine and manual households.

Tables 1.19-1.22

Cigar and pipe smoking

A decline in the prevalence of pipe and cigar smoking among men has been evident since the survey began, with most of the reduction occurring in the 1970s and 1980s.

In 2006, only 3 per cent of men smoked at least one cigar a month, compared with 34 per cent in 1974. Only a small number of women smoked cigars in 1974, and since 1978 the percentages have been scarcely measurable on the GHS. In previous years, cigar smoking has not been related to age, but in 2006 there was a clear age difference, with men aged 35 and over being more likely than younger men to say they had a cigar at least once a month.

Men were also asked whether or not they smoked a pipe 'at all nowadays'. Only 1 per cent of men in 2006 said they did, and they were almost all aged 50 and over.

Figure 1.4, Tables 1.23-1.24

Figure 1.4: Type of tobacco product smoked by men, 1974 to 2006

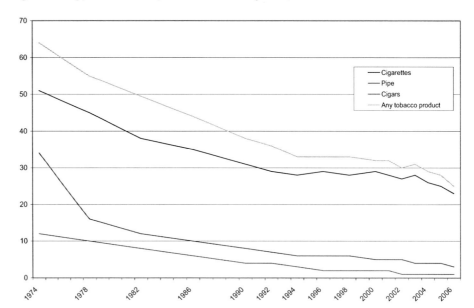

Age started smoking

The White Paper *Smoking Kills*[3] noted that people who start smoking at an early age are more likely than other smokers to smoke for a long period of time and more likely to die prematurely from a smoking-related disease.

About two thirds of respondents who were either current smokers or who had smoked regularly at some time in their lives had started smoking before they were 18. Indeed, almost two fifths had started smoking regularly before the age of 16, which was until recently the lowest age at which cigarettes could legally be bought[ix]. Men were more likely than women to have started smoking before they were 16 (41 per cent of men who had ever smoked regularly, compared with 36 per cent of women in 2006).

Since the early 1990s there appears to have been an increase in the proportion of women taking up smoking before the age of 16: in 1992, 28 per cent of women who had ever smoked had started before they were 16: this had risen to 36 per cent in 2005, but there was no further increase in 2006. There has been little change since 1992 in the proportion of men who had ever smoked who had started smoking regularly before the age of 16.

As the GHS has shown in previous years, there was an association between age started smoking regularly and socio-economic classification based on the current or last job of the household reference person. Of those in managerial and professional households, 31 per cent had started smoking before they were 16, compared with 45 per cent of those in routine and manual households.

Current heavy smokers were much more likely than light or ex-smokers to have started smoking at an early age. Of those smoking 20 or more cigarettes a day, 53 per cent started smoking regularly before they were 16, compared with only 33 per cent of those currently smoking fewer than 10 cigarettes a day.

Tables 1.25-1.27

Dependence on cigarette smoking

In order for the prevalence of cigarette smoking to reduce, young people have to be discouraged from starting to smoke and existing smokers have to be encouraged to stop. Since 1992, the GHS has asked three questions relevant to the likelihood of a smoker giving up. First, whether they would like to stop smoking, and then two indicators of dependence: whether they think they would find it easy or difficult not to smoke for a whole day; and how soon after waking they smoke their first cigarette. There has been very little change since 1992 in any of the three dependence measures used.

For an attempt to stop smoking to be successful, the smoker must want to stop. In 2006, 68 per cent of smokers said they would like to stop smoking altogether. The relationship between wanting to stop smoking and the number of cigarettes smoked is not straightforward. In every survey since the questions were first included in 1992, the proportion wanting to give up has been highest among those smoking on average 10-19 cigarettes a week, although, as in 2006, differences have not always been statistically significant.

It is interesting that it is not the heaviest smokers who are most likely to want to stop. This may be because they feel it would be too difficult or because they have been discouraged from wanting to stop by previous unsuccessful attempts. Furthermore, some previously heavy smokers who would like to give up may have cut down their consumption prior to an attempt to do so.

In 2006, 59 per cent of smokers felt that it would be either very or fairly difficult to go without smoking for a whole day. Not surprisingly, heavier smokers were more likely to say they would find it difficult – 82 per cent of those smoking 20 or more cigarettes a day did so, compared with only 26 per cent of those smoking fewer than 10 cigarettes a day.

Since women are less likely to be heavy smokers than men, it might be expected that women would be less likely to say they would find it hard to stop smoking for a day. As in almost every year shown in Table 1.29, however, this was not the case: in each of the three consumption categories shown, women were more likely than men to say they would find it hard not to smoke for a day, although the differences were not statistically significant in 2006. This difference between men and women smokers is not inconsistent with the overall similarity of the proportions saying they would find it hard not to smoke for a day (59 per cent of men and 60 per cent of women) because women are less likely than men to be heavy smokers, who are most likely to say they would find it difficult.

In 2006, 16 per cent of smokers had their first cigarette within five minutes of waking up. Heavy smokers were more likely than light smokers to smoke immediately on waking up: 36 per cent of those smoking 20 or more cigarettes did so, compared with only 2 per cent of those smoking fewer than 10 a day. Men were more likely than women to say they had their first cigarette within five minutes of waking - 18 per cent of men smokers, compared with 15 per cent of women smokers did so.

Women smokers are therefore more likely to perceive themselves as dependent despite the fact that on average they smoke fewer cigarettes a day than men, but appear to be less dependent in that they are less likely to smoke first thing when they wake up. There is no statistically significant difference between men and women smokers in the proportions wanting to give up.

Smokers in intermediate households were more likely than smokers in either managerial and professional or routine and manual households to say they would like to give up smoking altogether (73 per cent compared with 69 per cent and 66 per cent respectively). The difference was particularly marked among those smoking 20 or more cigarettes a day.

Overall, smokers in routine and manual households were more likely than others to say they would find it difficult to go without smoking for a whole day (63 per cent compared with 53 per cent among those in managerial and professional households, and 60 per cent among those in intermediate households). However, once amount smoked was taken into account (smokers in the routine and manual group smoke more on average than smokers in other social classes) the pattern of association was less clear.

Overall, smokers in managerial and professional households were less likely than other smokers to have had their first cigarette within five minutes of waking, and this was the case even when allowing for the fact that they smoked fewer cigarettes, on average. The differences between smokers in the other two groups of households – intermediate, and routine and manual – were smaller, and not consistently in the same direction.

Tables 1.28-1.33

Notes and references

[i] Goddard E, *General Household Survey 2005, Smoking and drinking among adults, 2005*, ONS 2006.

[ii] See *Chapter 4, General Household Survey 1992*, HMSO 1994. This includes a discussion of the differences found when smoking prevalence reported by young adults on the GHS was compared with prevalence among secondary school children.

[iii] See Appendix D, Living in Britain: results from the 2000 General Household Survey. The Stationery Office (London 2001).

[iv] *Smoking kills – a White Paper on tobacco*. The Stationery Office (London 1998) http://

[v] Available at www.hm-treasury.gov.uk/media/8/7/sr04_psa_ch3.pdf

[vi] *The NHS Cancer Plan*, Department of Health, 2000: available at www.dh.gov.uk/assetRoot/04/01/45/13/04014513.pdf

[vii] The results are published in Lader D et al, *Smoking-related behaviour and attitudes, 2006* ONS (London 2007)

[viii] An error was found in the automated procedure for coding the brand of cigarette smoked which was introduced when the GHS moved to computerised interviewing in April 1994. The net effect of this was that from 1994 to 2000, some brands were wrongly assigned to a low tar category. The coding procedure was revised for the 2001 survey. Corrected data for 1998 and 2000 are given in Tables 1.19 and 1.20.

[ix] The legal minimum age for the purchase of cigarettes and other tobacco has been 16 since 1908, but it was raised to 18 on 1 October 2007.

General Household Survey, 2006

Smoking and drinking among adults 2006

Table 1.1 Prevalence of cigarette smoking by sex and age: 1974 to 2006

Persons aged 16 and over *Great Britain*

Age	Unweighted							Weighted								Weighted base 2006 (000s) =100%[3]	Unweighted sample[3] 2006
	1974	1978	1982	1986	1990	1994	1998	1998	2000	2001	2002	2003	2004	2005[1]	2006[2]		
							Percentage smoking cigarettes										
Men																	
16-19	42	35	31	30	28	28	30	30	30	25	22	27	23	23	20	1,295	392
20-24	52	45	41	41	38	40	42	41	35	40	37	38	36	34	33	1,270	376
25-34	56	48	40	37	36	34	37	38	39	38	36	38	35	34	33	3,140	1053
35-49	55	48	40	37	34	31	32	33	31	31	29	32	31	29	26	5,636	2093
50-59	53	48	42	35	28	27	27	28	27	26	27	26	26	25	23	3,337	1374
60 and over	44	38	33	29	24	18	16	16	16	16	17	16	15	14	13	5,240	2389
All aged 16 and over	51	45	38	35	31	28	28	30	29	28	27	28	26	25	23	19,918	7677
Women																	
16-19	38	33	30	30	32	27	31	32	28	31	29	25	25	26	20	1,278	423
20-24	44	43	40	38	39	38	39	39	35	35	38	34	29	30	29	1,548	507
25-34	46	42	37	35	34	30	33	33	32	31	33	31	28	29	26	3,520	1320
35-49	49	43	38	34	33	28	28	29	27	28	27	28	28	26	25	6,392	2490
50-59	48	42	40	35	29	26	27	27	28	25	24	23	22	23	22	3,577	1513
60 and over	26	24	23	22	20	17	16	16	15	17	14	14	14	13	12	6,406	2752
All aged 16 and over	41	37	33	31	29	26	26	26	25	26	25	24	23	23	21	22,721	9005
Total																	
16-19	40	34	30	30	30	27	31	31	29	28	25	26	24	24	20	2,573	815
20-24	48	44	40	39	38	39	40	40	35	37	38	36	32	32	31	2,819	883
25-34	51	45	38	36	35	32	35	35	35	34	34	34	31	31	30	6,660	2373
35-49	52	45	39	36	34	30	30	31	29	29	28	30	29	27	25	12,027	4583
50-59	51	45	41	35	29	27	27	28	27	26	26	25	24	24	22	6,914	2887
60 and over	34	30	27	25	21	17	16	16	16	17	15	15	14	14	12	11,646	5141
All aged 16 and over	45	40	35	33	30	27	27	28	27	27	26	26	25	24	22	42,639	16682

1 2005 data includes last quarter of 2004/05 data due to survey change from financial year to calendar year.
2 Results for 2006 include longitudinal data (see Appendix B).
3 Trend tables show unweighted and weighted figures for 1998 to give an indication of the effect of the weighting. Bases for earlier years can be found in GHS reports for each year.

Office for National Statistics: January 2008

General Household Survey, 2006

Smoking and drinking among adults 2006

Table 1.2 Ex-regular cigarette smokers by sex and age: 1974 to 2006

Persons aged 16 and over Great Britain

Age	Unweighted							Weighted								Weighted base 2006 (000s) =100%[3]	Unweighted sample[3] 2006
	1974	1978	1982	1986	1990	1994	1998	1998	2000	2001	2002	2003	2004	2005[1]	2006[2]		
							Percentage of ex-regular cigarette smokers										
Men																	
16-19	3	4	4	5	4	5	5	5	3	4	3	5	4	3	4	1,295	392
20-24	9	9	9	11	8	7	8	9	7	9	7	7	8	7	11	1,270	376
25-34	18	18	20	20	16	16	13	13	12	15	13	13	15	14	16	3,140	1053
35-49	21	26	32	33	32	27	22	21	20	20	20	20	20	19	20	5,636	2093
50-59	30	35	38	38	42	40	41	40	36	36	35	32	34	34	31	3,337	1374
60 and over	37	43	47	52	52	55	54	54	52	47	51	50	50	51	49	5,240	2389
All aged 16 and over	23	27	30	32	32	31	31	29	27	27	28	27	28	27	27	19,918	7677
Women																	
16-19	4	5	6	7	6	6	7	8	6	6	5	6	4	4	4	1,278	423
20-24	9	8	9	9	8	10	8	8	11	12	10	10	8	9	11	1,548	507
25-34	12	14	15	16	14	14	14	14	13	16	16	16	14	15	17	3,520	1320
35-49	10	13	15	20	20	21	19	19	19	19	17	16	18	18	18	6,392	2490
50-59	13	18	19	18	20	22	25	25	24	24	26	27	27	25	25	3,577	1513
60 and over	11	16	20	23	27	29	29	29	29	29	30	29	28	29	30	6,406	2752
All aged 16 and over	11	14	16	18	19	21	21	20	20	21	21	21	20	21	21	22,721	9005

1 2005 data includes last quarter of 2004/5 data due to survey change from financial year to calendar year.
2 Results for 2006 include longitudinal data (see Appendix B).
3 Trend tables show unweighted and weighted figures for 1998 to give an indication of the effect of the weighting. Bases for earlier years can be found in GHS reports for each year.

Smoking and drinking among adults 2006

Table 1.3 Percentage who have never smoked cigarettes regularly by sex and age: 1974 to 2006

Persons aged 16 and over   Great Britain

Age	Unweighted							Weighted								Weighted base 2006 (000s) =100%[3]	Unweighted sample[3] 2006
	1974	1978	1982	1986	1990	1994	1998	1998	2000	2001	2002	2003	2004	2005[1]	2006[2]		
							Percentage who have never smoked regularly										
Men																	
16-19	56	61	65	65	68	67	64	65	67	71	75	68	72	74	77	1,295	392
20-24	38	46	50	47	54	53	49	50	58	51	55	54	55	59	56	1,270	376
25-34	26	33	39	43	48	50	50	49	49	47	51	49	50	53	51	3,140	1053
35-49	24	26	28	30	34	42	46	45	49	49	51	48	50	52	54	5,636	2093
50-59	16	17	20	26	31	33	32	32	37	38	38	41	40	41	46	3,337	1374
60 and over	18	18	20	19	24	27	30	30	32	36	32	34	35	35	38	5,240	2389
All aged 16 and over	25	29	32	34	37	40	41	42	44	45	46	45	46	47	50	19,918	7677
Women																	
16-19	58	62	64	62	62	67	62	61	66	63	66	69	70	70	76	1,278	423
20-24	47	49	51	54	53	52	53	53	54	53	52	55	62	61	61	1,548	507
25-34	42	44	48	48	52	55	53	53	54	53	51	53	58	56	57	3,520	1320
35-49	41	44	47	46	48	51	52	52	54	53	55	55	54	56	58	6,392	2490
50-59	38	39	41	47	51	52	48	48	48	51	50	50	51	51	53	3,577	1513
60 and over	63	60	57	55	54	54	55	56	56	54	55	57	58	58	58	6,406	2752
All aged 16 and over	49	49	51	51	52	54	53	53	54	53	54	55	57	57	58	22,721	9005

1 2005 data includes last quarter of 2004/5 data due to survey change from financial year to calendar year.
2 Results for 2006 include longitudinal data (see Appendix B).
3 Trend tables show unweighted and weighted figures for 1998 to give an indication of the effect of the weighting. Bases for earlier years can be found in GHS reports for each year.

General Household Survey, 2006

Smoking and drinking among adults 2006

Table 1.4 Cigarette-smoking status by sex and marital status

Persons aged 16 and over   *Great Britain: 2006[1]*

Marital status		Current cigarette smokers			Current non-smokers of cigarettes		Weighted base (000s)= 100%	Unweighted sample
		Light (under 20 per day)	Heavy (20 or more per day)	Total	Ex-regular cigarette smokers	Never or only occasionally smoked cigarettes		
Men								
Single	%	20	6	27	11	62	4,868	1502
Married/cohabiting	%	13	8	21	32	47	12,929	5357
Married couple	%	11	7	18	34	49	10,754	4536
Cohabiting couple	%	26	12	37	23	39	2,174	821
Widowed/divorced/separated	%	14	13	28	35	37	2,122	818
All aged 16 and over	%	15	8	23	27	50	19,918	7677
Women								
Single	%	22	5	26	10	63	4,280	1513
Married/cohabiting	%	0	0	18	22	59	13,682	5649
Married couple	%	12	4	16	23	62	11,392	4773
Cohabiting couple	%	26	7	33	21	46	2,290	876
Widowed/divorced/separated	%	16	7	23	25	52	4,758	1843
All aged 16 and over	%	16	5	21	21	58	22,720	9005
Total								
Single	%	21	6	27	11	63	9,147	3015
Married/cohabiting	%	14	6	20	27	53	26,611	11006
Married couple	%	11	5	17	28	55	22,147	9309
Cohabiting couple	%	26	9	35	22	43	4,464	1697
Widowed/divorced/separated	%	15	9	25	28	47	6,879	2661
All aged 16 and over	%	15	6	22	24	54	42,637	16682

1 Results for 2006 include longitudinal data (see Appendix B).

Smoking and drinking among adults 2006

Table 1.5 Cigarette-smoking status by age and marital status

Persons aged 16 and over *Great Britain: 2006[1]*

Marital status	Age					
	16-24	25-34	35-49	50-59	60 and over	Total
	Percentage smoking cigarettes					
Single	23	34	32	29	12	27
Married/cohabiting	35	26	22	20	11	20
Married couple	28	21	19	18	11	17
Cohabiting couple	38	35	37	34	16	35
Widowed/divorced/separated	42	48	39	32	15	25
All aged 16 and over	25	30	25	22	12	22
Weighted base (000s)= 100%						
Single	4,461	1,912	1,654	432	691	9,150
Married/cohabiting	906	4,484	8,887	5,300	7,033	26,610
Married couple	231	2,727	7,479	4,927	6,782	22,146
Cohabiting couple	674	1,758	1,408	374	250	4,464
Widowed/divorced/separated	26	264	1,485	1,181	3,924	6,880
All aged 16 and over	5,392	6,660	12,027	6,914	11,646	42,639
Unweighted sample						
Single	1397	641	539	160	278	3015
Married/cohabiting	293	1631	3503	2288	3291	11006
Married couple	75	997	2950	2115	3172	9309
Cohabiting couple	218	634	553	173	119	1697
Widowed/divorced/separated	8	101	541	439	1572	2661
All aged 16 and over	1698	2373	4583	2887	5141	16682

1 Results for 2006 include longitudinal data (see Appendix B).

General Household Survey, 2006

Smoking and drinking among adults 2006

Table **1.6** Prevalence of cigarette smoking by sex and whether household reference person is in a non-manual or manual socio-economic group: England 1992 to 2006[1,2]

Persons aged 16 and over   *England*

Socio-economic group of household reference person[3]	Unweighted				Weighted								Weighted base 2006 (000s) =100%[5]	Unweighted sample[5] 2006
	1992	1994	1996	1998	1998	2000	2001	2002	2003	2004	2005[4]	2006		
					Percentage smoking cigarettes									
Men														
Non-manual	22	21	21	21	22	24	22	21	22	22	19	18	9,181	3657
Manual	35	34	35	34	35	34	34	32	33	31	31	29	6,976	2618
Total[6]	29	28	28	28	29	29	28	27	27	26	25	23	17,163	6598
Women														
Non-manual	23	21	22	21	22	22	20	20	20	19	18	16	10,498	4275
Manual	30	30	33	31	31	29	31	30	29	28	28	27	6,949	2692
Total[6]	27	25	27	26	26	25	25	25	24	23	22	21	19,451	7693
All persons														
Non-manual	23	21	22	21	22	23	21	20	21	20	19	17	19,679	7932
Manual	33	32	34	32	33	31	32	31	31	30	29	28	13,925	5310
Total[6]	28	26	28	27	28	27	27	26	25	25	24	22	36,612	2618

1. Figures for 1992 to 1996 are taken from Department of Health bulletin Statistics on smoking: England, 1978 onwards. Figures for 2001 to 2006 are based on the NS-SEC classification recoded to produce SEG and should therefore be treated with caution.
2. Results for 2006 include longitudinal data (see Appendix B).
3. Head of household in years before 2000.
4. 2005 data includes last quarter of 2004/5 data due to survey change from financial year to calendar year.
5. Trend tables show unweighted and weighted figures for 1998 to give an indication of the effect of the weighting. Bases for earlier years can be found in GHS reports for each year.
6. Respondents whose head of household/household reference person was a full time student, in the Armed forces, had an inadequately described occupation, had never worked or were long-term unemployed are not shown as separate categories but are included in the total

General Household Survey, 2006

Smoking and drinking among adults 2006

Table **1.7** Prevalence of cigarette smoking by sex and socio-economic classification of the household reference person: England, 2001 to 2006[1]

Persons aged 16 and over — *England*

Socio-economic classification of household reference person	Weighted						Weighted base 2006 (000s) = 100%	Unweighted sample 2006
	2001	2002	2003	2004	2005[2]	2006		
	Percentage smoking cigarettes							
Men								
Managerial and professional	21	20	20	20	18	17	7,358	2961
Intermediate	29	27	28	26	24	22	3,053	1150
Routine and manual	34	32	34	32	32	32	5,982	2238
Total[3]	28	27	27	26	25	23	17,163	6598
Women								
Managerial and professional	17	17	17	17	16	14	7,826	3231
Intermediate	26	25	24	22	22	20	3,682	1435
Routine and manual	31	31	30	30	29	28	6,907	2664
Total[3]	25	25	24	23	22	21	19,451	7693
All persons								
Managerial and professional	19	19	18	19	17	15	15,184	6192
Intermediate	27	26	26	24	23	21	6,734	2585
Routine and manual	33	31	32	31	31	29	12,889	4902
Total[3]	27	26	25	25	24	22	36,612	14291

1 Results for 2006 include longitudinal data (see Appendix B).
2 2005 data includes last quarter of 2004/5 data due to survey change from financial year to calendar year.
3 Respondents whose household reference person was a full time student, had an inadequately described occupation, had never worked or was long-term unemployed these are not shown as separate categories but are included in the total.

General Household Survey, 2006

Smoking and drinking among adults 2006

Table 1.8 Prevalence of cigarette smoking by sex and socio-economic classification based on the current or last job of the household reference person

Persons aged 16 and over   *Great Britain: 2006[1]*

Socio-economic classification of household reference person[2]	Men		Women		Total	
	\multicolumn{6}{c}{Percentage smoking cigarettes}					
Managerial and professional						
Large employers and higher managerial	13		14		14	
Higher professional	13	17	9	14	11	15
Lower managerial and professional	20		17		18	
Intermediate						
Intermediate	22	21	19	21	20	21
Small employers and own account	21		22		22	
Routine and manual						
Lower supervisory and technical	25		25		25	
Semi-routine	33	31	29	28	31	29
Routine	35		29		32	
Total[2]		23		21		22
Weighted bases (000s) =100%						
Large employers and higher managerial		1,768		1,819		3,588
Higher professional		2,059		1,902		3,960
Lower managerial and professional		4,607		5,280		9,887
Intermediate		1,392		2,272		3,664
Small employers and own account		2,118		2,044		4,162
Lower supervisory and technical		2,385		2,225		4,610
Semi-routine		2,231		3,183		5,413
Routine		2,488		2,825		5,312
Total[2]		19,919		22,721		42,636
Unweighted sample						
Large employers and higher managerial		734		770		1504
Higher professional		829		799		1628
Lower managerial and professional		1831		2155		3986
Intermediate		527		878		1405
Small employers and own account		808		809		1617
Lower supervisory and technical		900		864		1764
Semi-routine		837		1248		2085
Routine		929		1070		1999
Total[2]		7677		9005		16682

1 Results for 2006 include longitudinal data (see Appendix B).
2 Respondents whose household reference person was a full time student, had an inadequately described occupation, had never worked or was long-term unemployed are not shown as separate categories but are included in the total.

General Household Survey, 2006

Smoking and drinking among adults 2006

Table **1.9** Prevalence of cigarette smoking by sex and socio-economic classification based on own current or last job, whether economically active or inactive, and, for economically inactive persons, age

Persons aged 16 and over
Great Britain: 2006[1]

Socio-economic classification	Men					Women					All persons				
	Active	Inactive 16-59	Inactive 60 and over	Total inactive	Total	Active	Inactive 16-59	Inactive 60 and over	Total inactive	Total	Active	Inactive 16-59	Inactive 60 and over	Total inactive	Total
	Percentage smoking cigarettes														
Managerial and professional	16	22	9	11	15	17	16	8	11	15	17	17	9	11	15
Intermediate	24	44	10	17	22	20	27	7	13	17	22	31	8	14	19
Routine and manual	34	61	16	29	32	32	40	15	25	28	33	46	16	26	30
Total[2]	24	37	12	20	23	23	28	12	18	21	24	30	12	19	22
Weighted bases (000s) =100%															
Managerial and professional	*5,702*	*231*	*1,526*	*1,756*	*7,463*	*4,780*	*675*	*1,191*	*1,867*	*6,648*	*10,482*	*905*	*2,717*	*3,623*	*14,110*
Intermediate	*2,484*	*153*	*630*	*782*	*3,266*	*3,063*	*603*	*1,445*	*2,049*	*5,111*	*5,547*	*757*	*2,074*	*2,830*	*8,377*
Routine and manual	*4,634*	*762*	*1,939*	*2,700*	*7,335*	*4,036*	*1,736*	*2,749*	*4,486*	*8,526*	*8,669*	*2498*	*4,688*	*7,186*	*15,861*
Total[2]	*13,760*	*1,996*	*4,154*	*6,146*	*19,918*	*12,786*	*4,242*	*5,682*	*9,926*	*22,719*	*26,546*	*6237*	*9,835*	*16,073*	*42,637*
Unweighted sample															
Managerial and professional	*2205*	*102*	*745*	*847*	*3052*	*1893*	*290*	*535*	*825*	*2718*	*4098*	*392*	*1280*	*1672*	*5770*
Intermediate	*928*	*57*	*282*	*339*	*1267*	*1202*	*244*	*632*	*876*	*2078*	*2130*	*301*	*914*	*1215*	*3345*
Routine and manual	*1678*	*264*	*846*	*1110*	*2788*	*1544*	*675*	*1143*	*1818*	*3362*	*3222*	*939*	*1989*	*2928*	*6150*
Total[2]	*5093*	*682*	*1898*	*2580*	*7673*	*4947*	*1630*	*2425*	*4055*	*9002*	*10040*	*2312*	*4323*	*6635*	*16675*

1 Results for 2006 include longitudinal data (see Appendix B).
2 Full time students, those who had never worked or were long-term unemployed, and those whose occupation was inadequately described are not shown as separate categories but are included in the total.

Office for National Statistics: January 2008

General Household Survey, 2006

Smoking and drinking among adults 2006

Table 1.10 Prevalence of cigarette smoking by sex and country: 1978 to 2006

Persons aged 16 and over   *Great Britain*

Country	Unweighted						Weighted								Weighted base 2006 (000s) =100%[3]	Unweighted sample[3] 2006
	1978	1982	1986	1990	1994	1998	1998	2000	2001	2002	2003	2004	2005[1]	2006[2]		
							Percentage smoking cigarettes									
Men																
England	44	37	34	31	28	28	29	29	28	27	27	26	25	23	*17,162*	*6599*
Wales	44	36	33	30	28	28	29	25	27	27	29	24	24	19	*1,021*	*410*
Scotland	48	45	37	33	31	33	35	30	32	29	35	29	28	25	*1,735*	*668*
Great Britain	45	38	35	31	28	28	30	29	28	27	28	26	25	23	*19,918*	*7677*
Women																
England	36	32	31	28	25	26	26	25	25	25	24	23	22	21	*19,451*	*7693*
Wales	37	34	30	31	27	26	27	24	26	27	26	22	21	20	*1,152*	*476*
Scotland	42	39	35	35	29	29	29	30	30	28	28	22	25	25	*2,116*	*836*
Great Britain	37	33	31	29	26	26	26	25	26	25	24	23	23	21	*22,719*	*9005*
All persons																
England	40	35	32	29	26	27	28	27	27	26	25	25	24	22	*36,613*	*14292*
Wales	40	35	31	31	27	27	28	25	27	27	27	23	22	20	*2,173*	*886*
Scotland	45	42	36	34	30	30	31	30	31	28	31	25	27	25	*3,852*	*1504*
Great Britain	40	35	33	30	27	27	28	27	27	26	26	25	24	22	*42,638*	*16682*

1 2005 data includes last quarter of 2004/5 data due to survey change from financial year to calendar year.
2 Results for 2006 include longitudinal data (see Appendix B).
3 Trend tables show unweighted and weighted figures for 1998 to give an indication of the effect of the weighting. Bases for earlier years can be found in GHS reports for each year.

Office for National Statistics: January 2008

General Household Survey, 2006

Smoking and drinking among adults 2006

Table **1.11** Prevalence of cigarette smoking by sex, country, and region of England: 1998 to 2006

Persons aged 16 and over *Great Britain*

Government Office Region	Weighted								Weighted base 2006 (000s) =100%[3]	Unweighted sample[3] 2006
	1998	2000	2001	2002	2003	2004	2005[1]	2006[2]		
Men				*Percentage smoking cigarettes*						
England										
North East	28	27	33	24	30	28	28	25	800	309
North West	29	29	28	28	30	27	26	26	2,219	901
Yorkshire and the Humber	30	29	30	27	25	30	27	24	1,805	723
East Midlands	27	27	28	24	31	27	25	21	1,699	686
West Midlands	32	27	27	25	26	26	23	25	1,739	674
East of England	26	27	27	25	28	26	25	22	1,982	784
London	34	31	29	29	28	26	25	24	2,237	662
South East	28	28	26	27	25	25	24	21	2,858	1115
South West	26	30	27	27	26	25	26	22	1,823	745
All England	29	29	28	27	27	26	25	23	17,162	6599
Wales	29	25	27	27	29	24	24	19	1,021	410
Scotland	35	30	32	29	35	29	28	25	1,735	668
Great Britain	30	29	28	27	28	26	25	23	19,918	7677
Women										
England										
North East	30	28	26	29	27	30	30	25	918	368
North West	32	30	29	28	30	28	23	23	2,667	1110
Yorkshire and the Humber	28	26	28	27	24	26	23	23	1,986	821
East Midlands	26	24	27	24	24	28	25	19	1,746	740
West Midlands	26	24	22	21	24	21	21	19	1,929	772
East of England	24	23	25	25	22	23	21	17	2,187	902
London	27	24	26	21	20	19	20	19	2,661	798
South East	21	23	23	25	22	20	21	19	3,233	1300
South West	25	24	22	24	22	21	25	23	2,121	882
All England	26	25	25	25	24	23	22	21	19,451	7693
Wales	27	24	26	27	26	22	21	20	1,152	476
Scotland	29	30	30	28	28	22	25	25	2,116	836
Great Britain	26	25	26	25	24	23	23	21	22,719	9005
All persons										
England										
North East	29	27	29	27	28	29	29	25	1,719	677
North West	31	30	29	28	30	28	24	25	4,885	2011
Yorkshire and the Humber	29	28	29	27	25	28	25	23	3,791	1544
East Midlands	27	25	28	24	27	27	25	20	3,444	1426
West Midlands	29	26	24	23	25	23	22	22	3,668	1446
East of England	25	25	26	25	25	24	23	19	4,170	1686
London	31	27	27	24	24	22	22	21	4,897	1460
South East	24	25	24	26	24	22	22	20	6,093	2415
South West	25	27	24	25	24	23	25	23	3,946	1627
All England	28	27	27	26	25	25	24	22	36,613	14292
Wales	28	25	27	27	27	23	22	20	2,173	886
Scotland	31	30	31	28	31	25	27	25	3,852	1504
Great Britain	28	27	27	26	26	25	24	22	42,638	16682

1 2005 data includes last quarter of 2004/5 data due to survey change from financial year to calendar year.
2 Results for 2006 include longitudinal data (see Appendix B).
3 Bases for earlier years can be found in GHS reports for each year.

Office for National Statistics: January 2008

General Household Survey, 2006

Smoking and drinking among adults 2006

Table 1.12 Cigarette-smoking status by sex, country and region: 2004-2006 combined

Persons aged 16 and over *Great Britain: 2004-2006*

Region		Current cigarette smokers			Current non-smokers of cigarettes			Weighted base (000s) =100%	Unweighted sample
		Heavy (20 or more per day)	Moderate (10-19 per day)	Light (fewer than 10 per day)	All current smokers	Ex-regular cigarette smokers	Never or only occasionally smoked cigarettes		
Men									
England									
North East	%	10	11	5	27	24	49	847	1048
North West	%	8	11	7	26	27	47	2,225	2873
Yorkshire and the Humber	%	10	10	6	27	25	48	1,749	2232
East Midlands	%	8	9	7	24	26	50	1,627	2101
West Midlands	%	9	10	6	24	29	47	1,670	2123
East of England	%	8	10	6	24	30	46	1,884	2431
London	%	6	9	9	25	23	53	2,287	2357
South East	%	7	10	6	23	31	46	2,826	3584
South West	%	6	9	8	24	32	43	1,758	2293
All England	%	8	10	7	25	28	48	16,875	21042
Wales	%	9	8	5	22	31	47	982	1240
Scotland	%	12	10	6	27	23	50	1,710	2175
Great Britain	%	8	10	7	25	27	48	19,567	24457
Women									
England									
North East	%	10	13	5	28	18	53	994	1266
North West	%	6	11	8	25	19	56	2,602	3434
Yorkshire and the Humber	%	7	10	6	24	21	55	1,997	2574
East Midlands	%	5	11	7	24	19	58	1,726	2293
West Midlands	%	5	10	6	20	19	60	1,891	2436
East of England	%	4	8	7	20	22	58	2,122	2805
London	%	4	7	8	20	16	64	2,676	2807
South East	%	4	9	7	20	24	56	3,156	4089
South West	%	6	10	8	23	22	55	1,991	2648
All England	%	5	10	7	22	20	58	19,152	24352
Wales	%	6	10	5	21	21	58	1,151	1470
Scotland	%	8	10	6	24	20	56	2,076	2693
Great Britain	%	6	10	7	22	20	57	22,379	28515
Total									
England									
North East	%	10	12	5	28	21	51	1,841	2314
North West	%	7	11	7	26	23	52	4,828	6307
Yorkshire and the Humber	%	8	10	6	25	23	52	3,744	4806
East Midlands	%	7	10	7	24	22	54	3,354	4394
West Midlands	%	7	10	6	22	24	54	3,561	4559
East of England	%	6	9	7	22	26	52	4,006	5236
London	%	5	8	9	22	19	59	4,963	5164
South East	%	6	9	7	21	27	51	5,982	7673
South West	%	6	10	8	24	27	49	3,750	4941
All England	%	7	10	7	23	24	53	36,028	45394
Wales	%	8	9	5	22	25	53	2,132	2710
Scotland	%	10	10	6	26	21	53	3,788	4868
Great Britain	%	7	10	7	23	24	53	41,946	52972

Office for National Statistics: January 2008

General Household Survey, 2006

Smoking and drinking among adults 2006

Table 1.13 Cigarette-smoking status by sex: 1974 to 2006

Persons aged 16 and over — *Great Britain*

	Unweighted							Weighted							
	1974	1978	1982	1986	1990	1994	1998	1998	2000	2001	2002	2003	2004	2005[1]	2006[2]
								Percentages							
Men															
Current cigarette smokers															
Light (under 20 per day)	25	22	20	20	17	17	18	19	18	19	17	18	18	17	15
Heavy (20 or more per day)	26	23	18	15	14	12	10	11	10	10	10	10	9	8	8
Total current cigarette smokers	51	45	38	35	31	28	28	30	29	28	27	28	26	25	23
Ex-regular cigarette smokers	23	27	30	32	32	31	31	29	27	27	28	27	28	27	27
Never or only occasionally	25	29	32	34	37	40	41	42	44	45	46	45	46	47	50
Weighted base (000s) =100%[3]								*19,229*	*20,350*	*19,913*	*19,561*	*19,187*	*19,561*	*19,496*	*19,918*
Unweighted sample[3]	*9852*	*10480*	*9199*	*8874*	*8106*	*7642*	*6579*		*6593*	*7055*	*6837*	*8097*	*6868*	*10038*	*7677*
Women															
Current cigarette smokers															
Light (under 20 per day)	28	23	22	21	20	18	19	19	19	19	18	18	17	17	16
Heavy (20 or more per day)	13	13	11	10	9	8	7	7	6	7	7	7	6	6	5
Total current cigarette smokers	41	37	33	31	29	26	26	26	25	26	25	24	23	23	21
Ex-regular cigarette smokers	11	14	16	18	19	21	21	20	20	21	21	21	20	21	21
Never or only occasionally	49	49	51	51	52	54	53	53	54	53	54	55	57	57	58
Weighted base (000s) =100%[3]								*21,654*	*22,044*	*21,987*	*22,236*	*21,842*	*22,396*	*22,315*	*22,721*
Unweighted sample 11480	*11480*	*12156*	*10641*	*10304*	*9445*	*9108*	*7830*		*7496*	*8299*	*7951*	*9327*	*8029*	*11627*	*9005*

1 2005 data includes last quarter of 2004/5 data due to survey change from financial year to calendar year.
2 Results for 2006 include longitudinal data (see Appendix B).
3 Trend tables show unweighted and weighted figures for 1998 to give an indication of the effect of the weighting. For the weighted data (1998 and 2000 to 2006) the weighted base (000s) is the base for percentages. Unweighted data (up to 1998) are based on the unweighted sample.

Office for National Statistics: January 2008

General Household Survey, 2006

Smoking and drinking among adults 2006

Table 1.14 Cigarette-smoking status by sex and age

Persons aged 16 and over *Great Britain: 2006[1]*

Age		Current cigarette smokers			Current non-smokers of cigarettes		Weighted base (000s) =100%	Unweighted sample
		Light (under 20 per day)	Heavy (20 or more per day)	All current smokers	Ex-regular cigarette smokers	Never or only occasionally smoked cigarettes		
Men								
16-19	%	17	2	20	4	77	*1,295*	*392*
20-24	%	26	7	33	11	56	*1,270*	*376*
25-34	%	25	8	33	16	51	*3,140*	*1053*
35-49	%	15	11	26	20	54	*5,636*	*2093*
50-59	%	14	9	23	31	46	*3,337*	*1374*
60 and over	%	7	6	13	49	38	*5,240*	*2389*
All aged 16 and over	%	15	8	23	27	50	*19,918*	*7677*
Women								
16-19	%	19	1	20	4	76	*1,278*	*423*
20-24	%	24	5	29	11	61	*1,548*	*507*
25-34	%	22	4	26	17	57	*3,520*	*1320*
35-49	%	18	7	25	18	58	*6,392*	*2490*
50-59	%	14	8	22	25	53	*3,577*	*1513*
60 and over	%	9	3	12	30	58	*6,406*	*2752*
All aged 16 and over	%	16	5	21	21	58	*22,721*	*9005*
Total								
16-19	%	18	2	20	4	76	*2,573*	*815*
20-24	%	25	6	31	11	59	*2,819*	*883*
25-34	%	23	6	30	16	54	*6,660*	*2373*
35-49	%	17	9	25	19	56	*12,027*	*4583*
50-59	%	14	8	22	28	50	*6,914*	*2887*
60 and over	%	8	4	12	38	49	*11,646*	*5141*
All aged 16 and over	%	15	6	22	24	54	*42,639*	*16682*

1 Results for 2006 include longitudinal data (see Appendix B).

Office for National Statistics: January 2008

Table 1.15 Average daily cigarette consumption per smoker by sex and age: 1974 to 2006

Current cigarette smokers aged 16 and over *Great Britain*

Age	Unweighted							Weighted								Weighted base 2006 (000s)= 100%[3]	Unweighted sample[3] 2006
	1974	1978	1982	1986	1990	1994	1998	1998	2000	2001	2002	2003	2004	2005[1]	2006[2]		
							Mean number of cigarettes per day										
Men																	
16-19	16	14	12	12	13	10	10	10	12	11	11	13	11	13	10	254	78
20-24	19	17	16	15	16	13	14	13	12	12	12	12	11	11	12	422	131
25-34	19	19	17	16	16	15	13	13	13	13	13	13	12	12	13	1,042	349
35-49	20	20	20	19	19	18	17	18	17	17	17	16	16	15	16	1,468	522
50-59	18	20	18	17	17	20	18	18	17	18	18	18	18	17	16	746	283
60 and over	14	15	16	15	15	14	16	16	15	15	16	15	14	15	18	674	294
All aged 16 and over	18	18	17	16	17	16	16	15	15	15	15	15	15	14	15	4,605	1657
Women																	
16-19	12	13	11	11	11	10	10	10	10	12	12	10	11	10	9	252	81
20-24	14	14	14	12	13	13	12	11	10	11	10	11	11	11	11	442	145
25-34	15	16	16	14	15	14	12	12	12	12	12	12	12	11	12	923	357
35-49	15	16	15	16	15	15	15	15	14	15	15	14	14	14	14	1,568	584
50-59	13	14	14	14	15	15	15	15	15	15	15	15	15	15	15	786	325
60 and over	10	11	11	12	12	13	12	12	12	12	13	13	13	13	13	768	322
All aged 16 and over	13	14	14	14	14	14	13	13	13	13	13	13	13	13	13	4,738	1814

1 2005 data includes last quarter of 2004/5 data due to survey change from financial year to calendar year.
2 Results for 2006 include longitudinal data (see Appendix B).
3 Trend tables show unweighted and weighted figures for 1998 to give an indication of the effect of the weighting. Bases for earlier years can be found in GHS reports for each year.

General Household Survey, 2006

Smoking and drinking among adults 2006

Table 1.16 Average daily cigarette consumption per smoker by sex, and socio-economic classification based on the current or last job of the household reference person

Current cigarette smokers aged 16 and over *Great Britain: 2006[1]*

Socio-economic classification of household reference person[1]	Men		Women		Total	
	\multicolumn{6}{c}{Mean number of cigarettes a day}					
Managerial and professional						
Large employers and higher managerial	13		10		11	
Higher professional	11	13	12	11	11	12
Lower managerial and professional	13		12		12	
Intermediate						
Intermediate	16	16	12	13	14	14
Small employers and own account	16		13		15	
Routine and manual						
Lower supervisory and technical	15		14		14	
Semi-routine	16	16	14	14	15	15
Routine	16		14		15	
Total[1]	15		13		14	

	Men	Women	Total
Weighted bases (000s) =100%			
Large employers and higher managerial	229	257	486
Higher professional	259	169	428
Lower managerial and professional	905	876	1,782
Intermediate	298	430	728
Small employers and own account	437	457	894
Lower supervisory and technical	603	560	1,163
Semi-routine	745	910	1,655
Routine	875	821	1,696
Total[2]	4,605	4,738	9,344
Unweighted sample			
Large employers and higher managerial	90	102	192
Higher professional	100	69	169
Lower managerial and professional	329	345	674
Intermediate	106	159	265
Small employers and own account	157	175	332
Lower supervisory and technical	213	211	424
Semi-routine	269	351	620
Routine	313	310	623
Total[2]	1657	1814	3471

1 Results for 2006 include longitudinal data (see Appendix B).

2 Respondents whose household reference person was a full time student, had an inadequately described occupation, had never worked or was long-term unemployed are not shown as separate categories but are included in the total.

Office for National Statistics: January 2008

General Household Survey, 2006

Smoking and drinking among adults 2006

Table 1.17 Type of cigarette smoked by sex: 1974 to 2006

Current cigarette smokers aged 16 and over *Great Britain*

Type of cigarette smoked	Unweighted							Weighted							
	1974	1978	1982	1986	1990	1994	1998	1998	2000	2001	2002	2003	2004	2005[1]	2006[2]
	%	%	%	%	%	%	%	%	%	%	%	%	%	%	%
Men															
Mainly filter	69	75	72	78	80	78	74	74	69	68	66	68	65	65	65
Mainly plain	18	11	7	4	2	2	1	1	1	1	1	1	1	1	1
Mainly hand-rolled	13	14	21	18	18	21	25	25	30	31	33	32	34	34	35
Weighted base (000s) =100%[3]								5,687	5,802	5,643	5,246	5,367	5,158	4,927	4,618
Unweighted sample[3]	4993	4646	3469	3072	2510	2150	1857		1796	1911	1765	2171	1748	2408	1661
	%	%	%	%	%	%	%	%	%	%	%	%	%	%	%
Women															
Mainly filter	91	95	94	96	97	96	92	92	89	87	86	87	85	84	83
Mainly plain	8	4	3	1	1	1	1	1	1	1	1	1	1	1	0
Mainly hand-rolled	1	1	3	2	2	4	7	8	10	12	13	12	14	16	16
Weighted base (000s) =100%[3]								5,735	5,619	5,635	5,560	5,287	5,156	5,060	4,743
Unweighted sample[3]	4600	4421	3522	3192	2748	2336	2044		1900	2101	1957	2226	1827	2579	1817

1 2005 data includes last quarter of 2004/5 data due to survey change from financial year to calendar year.
2 Results for 2006 include longitudinal data (see Appendix B).
3 Trend tables show unweighted and weighted figures for 1998 to give an indication of the effect of the weighting. For the weighted data (1998 and 2000 to 2006) the weighted base (000s) is the base for percentages. Unweighted data (up to 1998) are based on the unweighted sample.

Office for National Statistics: January 2008

Table 1.18 Type of cigarette smoked by sex and age

Current cigarette smokers aged 16 and over *Great Britain: 2006[1]*

Type of cigarette smoked	Age					
	16-24	25-34	35-49	50-59	60 and over	All aged 16 and over
	%	%	%	%	%	%
Men						
Mainly filter	75	72	61	57	61	65
Mainly plain	1	1	1	0	0	1
Mainly hand-rolled	24	27	39	43	39	35
Weighted base (000s) =100%	*675*	*1,044*	*1,473*	*752*	*674*	*4,618*
Unweighted sample	*209*	*350*	*524*	*284*	*294*	*1661*
Women						
Mainly filter	85	80	82	83	90	83
Mainly plain	0	1	0	0	1	0
Mainly hand-rolled	15	19	18	17	9	16
Weighted base (000s) =100%	*694*	*922*	*1,566*	*789*	*772*	*4,743*
Unweighted sample	*226*	*357*	*583*	*327*	*324*	*1817*
Total						
Mainly filter	80	76	71	70	77	74
Mainly plain	0	1	0	0	0	1
Mainly hand-rolled	20	23	28	30	23	25
Weighted base (000s) =100%	*1,369*	*1,967*	*3,039*	*1,541*	*1,446*	*9,362*
Unweighted sample	*435*	*707*	*1107*	*611*	*618*	*3478*

1 Results for 2006 include longitudinal data (see Appendix B).

Table 1.19 Grouped tar yield per cigarette: 1986 to 2006

Current smokers of manufactured cigarettes *Great Britain*

Tar yield	Unweighted					Weighted							
	1986	1988	1990	1992	1998	1998	2000	2001	2002	2003	2004	2005[1]	2006[2]
	%	%	%	%	%	%	%	%	%	%	%	%	%
<10mg	19	21	24	25	28	28	27	26	27	26	26	24	25
10<15mg	32	58	54	68	70	69	71	71	71	71	71	73	72
15+mg	40	17	19	4	0	0	0	0	0	0	0	0	0
No regular brand/dk tar	10	4	4	3	2	2	2	2	2	3	3	3	3
Weighted base (000s) =100%[3]						9,568	9,104	8,850	8,317	8,306	7,812	7,510	6,987
Unweighted sample[3]	5620	5363	4739	4662	3288		2955	3174	2870	3424	2716	3762	2606

1 2005 data includes last quarter of 2004/5 data due to survey change from financial year to calendar year.
2 Results for 2006 include longitudinal data (see Appendix B).
3 Trend tables show unweighted and weighted figures for 1998 to give an indication of the effect of the weighting. For the weighted data (1998 and 2000 to 2006) the weighted base (000s) is the base for percentages. Unweighted data (up to 1998) are based on the unweighted sample.

Table 1.20 Tar yield per cigarette: 1998 to 2006

Current smokers of manufactured cigarettes *Great Britain*

Tar yield	Weighted							
	1998	2000	2001	2002	2003	2004	2005[1]	2006[2]
	%	%	%	%	%	%	%	%
Less than 4mg	5	5	3	2	2	1	1	1
4<8mg	17	22	17	17	17	19	17	17
8<10mg	11	9	7	8	7	6	6	7
10<12mg	13	27	35	34	71	71	73	72
12<15mg	51	34	36	37	0	0	0	0
No regular brand/dk tar yield	2	2	2	2	3	3	3	3
Weighted base (000s) =100%	9,568	9,104	8,850	8,317	8,306	7,812	7,510	6,989
Unweighted sample	3288	2955	3174	2870	3424	2716	3762	2606

1 2005 data includes last quarter of 2004/5 data due to survey change from financial year to calendar year.
2 Results for 2006 include longitudinal data (see Appendix B).

General Household Survey, 2006

Smoking and drinking among adults 2006

Table 1.21 Tar yields by sex and age of smoker

Current smokers of manufactured[1] cigarettes aged 16 and over   *Great Britain: 2006[1]*

		Tar yield					Weighted base (000s) =100%	Unweighted sample
		Less than 4mg	4<8mg	8<10mg	10<12mg	No regular brand/dk tar yield		
Men								
16-19	%	0	6	4	81	9	197	60
20-24	%	3	12	3	79	3	315	97
25-34	%	1	23	6	68	2	759	253
35-49	%	2	15	2	76	4	902	325
50-59	%	0	10	3	82	4	429	163
60 and over	%	0	11	4	84	1	414	184
Total	%	1	15	4	76	3	3,016	1082
Women								
16-19	%	0	15	7	78	0	211	69
20-24	%	0	18	5	77	0	375	122
25-34	%	2	28	8	61	1	749	293
35-49	%	1	18	9	70	3	1,280	476
50-59	%	1	16	9	72	3	656	273
60 and over	%	3	18	12	65	2	700	291
Total	%	1	19	9	69	2	3,971	1524
Total								
16-19	%	0	11	6	79	4	409	129
20-24	%	2	15	4	78	1	691	219
25-34	%	1	26	7	64	2	1,508	546
35-49	%	2	17	6	72	3	2,182	801
50-59	%	1	13	7	76	4	1,085	436
60 and over	%	2	16	9	72	2	1,114	475
Total	%	1	17	7	72	3	6,989	2606

1 Results for 2006 include longitudinal data (see Appendix B).

Office for National Statistics: January 2008

Smoking and drinking among adults 2006

Table 1.22 Tar yields by sex and socio-economic classification based on the current or last job of the household reference person

Current smokers of manufactured[1] cigarettes aged 16 and over *Great Britain: 2006[2]*

Socio-economic class of household reference person[3]		Tar yields					Weighted base (000s) =100%	Unweighted sample
		Less than 4mg	4<8mg	8<10mg	10<12mg	No regular brand/dk tar yield		
Men								
Managerial and professional	%	2	23	4	68	3	1,061	395
Intermediate	%	2	15	6	75	2	505	178
Routine and manual	%	1	9	3	83	4	1,279	457
Total	%	1	15	4	76	3	3,015	1082
Women								
Managerial and professional	%	2	27	9	61	2	1,135	453
Intermediate	%	2	23	8	64	3	783	298
Routine and manual	%	1	13	9	75	2	1,851	702
Total	%	1	19	9	68	2	3,973	1524
All persons								
Managerial and professional	%	2	25	7	64	3	2,195	848
Intermediate	%	2	20	7	68	3	1,289	476
Routine and manual	%	1	12	7	78	2	3,130	1159
Total	%	1	17	7	72	3	6,987	2606

1 Thirty two per cent of male smokers and 12 per cent of female smokers said they mainly smoked hand-rolled cigarettes and have been excluded from this analysis.

2 Results for 2006 include longitudinal data (see Appendix B).

3 Respondents whose household reference person was a full time student, had an inadequately described occupation, had never worked or was long-term unemployed are not shown as separate categories but are included in the total.

General Household Survey, 2006

Smoking and drinking among adults 2006

Table 1.23 Prevalence of smoking by sex and type of product smoked: 1974 to 2006

Persons aged 16 and over — Great Britain

	Unweighted							Weighted							
	1974	1978	1982[1]	1986	1990	1994	1998	1998	2000	2001	2002	2003	2004	2005[2]	2006[3]
							Percentage smoking								
Men															
Cigarettes[4]	51	45	38	35	31	28	28	30	29	28	27	28	26	25	23
Pipe	12	10	..	6	4	3	2	2	2	2	1	1	1	1	1
Cigars[5]	34	16	12	10	8	6	6	6	5	5	5	4	4	4	3
All smokers[6]	64	55	45	44	38	33	33	34	32	32	30	31	29	28	25
Weighted base (000s) = 100%								*19,225*	*20,350*	*19,972*	*19,561*	*19,187*	*19,561*	*19,498*	*19,920*
Unweighted sample[6]	*9862*	*10439*	*9171*	*8884*	*8119*	*7662*	*6579*		*6593*	*7074*	*6835*	*8097*	*6868*	*10039*	*7678*
Women															
Cigarettes[4]	41	37	33	31	29	26	26	26	25	26	25	24	23	23	21
Cigars[5]	3	1	0	1	0	0	0	0	0	0	0	0	0	0	0
All smokers[6]	41	37	34	31	29	26	26	27	26	26	25	24	23	23	21
Weighted base (000s) = 100%[7]								*21,653*	*22,044*	*22,032*	*22,236*	*21,842*	*22,393*	*22,315*	*22,723*
Unweighted sample[6]	*11419*	*12079*	*10559*	*10312*	*9455*	*9137*	*7830*		*7496*	*8317*	*7951*	*9327*	*8028*	*11627*	*9006*

1 In 1982 and 1984 men were not asked about pipe smoking, and therefore the figures for all smokers exclude those who smoked only a pipe.
2 2005 data includes last quarter of 2004/5 data due to survey change from financial year to calendar year.
3 Results for 2006 include longitudinal data (see Appendix B).
4 Figures for cigarettes include all smokers of manufactured and hand-rolled cigarettes.
5 For 1974 the figures include occasional cigar smokers, that is, those who smoked less than one cigar a month.
6 The percentages for cigarettes, pipes and cigars add to more than the percentage for all smokers because some people smoked more than one type of product.
7 Trend tables show unweighted and weighted figures for 1998 to give an indication of the effect of the weighting. For the weighted data (1998 and 2000 to 2006) the weighted base (000s) is the base for percentages. Unweighted data (up to 1998) are based on the unweighted sample.

General Household Survey, 2006

Smoking and drinking among adults 2006

Table **1.24** Prevalence of smoking among men by age and type of product smoked[1]

Men aged 16 and over *Great Britain: 2006[1]*

Age	Cigarettes[2]	Pipe[3]	Cigars[3]	All smokers[4]	*Weighted base (000s) =100%*	*Unwgtd sample*
	Percentage smoking					
16-19	20	0	1	20	*1,298*	*393*
20-24	33	0	2	34	*1,270*	*376*
25-29	38	0	1	38	*1,396*	*458*
30-34	30	0	2	30	*1,744*	*595*
35-49	26	0	4	28	*5,635*	*2093*
50-59	23	2	4	26	*3,336*	*1374*
60 and over	13	2	3	16	*5,241*	*2389*
All aged 16 and over	23	1	3	25	*19,920*	*7678*

1 Results for 2006 include longitudinal data (see Appendix B).
2 Figures for cigarettes include all smokers of both manufactured and hand-rolled cigarettes.
3 Young people aged 16-17 were not asked about cigar or pipe-smoking.
4 The percentages for cigarettes, pipes and cigars add to more than the percentage for all smokers because some people smoked more than one type of product.

General Household Survey, 2006

Smoking and drinking among adults 2006

Table 1.25 Age started smoking regularly by sex: 1992 to 2006

Persons aged 16 and over who had ever smoked regularly   *Great Britain*

Age started smoking regularly	Unweighted				Weighted							
	1992	1994	1996	1998	1998	2000	2001	2002	2003	2004	2005[1]	2006[2]
	%	%	%	%	%	%	%	%	%	%	%	%
Men												
Under 16	40	41	41	43	42	43	42	42	42	42	41	41
16-17	27	27	27	26	26	27	26	28	26	26	26	26
18-19	17	16	17	17	17	15	16	16	16	16	17	17
20-24	12	11	11	10	11	11	11	11	11	10	11	12
25 and over	4	4	3	4	4	5	4	4	4	5	5	4
Weighted base (000s) =100%					*11,146*	*11,016*	*10,608*	*10,469*	*10,431*	*10,506*	*10,194*	*9,931*
Unweighted sample	*5143*	*4519*	*4295*	*3852*		*3625*	*3883*	*3696*	*4410*	*3700*	*5276*	*3902*
Women												
Under 16	28	30	32	31	32	33	35	33	35	35	36	36
16-17	28	28	28	29	28	27	27	28	26	18	27	28
18-19	19	19	17	18	17	19	17	18	19	18	17	17
20-24	15	13	13	14	14	12	12	13	13	12	12	12
25 and over	10	9	9	8	8	8	9	7	7	7	7	6
Weighted base (000s) =100%					*10,101*	*9,663*	*10,222*	*10,067*	*9,738*	*9,591*	*9,589*	*9,404*
Unweighted sample	*4640*	*4179*	*3991*	*3645*		*3302*	*3818*	*3589*	*4141*	*3446*	*4987*	*3733*
All persons												
Under 16	34	36	37	37	37	38	39	38	38	39	39	39
16-17	27	28	28	27	27	27	26	28	26	27	27	27
18-19	18	18	17	18	17	17	17	17	17	17	17	17
20-24	14	12	12	12	12	11	12	12	12	11	11	12
25 and over	7	7	6	6	6	6	6	5	6	6	6	5
Weighted base (000s) =100%					*21,247*	*20,679*	*20,830*	*20,537*	*20,169*	*20,097*	*19,783*	*19,337*
Unweighted sample	*9783*	*8698*	*8286*	*7497*		*6957*	*7701*	*7285*	*8551*	*7146*	*10263*	*7635*

1 2005 data includes last quarter of 2004/5 data due to survey change from financial year to calendar year.
2 Results for 2006 include longitudinal data (see Appendix B).

General Household Survey, 2006

Smoking and drinking among adults 2006

Table **1.26** Age started smoking regularly by sex and socio-economic classification based on the current or last job of the household reference person

Persons aged 16 and over who had ever smoked regularly *Great Britain: 2006[1]*

Age started smoking regularly	Socio-economic classification of household reference person[2]			
	Managerial & professional	Intermediate	Routine & manual	Total
	%	%	%	%
Men				
Under 16	33	42	48	41
16-17	30	23	25	26
18-19	20	17	15	17
20-24	13	14	9	12
25 and over	4	5	3	4
Weighted base (000s) =100%	*3,607*	*1,771*	*4,187*	*9,931*
Unweighted sample	*1479*	*689*	*1606*	*3902*
	%	%	%	%
Women				
Under 16	28	36	43	36
16-17	30	28	28	28
18-19	22	18	13	17
20-24	14	13	11	12
25 and over	6	6	6	6
Weighted base (000s) =100%	*3,141*	*1,809*	*4,025*	*9,404*
Unweighted sample	*1301*	*704*	*1571*	*3733*
	%	%	%	%
All persons				
Under 16	31	39	45	39
16-17	30	25	26	27
18-19	21	18	14	17
20-24	13	13	10	12
25 and over	5	5	4	5
Weighted base (000s) =100%	*6,749*	*3,579*	*8,212*	*19,337*
Unweighted sample	*2780*	*1393*	*3177*	*7635*

1 Results for 2006 include longitudinal data (see Appendix B).
2 Respondents whose household reference person was a full time student, had an inadequately described occupation, had never worked or was long-term unemployed are not shown as separate categories but are included in the total.

General Household Survey, 2006

Smoking and drinking among adults 2006

Table **1.27** Age started smoking regularly by sex, whether current smoker and if so, cigarettes smoked a day

Persons aged 16 and over who had ever smoked regularly *Great Britain: 2006[1]*

Age started smoking regularly	Current smoker				Ex-regular smoker	All who have ever smoked regularly
	20 or more a day	10-19 a day	0-9 a day	All current smokers[2]		
	%	%	%	%	%	%
Men						
Under 16	53	45	32	44	39	41
16-17	24	26	30	26	26	26
18-19	11	13	19	14	20	17
20-24	9	12	14	11	12	12
25 and over	4	5	5	4	3	4
Weighted base (000s) =100%	*1,582*	*1,753*	*1,225*	*4,571*	*5,361*	*9,931*
Unweighted sample	*580*	*636*	*425*	*1645*	*2257*	*3902*
	%	%	%	%	%	%
Women						
Under 16	52	41	35	42	31	36
16-17	25	28	30	28	28	28
18-19	11	16	13	14	20	17
20-24	6	10	16	11	14	12
25 and over	5	5	6	5	6	6
Weighted base (000s) =100%	*1,138*	*2,133*	*1,430*	*4,710*	*4,695*	*9,406*
Unweighted sample	*445*	*801*	*555*	*1805*	*1927*	*3732*
	%	%	%	%	%	%
All persons						
Under 16	53	43	33	43	35	39
16-17	25	27	30	27	27	27
18-19	11	15	16	14	20	17
20-24	8	11	15	11	13	12
25 and over	4	5	6	5	5	5
Weighted base (000s) =100%	*2,720*	*3,887*	*2,655*	*9,279*	*10,055*	*19,337*
Unweighted sample	*1025*	*1437*	*980*	*3450*	*4184*	*7634*

1 Results for 2006 include longitudinal data (see Appendix B).
2 Includes a few smokers who did not say how many cigarettes a day they smoked.

General Household Survey, 2006

Smoking and drinking among adults 2006

Table **1.28** Proportion of smokers who would like to give up smoking altogether, by sex and number of cigarettes smoked per day: 1992 to 2006

Current cigarette smokers aged 16 and over *Great Britain*

Number of cigarettes smoked a day	Unweighted				Weighted								Weighted base 2006 (000s) =100%[3]	Unweighted sample[3] 2006
	1992	1994	1996	1998	1998	2000	2001	2002	2003	2004	2005[1]	2006[2]		
					Percentage who would like to stop altogether									
Men														
20 or more	68	70	66	69	69	74	70	68	64	67	66	67	*1,597*	*585*
10-19	70	72	69	73	73	76	71	71	67	68	68	68	*1,769*	*641*
0-9	58	61	62	62	62	64	62	62	61	64	68	65	*1,240*	*431*
All smokers[4]	66	69	66	69	69	72	68	68	64	67	68	67	*4,620*	*1662*
Women														
20 or more	70	69	69	68	68	73	66	67	64	70	67	67	*1,142*	*447*
10-19	72	71	70	75	75	76	67	71	71	70	70	72	*2,136*	*802*
0-9	58	62	59	65	65	63	60	67	66	67	67	69	*1,461*	*565*
All smokers[4]	68	68	67	70	70	71	65	69	67	69	68	70	*4,747*	*1818*
Total														
20 or more	69	70	68	69	69	74	68	68	64	68	67	67	*2,738*	*1032*
10-19	71	71	70	74	74	76	69	71	69	69	69	70	*3,905*	*1443*
0-9	58	61	60	64	64	63	61	65	64	66	68	67	*2,701*	*996*
All smokers[4]	67	68	67	69	69	72	66	68	66	68	68	68	*9,366*	*3480*

1 2005 data includes last quarter of 2004/5 data due to survey change from financial year to calendar year
2 Results for 2006 include longitudinal data (see Appendix B).
3 Trend tables show unweighted and weighted figures for 1998 to give an indication of the effect of the weighting. Bases for earlier years can be found in GHS reports for each year.
4 Includes a few smokers who did not say how many cigarettes a day they smoked.

Smoking and drinking among adults 2006

Table 1.29 Proportion of smokers who would find it difficult to go without smoking for a day, by sex and number of cigarettes smoked per day: 1992 to 2006

Current cigarette smokers aged 16 and over Great Britain

Number of cigarettes smoked a day	Unweighted				Weighted								Weighted base 2006 (000s) =100%[3]	Unweighted sample[3] 2006
	1992	1994	1996	1998	1998	2000	2001	2002	2003	2004	2005[1]	2006[2]		

Percentage who would find it difficult not to smoke for a day

Men														
20 or more	76	78	78	78	78	78	74	77	78	77	77	80	1,592	583
10-19	54	57	54	54	54	56	55	57	53	57	60	64	1,764	639
0-9	20	17	20	25	23	14	21	23	19	16	23	24	1,231	428
All smokers[4]	55	56	56	56	56	53	52	56	53	52	55	59	4,602	1655
Women														
20 or more	86	86	87	87	86	88	87	86	83	82	84	84	1,137	446
10-19	68	68	66	66	65	67	65	66	64	67	65	70	2,123	799
0-9	23	20	24	24	25	22	24	21	22	21	27	27	1,452	562
All smokers[4]	61	60	61	59	59	58	58	59	56	57	58	60	4,720	1811
Total														
20 or more	80	82	83	82	82	82	80	81	80	79	80	82	2,729	1029
10-19	61	63	60	61	60	62	61	62	58	62	63	67	3,888	1438
0-9	21	19	23	24	24	18	22	22	21	19	25	26	2,683	990
All smokers[4]	58	59	58	58	57	56	55	57	55	55	56	59	9,322	3466

1 2005 data includes last quarter of 2004/5 data due to survey change from financial year to calendar year.
2 Results for 2006 include longitudinal data (see Appendix B).
3 Trend tables show unweighted and weighted figures for 1998 to give an indication of the effect of the weighting. Bases for earlier years can be found in GHS reports for each year.
4 Includes a few smokers who did not say how many cigarettes a day they smoked.

General Household Survey, 2006

Smoking and drinking among adults 2006

Table **1.30** Proportion of smokers who have their first cigarette within five minutes of waking, by sex and number of cigarettes smoked per day: 1992 to 2006

Current cigarette smokers aged 16 and over *Great Britain*

Number of cigarettes smoked a day	Unweighted				Weighted								Weighted base 2006 (000s) =100%[3]	Unweighted sample[3] 2006
	1992	1994	1996	1998	1998	2000	2001	2002	2003	2004	2005[1]	2006[2]		
					Percentage smoking within 5 minutes of waking									
Men														
20 or more	29	31	29	31	32	30	30	31	31	34	34	35	1,598	585
10-19	10	13	9	11	11	13	11	11	11	13	15	13	1,765	640
0-9	2	2	3	2	2	2	3	3	4	1	2	2	1,236	430
All smokers[4]	16	18	16	16	17	16	15	16	16	17	17	18	4,614	1660
Women														
20 or more	29	34	32	31	31	32	35	31	31	33	31	37	1,143	447
10-19	10	9	11	12	12	12	12	12	12	15	11	12	2,135	802
0-9	1	0	1	1	1	2	2	2	3	2	2	2	1,451	561
All smokers[4]	14	14	15	14	14	14	15	14	14	16	14	15	4,735	1813
Total														
20 or more	29	33	30	31	31	31	32	31	31	34	33	36	2,739	1032
10-19	10	11	10	12	12	13	11	11	11	14	13	12	3,899	1442
0-9	2	1	2	2	2	2	2	3	3	1	2	2	2,687	991
All smokers[4]	15	16	15	15	15	15	15	15	15	17	16	16	9,345	3473

1 2005 data includes last quarter of 2004/5 data due to survey change from financial year to calendar year.
2 Results for 2006 include longitudinal data (see Appendix B).
3 Trend tables show unweighted and weighted figures for 1998 to give an indication of the effect of the weighting. Bases for earlier years can be found in GHS reports for each year.
4 Includes a few smokers who did not say how many cigarettes a day they smoked.

General Household Survey, 2006

Smoking and drinking among adults 2006

Table 1.31 Proportion of smokers who would like to give up smoking altogether, by sex, socio-economic classification of household reference person, and number of cigarettes smoked a day

Current cigarette smokers aged 16 and over   *Great Britain: 2006[1]*

Number of cigarettes smoked a day	Socio-economic classification [2]			Total
	Managerial & professional	Intermediate	Routine & manual	
	Percentage who would like to stop altogether			
Men				
20 or more	66	80	61	67
10-19	69	70	67	68
0-9	64	69	65	65
All smokers[3]	67	74	64	67
Women				
20 or more	70	76	63	67
10-19	77	70	70	72
0-9	65	73	71	69
All smokers[3]	71	73	68	70
Total				
20 or more	67	79	62	67
10-19	73	70	69	70
0-9	64	72	68	67
All smokers[3]	69	73	66	68
Weighted base (000s)=100%				
Men				
20 or more	*344*	*304*	*863*	*1,597*
10-19	*546*	*265*	*851*	*1,769*
0-9	*504*	*168*	*509*	*1,240*
All smokers[3]	*1,397*	*744*	*2,228*	*4,620*
Women				
20 or more	*230*	*200*	*639*	*1,142*
10-19	*558*	*394*	*1,084*	*2,136*
0-9	*514*	*293*	*570*	*1,461*
All smokers[3]	*1,304*	*891*	*2,295*	*4,747*
Total				
20 or more	*575*	*503*	*1,502*	*2,738*
10-19	*1,103*	*659*	*1,935*	*3,905*
0-9	*1,019*	*461*	*1,078*	*2,701*
All smokers[3]	*2,702*	*1,634*	*4,522*	*9,366*
Unweighted sample				
Men				
20 or more	*136*	*106*	*315*	*585*
10-19	*203*	*99*	*306*	*641*
0-9	*180*	*58*	*174*	*431*
All smokers[3]	*520*	*265*	*797*	*1662*
Women				
20 or more	*100*	*74*	*246*	*447*
10-19	*216*	*143*	*409*	*802*
0-9	*200*	*117*	*217*	*565*
All smokers[3]	*517*	*336*	*873*	*1818*
Total				
20 or more	*236*	*180*	*561*	*1032*
10-19	*419*	*242*	*715*	*1443*
0-9	*380*	*175*	*391*	*996*
All smokers[3]	*1037*	*601*	*1670*	*3480*

1 Results for 2006 include longitudinal data (see Appendix B).
2 Respondents whose household reference person was a full time student, had an inadequately described occupation, had never worked or was long-term unemployed are not shown as separate categories but are included in the total.
3 Includes a few smokers who did not say how many cigarettes a day they smoked.

Smoking and drinking among adults 2006

Table 1.32 Proportion of smokers who would find it difficult to go without smoking for a day, by sex, socio-economic classification of household reference person, and number of cigarettes smoked a day

Current cigarette smokers aged 16 and over   *Great Britain: 2006[1]*

Number of cigarettes smoked a day	Socio-economic classification[2]			Total
	Managerial & professional	Intermediate	Routine & manual	
	Percentage who would find it difficult to stop for a day			
Men				
20 or more	79	79	80	80
10-19	62	65	65	64
0-9	22	17	26	24
All smokers[3]	52	60	62	59
Women				
20 or more	88	83	82	84
10-19	68	70	72	70
0-9	21	33	28	27
All smokers[3]	53	61	64	60
Total				
20 or more	83	80	81	82
10-19	65	68	69	67
0-9	21	28	27	26
All smokers[3]	53	60	63	59
Weighted base (000s)=100%				
Men				
20 or more	*344*	*304*	*858*	*1,592*
10-19	*546*	*262*	*849*	*1,764*
0-9	*502*	*161*	*509*	*1,231*
All smokers[3]	*1,395*	*734*	*2,221*	*4,602*
Women				
20 or more	*227*	*200*	*639*	*1,137*
10-19	*558*	*389*	*1079*	*2,123*
0-9	*512*	*294*	*568*	*1,452*
All smokers[3]	*1,299*	*887*	*2,288*	*4,720*
Total				
20 or more	*571*	*503*	*1,496*	*2,729*
10-19	*1,103*	*651*	*1,928*	*3,888*
0-9	*1,014*	*455*	*1,075*	*2,683*
All smokers[3]	*2,693*	*1,620*	*4,506*	*9,322*
Unweighted sample				
Men				
20 or more	*136*	*106*	*313*	*583*
10-19	*203*	*98*	*305*	*639*
0-9	*179*	*56*	*174*	*428*
All smokers[3]	*519*	*262*	*794*	*1655*
Women				
20 or more	*99*	*74*	*246*	*446*
10-19	*216*	*142*	*408*	*799*
0-9	*199*	*117*	*216*	*562*
All smokers[3]	*515*	*335*	*871*	*1811*
Total				
20 or more	*235*	*180*	*559*	*1029*
10-19	*419*	*240*	*713*	*1438*
0-9	*378*	*173*	*390*	*990*
All smokers[3]	*1034*	*597*	*1665*	*3466*

1 Results for 2006 include longitudinal data (see Appendix B).
2 Respondents whose household reference person was a full time student, had an inadequately described occupation, had never worked or was long-term unemployed are not shown as separate categories but are included in the total.
3 Includes a few smokers who did not say how many cigarettes a day they smoked.

General Household Survey, 2006

Smoking and drinking among adults 2006

Table **1.33** Proportion of smokers who have their first cigarette within five minutes of waking, by sex, socio-economic classification of household reference person, and number of cigarettes smoked a day

Current cigarette smokers aged 16 and over *Great Britain: 2006[1]*

Number of cigarettes smoked a day	Socio-economic classification[2]			
	Managerial & professional	Intermediate	Routine & manual	Total
	Percentage who smoke within 5 minutes of waking			
Men				
20 or more	28	37	37	35
10-19	9	12	16	13
0-9	1	4	1	2
All smokers[3]	11	20	21	18
Women				
20 or more	26	45	37	37
10-19	8	7	15	12
0-9	2	1	3	2
All smokers[3]	9	14	18	15
Total				
20 or more	27	40	37	36
10-19	9	9	15	12
0-9	1	2	2	2
All smokers[3]	10	17	19	16
Weighted base (000s)=100%				
Men				
20 or more	*344*	*303*	*865*	*1,598*
10-19	*546*	*265*	*846*	*1,765*
0-9	*500*	*167*	*508*	*1,236*
All smokers[3]	*1,393*	*742*	*2,224*	*4,614*
Women				
20 or more	*230*	*201*	*638*	*1,143*
10-19	*558*	*394*	*1,083*	*2,135*
0-9	*510*	*289*	*569*	*1,451*
All smokers[3]	*1,300*	*886*	*2,292*	*4,735*
Total				
20 or more	*574*	*504*	*1,502*	*2,739*
10-19	*1,102*	*658*	*1,930*	*3,899*
0-9	*1,010*	*456*	*1,078*	*2,687*
All smokers[3]	*2,691*	*1,627*	*4,517*	*9,345*
Unweighted sample				
Men				
20 or more	*136*	*106*	*315*	*585*
10-19	*203*	*99*	*305*	*640*
0-9	*179*	*58*	*174*	*430*
All smokers[3]	*519*	*265*	*796*	*1660*
Women				
20 or more	*100*	*74*	*246*	*447*
10-19	*216*	*143*	*409*	*802*
0-9	*198*	*115*	*217*	*561*
All smokers[3]	*515*	*333*	*873*	*1813*
Total				
20 or more	*236*	*180*	*561*	*1032*
10-19	*419*	*242*	*714*	*1442*
0-9	*377*	*173*	*391*	*991*
All smokers[3]	*1034*	*598*	*1669*	*3473*

1 Results for 2006 include longitudinal data (see Appendix B).
2 Respondents whose household reference person was a full time student, had an inadequately described occupation, had never worked or was long-term unemployed are not shown as separate categories but are included in the total.
3 Includes a few smokers who did not say how many cigarettes a day they smoked.

2 Drinking

Questions about drinking alcohol were included in the General Household Survey every two years from 1978 to 1998. Following the review of the GHS, the questions about drinking in the last seven days form part of the continuous survey, and have been included every year from 2000 onwards. Questions continuing the long-running series designed to measure average weekly alcohol consumption were included from 2000 to 2002 and again in 2005 and 2006. Before 1988 questions about drinking were asked only of those aged 18 and over, but since then respondents aged 16 and 17 have answered the questions using a self-completion questionnaire.

A key feature of this report is that it presents an updated method of converting what respondents say they drink into standard alcohol units. It also presents information on trends in alcohol consumption and on the association between consumption and characteristics of individuals such as sex, age, socio-economic position, and region.

Measuring alcohol consumption

Obtaining reliable information about drinking behaviour is difficult, and social surveys consistently record lower levels of consumption than would be expected from data on alcohol sales. This is partly because people may consciously or unconsciously under-estimate how much alcohol they consume. Drinking at home is particularly likely to be under-estimated because the quantities consumed are not measured and are likely to be larger than those dispensed in licensed premises.

There are different methods for obtaining survey information on drinking behaviour. One approach is to ask people to recall all episodes of drinking during a set period[x]. However, this is time-consuming and is not suitable for the GHS, where drinking is only one of a number of subjects covered.

The GHS currently uses two measures of alcohol consumption:

- average weekly alcohol consumption;
- maximum amount drunk on any one day in the previous seven days.

Average weekly alcohol consumption

Questions to establish average weekly alcohol consumption have been included on the GHS in their current form periodically since 1986. The measure was developed in response to earlier medical guidelines suggesting maximum recommended weekly amounts of alcohol. Its use continues to provide a consistent measure of alcohol consumption through which trends can be monitored. Respondents are asked how often over the last year they have drunk normal strength beer, strong beer (6% or greater ABV[xi]), wine, spirits, fortified wines and alcopops, and how much they have usually drunk on any one day. This information is combined to give an estimate of the respondent's weekly alcohol consumption (averaged over a year) in units of alcohol.

The method used for calculating usual weekly alcohol consumption is to multiply the number of units of each type drunk on a usual drinking day by the frequency with which it was drunk using the factors shown below, and then to total across all drinks.

Drinking frequency	Multiplying factor
Almost every day	7.0
5 or 6 days a week	5.5
3 or 4 days a week	3.5
Once or twice a week	1.5
Once or twice a month	0.375 (1.5 ÷ 4)
Once every couple of months	0.115 (6 ÷ 52)
Once or twice a year	0.029 (1.5 ÷ 52)

Maximum daily amount drunk last week

These questions have been included in the GHS since 1998, following the publication in 1995 of an inter-departmental review of the effects of drinking[xii]. This concluded that it was more appropriate to set benchmarks for daily than for weekly consumption of alcohol, partly because of concern about the health and social risks associated with single episodes of intoxication. The report considered that regular consumption of between three and four units[xiii] a day for men and two to three units a day for women does not carry a significant health risk, but that consistently drinking above these levels is not advised.

The government's advice on sensible drinking is now based on these daily benchmarks, and GHS data are used to monitor the extent to which people are following the advice given. Respondents are asked on how many days they drank alcohol during the previous week. They are then asked how much of each of six different types of drink (normal strength beer, strong beer, wine, spirits, fortified wines and alcopops) they drank on their heaviest drinking day during the previous week. These amounts are added to give an estimate of the maximum number of units the respondent had drunk on that day.

Updated method of converting volumes drunk to units

Estimates of alcohol consumption in surveys are given in standard units derived from assumptions about the alcohol content of different types of drink, combined with information from the respondent about the volume drunk.

In recent years, new types of alcoholic drink have been introduced, the alcohol content of some drinks has increased, and alcoholic drinks are now sold in more variable quantities than used to be the case. The GHS, in common with other surveys, has partially taken this into account: since 1998, alcopops and strong beer, lager and cider have been included as separate categories. However, it has recently also become necessary to reconsider the assumptions made in obtaining estimates of alcohol consumption, taking into account the following:

- increases in the size of glass in which wine is served on licensed premises;

- the increased alcoholic strength of wine;

- better estimates of the alcoholic strengths of beers, lagers and ciders.

For wine, it was decided to adopt a method which requires a question to be asked about glass size, which has the advantage that future changes in the average size of glass will be taken into account automatically. From 2008, the GHS will include additional questions to establish the size of wine glass, but in the interim a proxy conversion factor counting one glass of wine as 2 units will be used.

The changes in conversion factors are summarised in the table below, but are discussed in detail in a paper in the National Statistics Methodology series[xiv]. It was clear from the research undertaken for this paper that all surveys, including the GHS, have been undercounting the number of units in some types of drink - predominantly wine, but also, to a lesser degree, beer, lager and cider.

The change in method clearly doubles the units of wine consumed (since a glass is now taken as 2 units rather than 1). It also increases the units of strong beers by one third, and of normal strength beers by 12 per cent (the effect on normal strength beers is lower because the conversion factor of 1 unit for a half pint of normal strength beer remains unchanged, whereas all conversion factors for strong beers are increased). However, since strong beers account for only a small proportion of consumption, the 33 per cent increase in units does not have a major impact on the estimates of total consumption.

It should be noted, of course, that changing the way in which alcohol consumption estimates are derived does not in itself reflect a real change in drinking among the adult population.

Figure 2.1 Original and updated factors for converting alcohol volume to units

Type of drink	Usual volume (ml)	Original conversion factor (units)	Updated conversion factor (units)
Normal strength beer, lager, cider			
half pint	284	1.0	1.0
small can/bottle	330	1.0	1.5
large can/bottle	440	1.5	2.0
Strong beer, lager, cider (ABV = 6%)			
half pint	284	1.5	2.0
small can/bottle	330	1.5	2.0
large can/bottle	440	2.3	3.0
Table wine			
glass - 125ml	125		1.5
glass - 175ml	175		2.0
glass - 250ml/small can	250		3.0
glass - size unspecified	170	1.0	2.0
Fortified wine			
small glass	50	1.0	1.0
Spirits			
single	25	1.0	1.0
Alcopops/coolers			
bottle	275	1.5	1.5

Effect on GHS data of updated conversion factors

This section summarises the effect of the change in methodology on the data. More details, together with references to tables, are given in the following sections.

Average weekly alcohol consumption

Overall, the change in conversion factors increases average weekly alcohol consumption in 2006 by 32 per cent, from 10.2 units to 13.5 units. The effect of the change is not uniform across different subgroups of the adult population, because some groups are more likely than others to drink wine, and they are disproportionately affected. A higher proportion of women's than of men's consumption is wine, so the increase is proportionately greater for women: men's consumption increases by 27 per cent, from 14.8 to 18.7 units, women's by 45 per cent, from 6.2 to 9.0 units.

Figure 2.2 Average weekly alcohol units by sex: original and updated methods, 2006

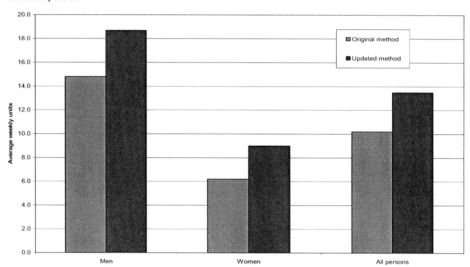

Other groups which are disproportionately affected because they are more likely to drink wine are as follows:

- those aged 25 and older;
- those in the managerial and professional socio-economic class (the first three categories of NS-SEC8);
- those in high income households, particularly those with a weekly income over £1,000;
- those living in England and Wales, rather than Scotland;
- those living in London, the South East and the South West of England.

Up to the publication in 1995 of the White Paper *Sensible Drinking* [xv] there was considerable interest in the percentages of men and women drinking above the then recommended weekly maximum consumption levels of 21 units for men and 14 for women, and to enable trends to continue to be monitored, the GHS still provides these data. Improving the estimates increases the proportion of men in

2006 drinking more than 21 units from 23 per cent to 31 per cent, and the proportion of women drinking more than 14 units from 12 per cent to 20 per cent.

The effect of updating the estimates is even more marked in relation to the percentages of men and women with very high levels of alcohol consumption - more than 50 units a week for men, and more than 35 units for women. These increase from 5 per cent to 8 per cent for men, and from 2 per cent to 5 per cent for women.

Maximum drunk on any one day in the previous week

Looking first at the effect of the updated conversion factors on the actual amounts drunk on the heaviest drinking day in the previous week, it can be seen that the magnitude of the increases is very similar to those for average weekly consumption.

However, the effect on the proportions of respondents drinking more than the recommended levels of four units (men) and three units (women) is somewhat different, the increase being much more marked among women than among men: the proportion of men drinking more than four units on at least one day in the previous week increases from 33 per cent to 40 per cent, but the proportion of women drinking more than three units increases from 20 per cent to 33 per cent. This is because of where the thresholds fall in relation to wine consumption. On the previous definition, two glasses of wine equals 2 units, which is below the threshold levels for both men and women. Using the updated conversion factors, however, two glasses of wine equals four units, and whereas men drinking this amount are at their threshold, women are over theirs.

Figure 2.3 Maximum drunk on any one day last week, by sex: original and updated methods, 2006

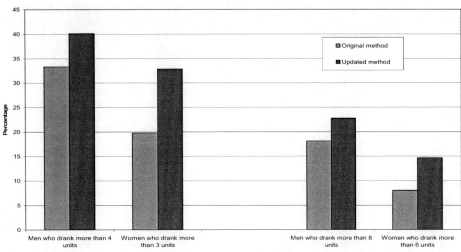

Apart from this greater relative effect on women's drinking in relation to daily guidelines, the associations are similar to those noted above in relation to average weekly consumption. The proportions drinking more than the daily amounts of four units for men and three units for women on at least one day in the previous week are disproportionately increased in the following groups:

- those aged 45 and over;

- those in the managerial and professional socio-economic class of the three-category version of NS-SEC (and in particular, among women, the first two categories of NS-SEC8);

- those in households with a weekly income over £1,000;

- those living in London and the South East of England.

Trends in alcohol consumption

To be comparable with previous years, the 2006 data discussed in this section are those derived using the original method of conversion to units.

Trends in average weekly alcohol consumption

Consideration of trends is complicated by the introduction of weighting. This increased the proportion of men drinking more than 21 units a week in 1998 by about one percentage point. The comparison of weighted and unweighted figures for later years, although not shown in the tables, is similar.

During the 1990s the GHS showed a slight increase in overall weekly alcohol consumption among men and a much more marked one among women. Following an increase between 1998 and 2000, there has been a decline since 2002 in the proportion of men drinking more than 21 units a week, on average, and in the proportion of women drinking more than 14 units. At first sight, the fall appears to have been most marked between 2002 and 2005, but this is largely because of the longer time interval between surveys. The proportion of men drinking more than 21 units a week on average fell from 29 per cent in 2000 to 23 per cent in 2006. There was also a fall in the proportion of women drinking more than 14 units a week (from 17 per cent in 2000 to 12 per cent in 2006).

Figure 2.4 Percentage of men drinking more than 21 units a week, and women drinking more than 14 units a week: original method, 1998 to 2006

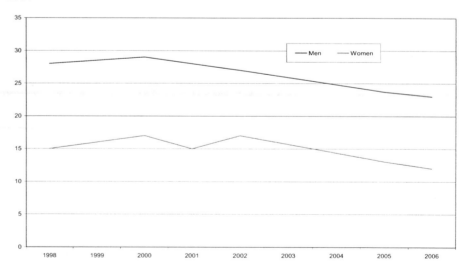

The fall in consumption occurred among men and women in all age groups, but was most evident among those aged 16 to 24. Among young men in this age group, the proportion drinking more than 21 units a week fell from 41 per cent in 2000 to 26 per cent in 2006, and among young women, the proportion drinking more than 14 units a week fell from 33 per cent to 19 per cent over that period.

There is a suggestion of a slight decline since 2002 in the proportion of men drinking more than 50 units a week on average, but no significant change in the proportion of women drinking more than 35 units.

In 2006, men drank an average of 14.8 units a week (equivalent to about seven and a half pints of beer), about 2.5 units less than they were drinking from 1998 to 2002. Women drank an average of 6.2 units a week in 2006. The decrease of about 1.5 units since 2002 reverses the steady rise in women's consumption seen over the previous decade. Similarly, the average consumption of young men and women aged 16 to 24 was lower in 2006 than in 2002. Among young men, consumption fell from 21.5 to 16.4 units, and among young women, from 14.1 to 9.0 units.

Figure 2.5 Average weekly alcohol units, by sex: original method, 1998 to 2006

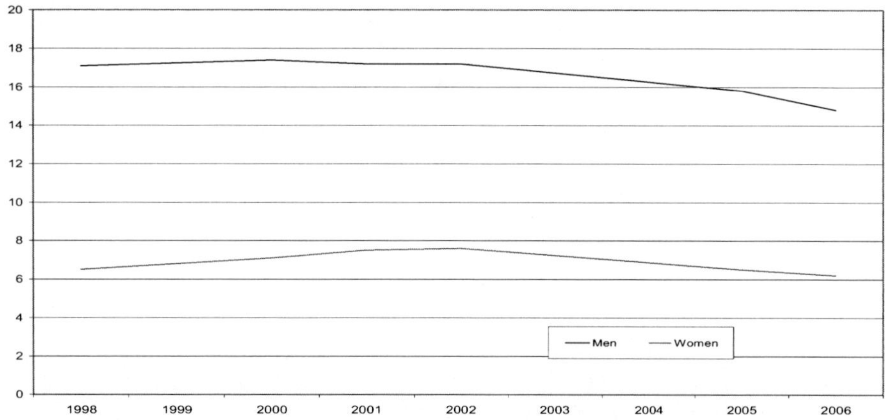

The British Beer and Pub Association (BBPA) makes annual estimates of per capita alcohol consumption using data provided by HM Revenue and Customs[xvi]. These show a steady increase in consumption from 1998 to 2004, followed by a decline of about 5 per cent over the next couple of years. The decline measured by the GHS is much greater, at about 15 per cent between 2002 and 2006. There may have been a small fall in consumption in recent years, but two factors are likely to account for the lack of a consistent pattern in the two sources:

1. There may be an increased tendency among respondents to under-report consumption. Recent extensive publicity about the dangers of drinking, and in particular binge drinking, may have led some people to moderate their behaviour, but might equally have made others less inclined to admit to how much they have drunk.

2. The introduction of improved methods of estimating units occurs as a step change, whereas the actual changes which the updated method takes into account happened over a number of years. It is likely, therefore, that progressive underestimation of wine consumption has occurred as glass sizes and alcoholic strength have increased. Although the increase in alcoholic strength of wines has probably been gradual over a long period, the increase in average wine glass size appears to have happened in the last few years, and this might account for the recent decline in consumption as measured by the GHS.

The fall in consumption among young people is unlikely to be due to progressive under-estimation of consumption, since they drink little wine. However, there is some indication that it is becoming more difficult to persuade young people to take part in the GHS (they form a decreasing proportion of respondents). Although this is partially corrected by weighting the data to known population totals, such reweighting cannot compensate if those lost to the survey have heavier consumption on average than those who do take part.

Tables 2.1-2.2

Trends in last week's drinking

Questions about the maximum daily amount drunk last week were first included on the GHS in 1998, so these data can provide evidence only on recent trends. Table 2.3 shows that following a period of little change between 1998 and 2004, there appears to have been a slight fall in the proportions of men and women who say that they had an alcoholic drink in the previous week, and also in the proportion of men who had drunk on five or more days in the week.

Figure 2.6 Percentage of men drinking more than four units, and of women drinking more than three units, on at least one day in the previous week: original method, Great Britain 1998 to 2006

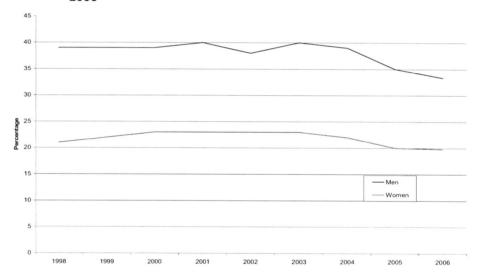

The proportion of men exceeding the daily benchmarks, 33 per cent in 2006, continued the fall from 39 per cent in 2004 to 35 per cent in 2005. There was no comparable fall among women in 2006, among whom the proportion remained at 20 per cent, but this is nonetheless lower than in the early years of the decade. There was little change between 2005 and 2006 in the proportions of men and women drinking heavily (more than 8 units, and 6 units respectively) on at least one day in the previous week.

There is considerable fluctuation in the trends within age groups which makes it difficult to be confident about the overall pattern. The proportion of young men aged 16 to 24 drinking more than four and more than eight units changed little between 1998 and 2003, but then fell in 2004 and again in 2005 and 2006. On the face of it, data for 2006 support previous indications that the recent upward trend in heavy drinking among young women may have peaked. The proportion of 16-24 year old women who had drunk more than six units on at least one day in the previous week increased from 24 per cent to 28 per cent between 1998 and 2002 but had fallen to 20 per cent in 2006. As with young men, however, the apparent decline may be due to factors other than a real change in behaviour. Among older women, there was no discernible trend in the proportions drinking heavily.

Tables 2.3-2.4

Alcohol consumption in 2006

The results discussed in this section are those based on the updated method of converting volumes drunk to units, though tables show the original method also. The introduction of the updated estimates has altered the association of alcohol with various personal characteristics, mainly because some groups are more likely to be wine drinkers, and are therefore disproportionately affected by the change in method.

Average weekly alcohol consumption in 2006

Weekly alcohol consumption and sex and age

With the original method, alcohol consumption among both men and women was highest among those aged 16 to 24, and then declined with increasing age. With the updated estimates, the age difference has almost disappeared, because older people are more likely to drink wine. It is still the case, however, that consumption is considerably lower among those aged 65 and over than among other age groups.

Overall, in 2006 men's consumption was about twice that of women (18.7 units compared with 9.0 units) but the difference was less marked among younger than older people. This reflects the trend that has occurred in recent years for women's consumption to increase relative to that of men, particularly among younger age groups.

Table 2.5

Weekly alcohol consumption and household socio-economic class

A review of information on inequalities in health, undertaken by the Department of Health[xvii], noted that both mortality and morbidity show a clear association with socio-economic position, with death rates much higher among unskilled men than among those in professional households (overall, up to 22,000 premature deaths a year are thought to be attributable to alcohol misuse[xviii]). Over many years, the GHS has shown little difference in usual weekly alcohol consumption between those in non-manual and manual households. Where differences do exist, it has been those in the non-manual categories who tend to have the higher weekly consumption. The updated method of estimating consumption increases this difference, because the effect of the change is greatest for those in non-manual groups since they are more likely to drink wine.

Previously, there was no clear socio-economic gradient in relation to alcohol consumption among men, but with the updated estimates, and using the three-category classification, men in the routine and manual group were drinking on average 16.7 units a week, considerably less than the 19.9 units of men in the other two groups.

The pattern among women was clearer, and similar to that shown in previous reports. Average weekly consumption in 2006 was highest, at 10.7 units, in the managerial and professional group, and lowest, at 7.1 units, among those in routine and manual worker households.

Table 2.6

Weekly alcohol consumption, income and economic activity status

Average weekly alcohol consumption was higher among men and women in high income households than among other men and women. Among those living in households with a gross income of more than £1,000 a week, men drank on average 22.1 units as week, and women 12.2 units. These levels compared with 17.8 units and 6.1 units respectively among those in households with an income of £200 or less.

Among those in full-time employment, however, there was no significant variation in average weekly alcohol consumption according to earnings.

Tables 2.7-2.9

Regional variation in average weekly alcohol consumption

Average weekly consumption was higher in England (13.7 units) and Wales (13.5 units) than in Scotland (11.6 units), but the overall figures mask some differences between men and women. Among men, consumption was lower in Scotland than in the other two countries, but among women, consumption in both Wales and Scotland was lower than in England.

Some differences between men and women were also evident in the variation of consumption in the English regions. Among men, consumption was highest in the three northern regions (North East, North West and Yorkshire and the Humber) and in the South West. The pattern was similar among women, except that consumption was not particularly high in the North East.

It should be noted that sample sizes in some regions are small and some fluctuation in results from year to year is to be expected. This can affect whether a particular region or country appears to have a high or low consumption level relative to that of other areas, and this may not be due to real differences in the population from which the sample is drawn. It is therefore unwise to give undue weight to data for a single year only.

Table 2.10

Last week's drinking in 2006

Frequency of drinking during the last week

Patterns of drinking behaviour in 2006 were broadly the same as those described in earlier GHS reports. Men were more likely than women to have had an alcoholic drink in the previous week: 71 per cent of men and 56 per cent of women had had a drink on at least one day during the previous week. Men also drank on more days of the week than women. More than one in five men (21 per cent) compared with just over one in ten women (11 per cent) had drunk on at least five of the preceding seven days. Similarly, men were much more likely than women to have drunk alcohol every day during the previous week (12 per cent compared with 7 per cent).

The proportions drinking last week also varied between age groups. Those in the youngest and oldest age groups (16 to 24 and 65 and over) were less likely than those in the middle age range to report drinking alcohol during the previous week. The proportion who had drunk alcohol in the previous week was particularly low among women aged 65 and over, 44 per cent of whom had done so, compared with 67 per cent of men in the same age group, and around 60 per cent of women aged 25 to 64.

However, although they were less likely to have had a drink at all in the previous week, older people drank more frequently than younger people. For example, 20 per cent of men and 11 per cent of women aged 65 and over had drunk every day during the previous week, compared with only 4 per cent of men and 1 per cent of women aged 16 to 24.

Table 2.11

Maximum daily amount drunk last week

Two measures of daily consumption are shown in the tables. The first is the proportion exceeding the recommended daily benchmarks (men drinking more than four units and women drinking more than three units in one day). The second measure is intended to indicate heavy drinking that would be likely to lead to intoxication. Although people vary in their susceptibility to the effect of alcohol, this level is taken as a rough guide to be more than eight units on one day for men and more than six units for women. For 2006, two sets of data are shown: the first uses the original method of converting volumes of alcohol into units, and the second using the updated method. The earlier method is used to provide trend data, and the updated estimates are used in discussion of the association of alcohol consumption with respondents' characteristics.

Men were more likely than women to have exceeded the daily benchmarks on at least one day during the previous week, although the updated estimates have reduced the difference from 13 percentage points to seven. Men were also more likely than women to have drunk heavily.

It was noted earlier that older people drink more frequently than younger people. However, among both men and women, those aged 65 and over were significantly less likely than respondents in other age groups to have exceeded the recommended number of daily units on at least one day. For example, 42 per cent of young men aged 16 to 24 had exceeded four units on at least one day during the previous week, compared with only 21 per cent of men aged 65 and over. Among women, 39 per cent of those in the youngest age group had exceeded three units on at least one day compared with only 14 per cent of those aged 65 and over.

Similar patterns were evident for heavy drinking: 30 per cent of men aged 16 to 24, but only 7 per cent of those aged 65 and over, had drunk more than eight units on at least one day during the previous week. Among young women aged 16 to 24, 25 per cent had drunk heavily on at least one day during the preceding week, compared with only 2 per cent of women in the oldest age group.

Table 2.12

Drinking last week and socio-economic characteristics

Looking first at the frequency of drinking alcohol, men and women in large employer/higher managerial households were the most likely to have drunk alcohol in the previous week, while those in households where the reference person was in a semi-routine or routine occupation were the least likely. A similar pattern was apparent in the proportions drinking on five or more days in the previous week. For example, 72 per cent of women in large employer/higher managerial households had had a drink in the last week and 18 per cent had done so on five or more days. Among women in households where the reference person was in a routine occupation, these proportions were much lower, at 42 per cent and 6 per cent respectively.

Variations in amounts drunk were also marked, particularly for women: those in large employer/higher managerial households were much more likely than those in the routine group to have drunk more than the recommended three units on any one day (47 per cent compared with 23 per cent), and also more likely to have drunk heavily on at least one day in the previous week (19 per cent compared with 11 per cent).

Tables 2.13-2.14

Drinking last week and household income

In general, the higher the level of gross weekly household income, the more likely both men and women were to have drunk alcohol in the previous week and to have exceeded the daily benchmarks. Among men in households with a gross weekly income of over £1,000, 83 per cent had had a drink in the previous week, and 51 per cent had drunk more than four units on at least one day. Among men in households with an income of £200 or less, only 61 per cent had had a drink and only 32 per cent had drunk more than four units on any one day. A similar pattern occurred for women and for the proportions drinking heavily. Thus, for example, adults in households with a gross weekly income of over £1,000 were about twice as likely as those in households with a gross weekly household income of £200 or less to have drunk more than eight and six units respectively on at least one day in the previous week.

Tables 2.15-2.16

Drinking last week, economic activity status and earnings from employment

Variations in alcohol consumption by economic status reflect differences in both the income and age profiles of the groups and also, probably, differences in health. Among men aged 16 to 64, those in employment were most likely to have drunk alcohol during the previous week - 76 per cent had done so compared with 54 per cent of the unemployed and 59 per cent of those who were economically inactive. In terms of quantity, working men were more likely than the unemployed and the economically inactive to have drunk more than 4 units on one day - 47 per cent, compared with 37 per cent and 35 per cent respectively. As noted above, lower levels of drinking among economically inactive men are partly due to the large proportion of this group who are aged 60 to 64.

Among women aged 16 to 64, 65 per cent of those who were working, 54 per cent of those who were unemployed, and 47 per cent of those who were economically inactive had drunk alcohol in the previous week. Working women were almost twice as likely as those who were economically inactive to have drunk heavily on at least one day in the previous week.

Among those working full time, variations in the frequency of drinking in relation to earnings from employment showed the same pattern of association as that with household income. Men and women who were high earners were more likely than the lower paid both to have drunk alcohol at all and to have drunk on five or more days. For example, among full-time workers aged 16 to 64 who were earning more than £800 per week, 29 per cent of men and 16 per cent of women had drunk on five or more days in the previous week, compared with 23 per cent of men and 10 per cent of women earning £200 or less per week, The two measures of consumption, however, showed no consistent pattern of association with earnings, either for men or for women.

Tables 2.17-2.20

Regional variation in drinking last week

As with average weekly alcohol consumption, care should be taken in interpreting the results for any one year. This is because sample sizes in some regions are small, making them subject to relatively high levels of sampling error.

In 2006, men and women in Scotland were less likely to have drunk on at least 5 days in the previous week than those living in England or Wales (14 per cent compared with 21 per cent and 23 per cent respectively for men; 9 per cent compared with 12 per cent and 11 per cent for women). However, men in Scotland were no less likely than those in England and Wales to have drunk more than the recommended maximum of four units nor to have drunk heavily. Differences among women in the proportions drinking more than the recommended amount or drinking heavily were not statistically significant.

Looking at the English regions, the daily benchmarks were most likely to be exceeded in the North West (47 per cent of men and 40 per cent of women) and Yorkshire and the Humber (48 per cent and 40 per cent respectively). The lowest proportions doing so were in London, where 35 per cent of men had drunk more than four units, and 27 per cent of women had drunk more than three units.

The relatively low proportions in London of men and women drinking more than the daily guideline amounts are largely explained by the high proportion of people in ethnic minority groups: the 2005 GHS report showed that 28 per cent of adults in London based on a sample of five years data combined were from non-white categories. Even the large sample was too small to permit detailed analysis by region and ethnic group, but an analysis based on the white population only showed that the proportions in London who had exceeded the recommended daily amounts, although still a little lower than average, were more similar to those in other regions (no table shown).

The same broad pattern of regional variation in daily drinking has been evident since these questions were first included in 1998. As noted above, however, sample sizes in some regions are small and some fluctuation in results from year to year is to be expected. This can affect whether a particular region or country appears to have a high or low consumption-level relative to other areas.

Tables 2.21-2.22

Notes and references

[x] Goddard E. *Obtaining information about drinking through surveys of the general population*. National Statistics Methodology Series NSM 24 (ONS 2001)

[xi] ABV is the percentage alcohol by volume.

[xii] *Sensible drinking: the report of an inter-departmental group.* (Department of Health 1995)

[xiii] Assuming one unit of alcohol to be a half pint of normal strength beer, lager or cider, a single measure of spirits, a glass of wine, or a small glass of sherry or other fortified wine.

[xiv] Goddard E *Estimating alcohol consumption from survey data: updated method of converting volumes to units* National Statistics Methodology Series NSM 37 (ONS 2007), also available at http://www.statistics.gov.uk/statbase/product.asp?vlnk=15067

[xv] *Sensible drinking:* op.cit.

[xvi] *BBPA Statistical Handbook 2007: a compilation of drinks industry statistics.* (British Beer & Pub Association, Brewing Publications Ltd, 2007).

[xvii] *Inequalities in Health*, (The Stationery Office 1998) also available at http://www.archive.official-documents.co.uk/doh/ih.htm

[xviii] *Safe. Sensible. Social. The next steps in the National Alcohol Strategy.* (Department of Health 2007), available at http://www.dh.gov.uk/en/Publicationsandstatistics/publications/PublicationsPolicyand Guidance/DH_075218

General Household Survey, 2006

Smoking and drinking among adults 2006

Table 2.1 Average weekly alcohol consumption (units), by sex and age: 1992-2006

Persons aged 16 and over   *Great Britain*

Age	Unweighted				Weighted							*Weighted base 2006 (000's) =100%*	*Unwtd sample[3] 2006*
	1992	1994	1996	1998	1998	2000	2001	2002	2005[1]	2006[2] original method	2006[2] improved method		
Men													
16-24	19.1	17.4	20.3	23.6	25.5	25.9	24.8	21.5	18.2	16.4	18.6	*2,607*	*781*
25-44	18.2	17.5	17.6	16.5	17.1	17.7	18.4	18.7	16.2	15.6	19.7	*7,057*	*2468*
45-64	15.6	15.5	15.6	17.3	17.4	16.8	16.1	17.5	17.7	16.0	20.8	*6,450*	*2671*
65 and over	9.7	10.0	11.0	10.7	10.6	11.0	10.8	10.7	10.4	10.4	13.5	*3,836*	*1767*
Total	15.9	15.4	16.0	16.4	17.1	17.4	17.2	17.2	15.8	14.8	18.7	*19,950*	*7687*
Women													
16-24	7.3	7.7	9.5	10.6	11.0	12.6	14.1	14.1	10.9	9.0	10.8	*2,863*	*944*
25-44	6.3	6.2	7.2	7.1	7.1	8.1	8.3	8.4	7.1	6.8	10.1	*7,875*	*3006*
45-64	5.3	5.3	5.9	6.4	6.4	6.2	6.8	6.7	6.3	6.2	9.8	*7,095*	*3014*
65 and over	2.7	3.2	3.5	3.3	3.2	3.5	3.6	3.8	3.5	3.5	5.1	*4,911*	*2050*
Total	5.4	5.4	6.3	6.4	6.5	7.1	7.5	7.6	6.5	6.2	9.0	*22,744*	*9014*
All persons													
16-24	12.9	12.3	14.7	16.6	18.0	19.3	19.4	17.6	14.3	12.5	14.6	*5,470*	*1725*
25-44	11.8	11.4	11.9	11.4	12.0	12.9	13.3	13.3	11.3	11.0	14.6	*14,932*	*5474*
45-64	10.2	10.2	10.5	11.6	11.7	11.4	11.3	11.9	11.7	10.9	15.0	*13,545*	*5685*
65 and over	5.6	6.0	6.8	6.5	6.3	6.7	6.6	6.8	6.5	6.5	8.7	*8,747*	*3817*
Total	10.2	10.0	10.7	11.0	11.5	12.0	12.1	12.1	10.8	10.2	13.5	*42,694*	*16701*

1 2005 data includes last quarter of 2004/5 data due to survey change from financial year to calendar year.

2 Results for 2006 include longitudinal data (see Appendix B).

3 Trend tables show unweighted and weighted figures for 1998 to give an indication of the effect of the weighting. Bases for earlier years can be found in GHS reports for each year.

Office for National Statistics: January 2008

General Household Survey, 2006

Smoking and drinking among adults 2006

Table **2.2** Weekly alcohol consumption level: percentage exceeding specified amounts by sex and age: 1988-2006

Persons aged 16 and over   *Great Britain*

Age	Unweighted data					Weighted data							Weighted base 2006 (000's) =100%	Unwgtd sample[3] 2006
	1988	1992	1994	1996	1998	1998	2000	2001	2002	2005[1]	2006[2] original method	2006[2] improved method		
Men														
Percentage of men who drank more than 21 units														
16-24	31	32	29	35	36	38	41	40	37	27	26	30	2,607	781
25-44	34	31	30	30	27	28	30	30	29	26	24	33	7,058	2468
45-64	24	25	27	26	30	30	28	26	28	25	24	34	6,449	2671
65 and over	13	15	17	18	16	16	17	15	15	14	14	21	3,836	1767
Total	26	26	27	27	27	28	29	28	27	24	23	31	19,950	7687
Percentage of men who drank more than 50 units														
16-24	10	9	9	10	13	14	14	15	12	9	7	9	2,607	781
25-44	9	8	7	6	6	6	7	7	8	5	5	9	7,058	2468
45-64	6	6	6	5	6	7	6	5	6	6	6	10	6,449	2671
65 and over	2	2	3	3	3	3	3	2	3	3	2	5	3,836	1767
Total	7	6	6	6	6	7	7	7	7	6	5	8	19,950	7687
Women														
Percentage of women who drank more than 14 units														
16-24	15	17	19	22	25	25	33	32	33	24	19	24	2,863	944
25-44	14	14	15	16	16	16	19	17	19	14	14	23	7,874	3006
45-64	9	11	12	13	16	15	14	14	14	13	12	21	7,096	3014
65 and over	4	5	7	7	6	6	7	6	7	5	5	10	4,910	2050
Total	10	11	13	14	15	15	17	15	17	13	12	20	22,743	9014
Percentage of women who drank more than 35 units														
16-24	3	4	4	5	6	7	9	10	10	6	5	7	2,863	944
25-44	2	2	2	2	2	2	3	3	3	2	2	6	7,874	3006
45-64	1	1	2	2	2	2	2	2	2	2	2	6	7,096	3014
65 and over	0	0	1	1	1	1	1	1	1	1	1	2	4,910	2050
Total	2	2	2	2	2	2	3	3	3	2	2	5	22,743	9014

1 2005 data includes last quarter of 2004/5 data due to survey change from financial year to calendar year.
2 Results for 2006 include longitudinal data (see Appendix B).
3 Trend tables show unweighted and weighted figures for 1998 to give an indication of the effect of the weighting. Bases for earlier years can be found in GHS reports for each year.

General Household Survey, 2006

Smoking and drinking among adults 2006

Table 2.3 Drinking last week, by sex and age: 1998 to 2006

Persons aged 16 and over *Great Britain*

Drinking last week	1998	2000	2001	2002	2003	2004	2005[1]	2006[2]	Weighted base 2006 (000s)=100%	Unweighted sample[3] 2006
					Percentages					
Men										
Drank last week										
16-24	70	70	70	69	70	66	64	60	2,597	777
25-44	79	78	78	77	77	76	74	73	7,058	2468
45-64	77	77	76	76	78	76	77	76	6,447	2669
65 and over	65	67	68	67	69	68	66	67	3,837	1767
Total	75	75	75	74	75	73	72	71	19,939	7681
Drank on 5 or more days last week										
16-24	13	11	14	11	14	8	10	8	2,597	777
25-44	21	19	19	19	20	20	18	17	7,058	2468
45-64	29	26	25	26	26	28	28	26	6,447	2669
65 and over	25	28	27	28	29	28	26	27	3,837	1767
Total	23	22	22	22	23	23	22	21	19,939	7681
Women										
Drank last week										
16-24	62	64	59	61	61	60	56	53	2,864	944
25-44	65	67	66	65	65	62	62	60	7,876	3006
45-64	61	61	61	63	64	62	61	61	7,094	3013
65 and over	45	43	45	46	45	45	43	44	4,910	2050
Total	59	60	59	59	60	58	57	56	22,744	9013
Drank on 5 or more days last week										
16-24	8	7	8	7	4	5	5	3	2,864	944
25-44	12	11	11	11	10	9	11	9	7,876	3006
45-64	15	15	17	17	17	18	17	15	7,094	3013
65 and over	14	14	15	15	16	16	14	15	4,910	2050
Total	13	13	13	13	13	13	13	11	22,744	9013

1 2005 data includes last quarter of 2004/5 data due to survey change from financial year to calendar year.
2 Results for 2006 include longitudinal data (see Appendix B).
3 Trend tables show unweighted and weighted figures for 1998 to give an indication of the effect of the weighting.
  Bases for earlier years can be found in GHS reports for each year.

General Household Survey, 2006

Smoking and drinking among adults 2006

Table 2.4 Maximum drunk on any one day last week by sex and age: 1998 to 2006

Persons aged 16 and over *Great Britain*

Maximum daily amount	1998	2000	2001	2002	2003	2004	2005[1]	2006[2] original method	2006[2] improved method	Weighted base 2006 (000s)=100%	Unweighted sample[3] 2006
					Percentages						
Men											
Drank more than 4 units on at least one day											
16-24	52	50	50	49	51	47	42	39	42	2,586	774
25-44	48	45	49	46	47	48	42	42	48	7,046	2463
45-64	37	38	37	38	41	37	35	33	42	6,450	2670
65 and over	16	16	18	16	19	20	16	14	21	3,836	1767
Total	39	39	40	38	40	39	35	33	40	19,918	7674
Drank more than 8 units on at least one day											
16-24	39	37	37	35	37	32	30	27	30	2,586	774
25-44	29	27	30	28	30	31	25	25	31	7,046	2463
45-64	17	17	17	18	20	18	16	15	21	6,450	2670
65 and over	4	5	5	5	6	7	4	4	7	3,836	1767
Total	22	21	22	21	23	22	19	18	23	19,918	7674
Women											
Drank more than 3 units on at least one day											
16-24	42	42	40	42	40	39	36	34	39	2,859	943
25-44	28	31	31	31	30	28	26	27	40	7,877	3007
45-64	17	19	19	19	20	20	18	17	35	7,096	3014
65 and over	4	4	5	5	4	5	4	4	14	4,908	2049
Total	21	23	23	23	23	22	20	20	33	22,740	9013
Drank more than 6 units on at least one day											
16-24	24	27	27	28	26	24	22	20	25	2,859	943
25-44	11	13	14	13	13	13	11	12	21	7,877	3007
45-64	5	5	5	5	5	6	4	4	12	7,096	3014
65 and over	1	1	1	1	1	1	1	0	2	4,908	2049
Total	8	10	10	10	9	9	8	8	15	22,740	9013

1 2005 data includes last quarter of 2004/5 data due to survey change from financial year to calendar year.
2 Results for 2006 include longitudinal data (see Appendix B).
3 Trend tables show unweighted and weighted figures for 1998 to give an indication of the effect of the weighting.
  Bases for earlier years can be found in GHS reports for each year.

Smoking and drinking among adults 2006

Table 2.5 Average weekly alcohol consumption (units), by sex and age

Persons aged 16 and over   *Great Britain: 2006[1]*

Age	Original method Average weekly alcohol consumption			Improved method Average weekly alcohol consumption			Weighted base (000's) =100%			Unweighted sample		
	Men	Women	Total	Men	Women	Total	Men	Women	Total	Men	Women	Total
16-24	16.4	9.0	12.5	18.6	10.8	14.6	2,607	2,863	5,470	781	944	1725
25-44	15.6	6.8	11.0	19.7	10.1	14.6	7,057	7,875	14,932	2468	3006	5474
45-64	16.0	6.2	10.9	20.8	9.8	15.0	6,450	7,095	13,545	2671	3014	5685
65 and over	10.4	3.5	6.5	13.5	5.1	8.7	3,836	4,911	8,747	1767	2050	3817
Total	14.8	6.2	10.2	18.7	9.0	13.5	19,950	22,744	42,694	7687	9014	16701

1 Results for 2006 include longitudinal data (see Appendix B).

General Household Survey, 2006

Smoking and drinking among adults 2006

Table **2.6** Average weekly alcohol consumption (units), by sex and socio-economic class based on the current or last job of the household reference person

Persons aged 16 and over *Great Britain: 2006[2]*

Socio-economic classification of household reference person[1]	Original method					Improved method				
	Men		Women		Total	Men		Women		Total
Managerial and professional										
Large employer and higher managerial	13.2 ⎤		7.3 ⎤		10.4 ⎤	22.9 ⎤		12.5 ⎤		17.6 ⎤
Higher professional	14.9 ⎥	14.9	6.6 ⎥	7.0	10.5 ⎥ 10.8	17.8 ⎥	19.9	11.2 ⎥	10.7	14.6 ⎥ 15.1
Lower managerial and professional	15.2 ⎦		5.6 ⎦		9.2 ⎦	19.6 ⎦		9.9 ⎦		14.4 ⎦
Intermediate										
Intermediate	16.7 ⎤	16.1	6.9 ⎤	6.2	11.9 ⎤ 10.6	19.3 ⎤	19.9	8.3 ⎤	9.1	12.5 ⎤ 14.0
Small employers/own account workers	14.4 ⎦		5.8 ⎦		10.2 ⎦	20.4 ⎦		10.1 ⎦		15.3 ⎦
Routine and manual										
Lower supervisory and technical	13.4 ⎤		5.5 ⎤		8.7 ⎤	17.2 ⎤		7.9 ⎤		12.7 ⎤
Semi-routine	14.5 ⎥	14.1	4.7 ⎥	5.3	9.3 ⎥ 9.4	16.0 ⎥	16.7	7.5 ⎥	7.1	11.0 ⎥ 11.6
Routine	8.9 ⎦		4.6 ⎦		6.3 ⎦	17.0 ⎦		6.1 ⎦		11.2 ⎦
Total[1]	14.8		6.2		10.2	18.7		9.0		13.5
Weighted bases (000's) =100%										
Large employer and higher managerial		*2,064*		*1,916*	*3,979*		*2,064*		*1,916*	*3,979*
Higher professional		*4,610*		*5,277*	*9,887*		*4,610*		*5,277*	*9,887*
Lower managerial and professional		*1,393*		*2,268*	*3,661*		*1,393*		*2,268*	*3,661*
Intermediate		*2,139*		*2,046*	*4,185*		*2,139*		*2,046*	*4,185*
Small employers/own account workers		*2,380*		*2,233*	*4,613*		*2,380*		*2,233*	*4,613*
Lower supervisory and technical		*2,234*		*3,182*	*5,416*		*2,234*		*3,182*	*5,416*
Semi-routine		*2,498*		*2,830*	*5,327*		*2,498*		*2,830*	*5,327*
Routine		*319*		*507*	*827*		*319*		*507*	*827*
Total[1]		*19,950*		*22,744*	*42,694*		*19,950*		*22,744*	*42,694*
Unweighted sample										
Large employer and higher managerial		*733*		*770*	*1503*		*733*		*770*	*1503*
Higher professional		*831*		*805*	*1636*		*831*		*805*	*1636*
Lower managerial and professional		*1832*		*2154*	*3986*		*1832*		*2154*	*3986*
Intermediate		*527*		*877*	*1404*		*527*		*877*	*1404*
Small employers/own account workers		*814*		*810*	*1624*		*814*		*810*	*1624*
Lower supervisory and technical		*899*		*866*	*1765*		*899*		*866*	*1765*
Semi-routine		*838*		*1248*	*2086*		*838*		*1248*	*2086*
Routine		*932*		*1072*	*2004*		*932*		*1072*	*2004*
Total[1]		*7687*		*9014*	*16701*		*7687*		*9014*	*16701*

1 Full-time students, members of the Armed Forces, the long term unemployed and those who have never worked are not shown as separate categories but are included in the totals.
2 Results for 2006 include longitudinal data (see Appendix B).

General Household Survey, 2006

Smoking and drinking among adults 2006

Table 2.7 Average weekly alcohol consumption (units), by sex and usual gross weekly household income (£)

Persons aged 16 and over *Great Britain: 2006[1]*

Usual gross weekly household income (£)	Original method Average weekly alcohol consumption			Improved method Average weekly alcohol consumption			Weighted base (000's) =100%			Unweighted sample		
	Men	Women	Total	Men	Women	Total	Men	Women	Total	Men	Women	Total
Up to 200.00	15.0	4.4	8.5	17.8	6.1	10.6	2,785	4,467	7,252	1067	1733	2800
200.01 - 400.00	11.6	5.4	8.3	14.4	7.7	10.7	3,535	4,231	7,767	1418	1740	3158
400.01 - 600.00	14.5	6.5	10.4	17.9	9.3	13.6	3,138	3,205	6,343	1205	1291	2496
600.01 - 800.00	15.1	7.0	11.0	19.1	10.0	14.5	2,808	2,905	5,713	1078	1141	2219
800.01 - 1000.00	17.4	7.1	12.2	22.0	10.3	16.1	1,966	1,967	3,933	730	757	1487
1000.01 or more	16.6	7.9	12.2	22.1	12.2	17.1	3,979	4,013	7,992	1501	1565	3066
Total	14.8	6.2	10.2	18.7	9.0	13.5	19,950	22,744	42,694	7687	9014	16701

1 Results for 2006 include longitudinal data (see Appendix B).

Table 2.8 Average weekly alcohol consumption (units), by sex and economic activity status

Persons aged 16-64 *Great Britain: 2006[1]*

Economic activity status	Original method Average weekly alcohol consumption			Improved method Average weekly alcohol consumption			Weighted base (000's) =100%			Unweighted sample		
	Men	Women	Total	Men	Women	Total	Men	Women	Total	Men	Women	Total
Working	16.0	7.4	11.8	20.3	10.9	15.7	12,732	12,155	24,887	4704	4709	9413
Unemployed	15.0	7.8	12.3	17.8	9.8	14.8	678	408	1,085	229	140	369
Economically inactive	15.4	5.7	9.0	18.9	8.0	11.7	2,696	5,263	7,959	984	2112	3096
Total	15.9	6.9	11.2	20.0	10.1	14.8	16,106	17,826	33,931	5917	6961	12878

1 Results for 2006 include longitudinal data (see Appendix B).

Smoking and drinking among adults 2006

Table 2.9 Average weekly alcohol consumption (units), by sex and usual gross weekly earnings (£)

Persons aged 16-64 in full-time employment *Great Britain: 2006[1]*

Usual gross weekly earnings (£)	Original method Average weekly alcohol consumption			Improved method Average weekly alcohol consumption			Weighted base (000's) =100%			Unweighted sample		
	Men	Women	Total	Men	Women	Total	Men	Women	Total	Men	Women	Total
Up to 200.00	19.4	8.5	14.8	24.2	11.6	18.9	834	606	1,440	301	222	523
200.01 - 300.00	16.4	7.7	11.8	19.3	10.7	14.7	1,397	1,580	2,978	493	580	1073
300.01 - 400.00	15.7	7.4	12.3	19.0	10.9	15.7	2,087	1,427	3,514	744	544	1288
400.01 - 600.00	16.8	7.6	13.7	21.1	11.8	18.0	2,929	1,476	4,406	1084	557	1641
600.01 - 800.00	16.5	8.2	13.8	21.5	13.5	18.9	1,344	635	1,979	517	255	772
800.01 or more	15.5	7.5	13.8	21.7	12.1	19.6	1,647	458	2,105	657	180	837
Total	16.4	7.7	13.2	20.7	11.4	17.3	11,207	6,647	17,855	4178	2521	6699

1 Results for 2006 include longitudinal data (see Appendix B).

Table 2.10 Average weekly alcohol consumption (units), by sex and Government Office Region

Persons aged 16 and over *Great Britain: 2006[1]*

Government Office Region	Original method Average weekly alcohol consumption			Improved method Average weekly alcohol consumption			Weighted base (000's) =100%			Unweighted sample		
	Men	Women	Total	Men	Women	Total	Men	Women	Total	Men	Women	Total
North East	17.9	6.4	11.8	21.4	8.7	14.6	803	919	1,722	310	368	678
North West	17.4	6.9	11.7	21.7	9.9	15.3	2,220	2,669	4,889	901	1110	2011
Yorkshire and the Humber	17.3	8.0	12.4	21.4	11.2	16.0	1,805	1,980	3,785	723	819	1542
East Midlands	15.4	6.5	10.9	19.2	9.4	14.2	1,703	1,749	3,452	688	741	1429
West Midlands	12.5	6.4	9.3	15.6	8.8	12.0	1,748	1,932	3,681	677	773	1450
East of England	14.4	5.6	9.8	18.2	8.3	13.0	1,990	2,201	4,191	786	907	1693
London	13.3	4.5	8.5	16.9	6.8	11.4	2,237	2,657	4,893	662	796	1458
South East	13.2	6.3	9.5	17.5	9.5	13.3	2,858	3,236	6,093	1115	1301	2416
South West	15.7	6.9	11.0	20.0	10.2	14.7	1,824	2,125	3,950	745	884	1629
England	14.9	6.3	10.4	18.9	9.2	13.7	17,189	19,468	36,657	6607	7,699	14306
Wales	16.1	5.3	10.4	19.9	7.8	13.5	1,023	1,152	2,176	411	477	888
Scotland	13.0	5.5	8.9	16.3	7.8	11.6	1,738	2,123	3,861	669	838	1507
Great Britain	14.8	6.2	10.2	18.7	9.0	13.5	19,950	22,744	42,694	7687	9014	16701

1 Results for 2006 include longitudinal data (see Appendix B).

General Household Survey, 2006

Smoking and drinking among adults 2006

Table 2.11 Whether drank last week and number of drinking days by sex and age

Persons aged 16 and over *Great Britain: 2006[1]*

Drinking days last week	Age				
	16-24	25-44	45-64	65 and over	Total
	%	%	%	%	%
Men					
0	40	27	24	33	29
1	21	20	17	16	18
2	17	17	14	11	15
3	9	12	11	8	10
4	5	8	8	5	7
5	2	5	6	4	5
6	2 ⎱ 8	3 ⎱ 17	4 ⎱ 26	3 ⎱ 27	3 ⎱ 21
7	4	8	15	20	12
% who drank last week	60	73	76	67	71
Weighted base (000's) =100%	*2,597*	*7,058*	*6,447*	*3,837*	*19,939*
Unweighted sample	*777*	*2468*	*2669*	*1767*	*7682*
	%	%	%	%	%
Women					
0	47	40	39	56	44
1	24	22	19	16	20
2	14	15	13	7	12
3	8	8	9	4	8
4	4	6	6	3	5
5	1	3	4	3	3
6	1 ⎱ 3	1 ⎱ 9	3 ⎱ 15	1 ⎱ 15	2 ⎱ 11
7	1	5	8	11	7
% who drank last week	53	60	61	44	56
Weighted base (000's) =100%	*2,864*	*7,876*	*7,094*	*4,910*	*22,744*
Unweighted sample	*944*	*3006*	*3013*	*2050*	*9013*
	%	%	%	%	%
All persons					
0	44	34	32	46	37
1	22	21	18	16	19
2	15	16	13	9	14
3	9	10	10	6	9
4	4	7	7	3	6
5	2	4	5	3	4
6	2 ⎱ 6	2 ⎱ 13	3 ⎱ 20	2 ⎱ 20	3 ⎱ 16
7	2	6	12	15	9
% who drank last week	56	66	68	54	63
Weighted base (000's) =100%	*5,460*	*14,932*	*13,539*	*8,748*	*42,679*
Unweighted sample	*1,721*	*5,474*	*5,682*	*3,817*	*16,695*

1 Results for 2006 include longitudinal data (see Appendix B).

General Household Survey, 2006

Smoking and drinking among adults 2006

Table 2.12 Maximum drunk on any one day last week, by sex and age

Persons aged 16 and over   *Great Britain: 2006[2]*

Maximum daily amount	Age				
	16-24	25-44	45-64	65 and over	Total
Original method	%	%	%	%	%
Men					
Drank nothing last week	40	27	24	33	29
Up to 4 units	21	31	43	53	38
More than 4, up to 8 units	11 ⎤ 39	17 ⎤ 42	18 ⎤ 33	10 ⎤ 14	15 ⎤ 33
More than 8 units	27 ⎦	25 ⎦	15 ⎦	4 ⎦	18 ⎦
	%		%	%	%
Women					
Drank nothing last week	47	40	40	56	44
Up to 3 units	19	34	43	40	36
More than 3, up to 6 units	14 ⎤ 34	15 ⎤ 27	13 ⎤ 17	4 ⎤ 4	12 ⎤ 20
More than 6 units	20 ⎦	12 ⎦	4 ⎦	0 ⎦	8 ⎦
	%	%	%	%	%
All persons[1]					
Drank nothing last week	44	34	32	46	37
Up to 4/3 units	20	33	43	46	37
More than 4/3, up to 8/6 units	13 ⎤ 36	16 ⎤ 34	15 ⎤ 25	6 ⎤ 9	13 ⎤ 26
More than 8/6 units	24 ⎦	18 ⎦	10 ⎦	2 ⎦	13 ⎦
Improved method	%	%	%	%	%
Men					
Drank nothing last week	40	27	24	33	29
Up to 4 units	18	25	33	46	31
More than 4, up to 8 units	12 ⎤ 42	17 ⎤ 48	21 ⎤ 42	14 ⎤ 21	17 ⎤ 40
More than 8 units	30 ⎦	31 ⎦	21 ⎦	7 ⎦	23 ⎦
	%	%	%	%	%
Women					
Drank nothing last week	47	40	40	56	44
Up to 3 units	14	20	25	30	23
More than 3, up to 6 units	14 ⎤ 39	19 ⎤ 40	23 ⎤ 35	12 ⎤ 14	18 ⎤ 33
More than 6 units	25 ⎦	21 ⎦	12 ⎦	2 ⎦	15 ⎦
	%	%	%	%	%
All persons[1]					
Drank nothing last week	44	34	32	46	37
Up to 4/3 units	16	23	29	37	27
More than 4/3, up to 8/6 units	13 ⎤ 40	18 ⎤ 44	22 ⎤ 39	13 ⎤ 17	18 ⎤ 36
More than 8/6 units	27 ⎦	25 ⎦	16 ⎦	4 ⎦	18 ⎦
Weighted base (000's) =100%					
Men	2,586	7,046	6,450	3,836	19,918
Women	2,859	7,877	7,096	4,908	22,740
All persons	5,446	14,923	13,544	8,744	42,657
Unweighted sample					
Men	774	2464	2670	1767	7675
Women	943	3007	3014	2049	9013
All persons	1717	5471	5684	3816	16688

1 The first of each pair of figures shown relates to men, and the second, to women.
2 Results for 2006 include longitudinal data (see Appendix B).

General Household Survey, 2006

Smoking and drinking among adults 2006

Table 2.13 Drinking last week, by sex, and socio-economic classification based on the current or last job of the household reference person

Persons aged 16 and over *Great Britain: 2006[2]*

Socio-economic classification of household reference person[1]	Men		Women		All persons	
	\multicolumn{6}{l}{Percentage who drank last week}					
Managerial and professional						
Large employer and higher managerial	84		72		78	
Higher professional	79	79	67	66	73	72
Lower managerial and professional	76		63		69	
Intermediate						
Intermediate	73	71	55	55	62	63
Small employers/own account workers	71		56		63	
Routine and manual						
Lower supervisory and technical	70		53		62	
Semi-routine	62	64	46	47	53	55
Routine	61		42		51	
Total[1]	71		56		63	
	\multicolumn{6}{l}{Percentage who drank on 5 or more days last week}					
Managerial and professional						
Large employer and higher managerial	30		18		24	
Higher professional	24	25	16	15	20	20
Lower managerial and professional	23		14		18	
Intermediate						
Intermediate	21	23	11	12	15	17
Small employers/own account workers	24		13		19	
Routine and manual						
Lower supervisory and technical	16		9		13	
Semi-routine	14	15	8	8	10	11
Routine	16		6		11	
Total[1]	21		11		16	
Weighted bases (000's) =100%						
Large employer and higher managerial	1,767		1,821		3,588	
Higher professional	2,064		1,916		3,980	
Lower managerial and professional	4,608		5,277		9,886	
Intermediate	1,390		2,268		3,659	
Small employers/own account workers	2,139		2,046		4,185	
Lower supervisory and technical	2,378		2,231		4,609	
Semi-routine	2,232		3,182		5,413	
Routine	2,493		2,830		5,323	
Total	19,939		22,741		42,680	
Unweighted sample						
Large employer and higher managerial	733		770		1503	
Higher professional	831		805		1636	
Lower managerial and professional	1731		2154		3885	
Intermediate	526		877		1403	
Small employers/own account workers	814		810		1624	
Lower supervisory and technical	898		865		1763	
Semi-routine	837		1248		2085	
Routine	930		1072		2002	
Total	7681		9013		16694	

1 Full-time students, members of the Armed Forces, the long term unemployed and those who have never worked are not shown as separate categories but are included in the totals.
2 Results for 2006 include longitudinal data (see Appendix B).

General Household Survey, 2006

Smoking and drinking among adults 2006

Table 2.14 Maximum number of units drunk on at least one day last week, by sex and socio-economic classification based on the current or last job of the household reference person

Persons aged 16 and over *Great Britain: 2006[3]*

Socio-economic classification of household reference person[1]	Original method			Improved method		
	Men	Women	All persons	Men	Women	All persons
	Percentage who drank more than 4/3 units on at least one day last week: original method[2]			Percentage who drank more than 4/3 units on at least one day last week: improved method[2]		
Managerial and professional						
Large employer and higher managerial	37	24	30	47	47	47
Higher professional	32 } 34	21 } 22	27 } 28	42 } 44	41 } 40	41 } 42
Lower managerial and professional	35	21	27	44	37	40
Intermediate						
Intermediate	33 } 34	17 } 18	23 } 26	41 } 41	30 } 32	34 } 36
Small employers/own account workers	35	20	28	41	34	37
Routine and manual						
Lower supervisory and technical	34	20	28	38	30	34
Semi-routine	29 } 32	17 } 18	22 } 24	33 } 35	26 } 26	29 } 30
Routine	32	17	24	35	23	29
Total[1]	33	20	26	40	33	36
	Percentage who drank more than 8/6 units on at least one day last week: original method[2]			Percentage who drank more than 8/6 units on at least one day last week: original method[2]		
Managerial and professional						
Large employer and higher managerial	20	8	14	27	19	23
Higher professional	15 } 18	8 } 8	11 } 13	21 } 24	16 } 17	19 } 21
Lower managerial and professional	18	8	13	24	17	20
Intermediate						
Intermediate	17 } 18	7 } 8	11 } 12	22 } 23	13 } 13	16 } 18
Small employers/own account workers	19	8	14	23	14	19
Routine and manual						
Lower supervisory and technical	20	7	14	24	13	19
Semi-routine	15 } 19	9 } 8	11 } 13	18 } 21	13 } 12	15 } 16
Routine	20	7	13	22	11	16
Total[1]	18	8	13	23	15	18
Weighted bases (000's) =100%						
Large employer and higher managerial	1,763	1,821	3,584	1,763	1,821	3,584
Higher professional	2,062	1,916	3,977	2,062	1,916	3,977
Lower managerial and professional	4,598	5,273	9,871	4,598	5,273	9,871
Intermediate	1,390	2,268	3,658	1,390	2,268	3,658
Small employers/own account workers	2,138	2,046	4,185	2,138	2,046	4,185
Lower supervisory and technical	2,383	2,234	4,617	2,383	2,234	4,617
Semi-routine	2,231	3,179	5,410	2,231	3,179	5,410
Routine	2,490	2,829	5,321	2,490	2,829	5,321
Total[1]	19,917	22,739	42,656	19,917	22,739	42,656
Unweighted sample						
Large employer and higher managerial	731	770	1501	731	770	1501
Higher professional	830	805	1635	830	805	1635
Lower managerial and professional	1828	2153	3981	1828	2153	3981
Intermediate	526	877	1403	526	877	1403
Small employers/own account workers	814	810	1624	814	810	1624
Lower supervisory and technical	900	866	1766	900	866	1766
Semi-routine	836	1247	2083	836	1247	2083
Routine	930	1072	2002	930	1072	2002
Total[1]	7674	9013	16687	7674	9013	16687

1 Full-time students, members of the Armed Forces, the long term unemployed and those who have never worked are not shown as separate categories but are included in the totals.
2 The first of each pair of figures shown relates to men, and the second, to women.
3 Results for 2006 include longitudinal data (see Appendix B).

Smoking and drinking among adults 2006

Table 2.15 Drinking last week, by sex and usual gross weekly household income

Persons aged 16 and over *Great Britain: 2006[2]*

Drinking last week	Usual gross weekly household income (£)						Total[1]
	Up to 200.00	200.01 - 400.00	400.01 - 600.00	600.01 - 800.00	800.01 - 1000.00	1000.01 or more	
	Percentages						
Drank last week							
Men	61	62	71	74	79	83	71
Women	40	51	56	63	64	70	56
All persons	48	56	63	68	71	76	63
Drank on 5 or more days							
Men	18	19	19	18	22	26	21
Women	9	11	12	12	11	14	11
All persons	12	15	16	15	16	20	16
Weighted base (000's) =100%							
Men	2,780	3,535	3,131	2,808	1,966	3,979	19,936
Women	4,467	4,231	3,205	2,905	1,967	4,013	22,742
All persons	7,247	7,768	6,335	5,713	3,933	7,992	42,678
Unweighted sample							
Men	1065	1418	1202	1078	730	1501	7681
Women	1733	1740	1291	1141	757	1565	9013
All persons	2798	3158	2493	2219	1487	3066	16694

1 The total includes those for whom household income was not available
2 Results for 2006 include longitudinal data (see Appendix B).

General Household Survey, 2006

Smoking and drinking among adults 2006

Table **2.16** Maximum drunk on any one day last week by sex and usual gross weekly household income

Persons aged 16 and over — *Great Britain: 2006[3]*

Drinking last week	Usual gross weekly household income (£)						
	Up to 200.00	200.01 - 400.00	400.01 - 600.00	600.01 - 800.00	800.01 - 1000.00	1000.01 or more	Total[1]
				Percentages			
Original method							
Drank more than 4/3 units on at least one day last week[2]							
Men	27	23	33	37	43	40	33
Women	12	16	20	23	27	27	20
All persons	18	19	26	30	35	34	26
Drank more than 8/6 units on at least one day last week[2]							
Men	14	11	19	22	24	22	18
Women	5	7	9	11	10	9	8
All persons	8	9	14	16	17	16	13
Improved method							
Drank more than 4/3 units on at least one day last week[2]							
Men	32	28	38	43	51	51	40
Women	20	27	33	36	41	47	33
All persons	25	28	35	39	46	49	36
Drank more than 8/6 units on at least one day last week[2]							
Men	16	14	23	27	30	29	23
Women	7	11	14	18	20	21	15
All persons	11	12	19	22	25	25	18
Weighted base (000's) =100%							
Men	2,782	3,531	3,134	2,807	1,962	3,976	19,913
Women	4,465	4,230	3,205	2,905	1,967	4,010	22,739
All persons	7,247	7,764	6,340	5,713	3,928	7,986	42,659
Unweighted sample							
Men	1066	1417	1204	1077	728	1500	7674
Women	1732	1740	1291	1141	757	1564	9013
All persons	2798	3157	2495	2218	1485	3064	16687

1 The total includes those for whom household income was not available
2 The first of each pair of figures shown relates to men, and the second, to women.
3 Results for 2006 include longitudinal data (see Appendix B).

Smoking and drinking among adults 2006

Table 2.17 Drinking last week, by sex and economic activity status

Persons aged 16-64 *Great Britain: 2006[1]*

Drinking last week	Economic activity status			
	Working	Unemployed	Economically inactive	Total
	Percentages			
Drank last week				
Men	76	54	59	72
Women	65	54	47	59
All persons	70	54	51	65
Drank on 5 or more days last week				
Men	20	10	17	19
Women	11	6	9	10
All persons	16	9	12	15
Weighted base (000's) =100%				
Men	*12,726*	*678*	*2,688*	*16,092*
Women	*12,155*	*407*	*5,261*	*17,823*
All persons	*24,880*	*1,085*	*7,949*	*33,914*
Unweighted sample				
Men	*4701*	*229*	*981*	*5911*
Women	*4709*	*140*	*2111*	*6960*
All persons	*9410*	*369*	*3092*	*12871*

1 Results for 2006 include longitudinal data (see Appendix B).

Smoking and drinking among adults 2006

Table 2.18 Maximum drunk on any one day last week, by sex and economic activity status

Persons aged 16-64 Great Britain: 2006[2]

Drinking last week	Economic activity status			
	Working	Unemployed	Economically inactive	Total
	Percentages			
Original method				
Drank more than 4/3 units on at least one day[1]				
Men	39	34	31	38
Women	27	25	17	24
All persons	34	30	22	31
Drank more than 8/6 units on at least one day[1]				
Men	23	18	16	21
Women	12	10	7	10
All persons	17	15	10	15
Improved method				
Drank more than 4/3 units on at least one day[1]				
Men	47	37	35	45
Women	43	35	27	38
All persons	45	36	30	41
Drank more than 8/6 units on at least one day[1]				
Men	28	22	20	27
Women	21	16	12	18
All persons	25	20	15	22
Weighted base (000's) =100%				
Men	*12,714*	*674*	*2,686*	*16,074*
Women	*12,157*	*407*	*5,260*	*17,824*
All persons	*24,871*	*1,081*	*7,945*	*33,897*
Unweighted sample				
Men	*4696*	*228*	*980*	*5904*
Women	*4710*	*140*	*2111*	*6961*
All persons	*9406*	*368*	*3091*	*12865*

1 The first of each pair of figures shown relates to men, and the second, to women.
2 Results for 2006 include longitudinal data (see Appendix B).

Table 2.19 Drinking last week, by sex and usual gross weekly earnings

Persons aged 16-64 in full-time employment *Great Britain: 2006[2]*

Drinking last week	Usual gross weekly earnings[1] (£)						
	Up to 200.00	200.01 -300.00	300.01 -400.00	400.01 -600.00	600.01 -800.00	800.01 or more	Total
	Percentages						
Drank last week							
Men	74	69	73	79	85	86	78
Women	61	64	66	67	77	76	67
All persons	68	66	70	75	82	84	74
Drank on 5 or more days							
Men	23	17	16	19	25	29	20
Women	10	9	12	11	19	16	12
All persons	17	13	14	16	23	26	17
Weighted base (000's) =100%							
Men	*834*	*1,397*	*2,087*	*2,929*	*1,344*	*1,647*	*11,205*
Women	*606*	*1,580*	*1,427*	*1,476*	*635*	*458*	*6,646*
All persons	*1,440*	*2,978*	*3,515*	*4,406*	*1,979*	*2,104*	*17,853*
Unweighted sample							
Men	*301*	*493*	*744*	*1084*	*517*	*657*	*4177*
Women	*222*	*580*	*544*	*557*	*255*	*180*	*2521*
All persons	*523*	*1073*	*1288*	*1641*	*772*	*837*	*6698*

1 Usual gross weekly earnings for the respondent
2 Results for 2006 include longitudinal data (see Appendix B).

Table 2.20 Maximum drunk on any one day last week, by sex and usual gross weekly earnings

Persons aged 16-64 in full-time employment *Great Britain: 2006[3]*

Drinking last week	Usual gross weekly earnings[1] (£)						
	Up to 200.00	200.01 -300.00	300.01 -400.00	400.01 -600.00	600.01 -800.00	800.01 or more	Total
	Percentages						
Original method							
Drank more than 4/3 units on at least one day[2]							
Men	39	37	43	46	41	39	41
Women	31	31	30	30	27	34	30
All persons	36	33	38	41	36	38	37
Drank more than 8/6 units on at least one day[2]							
Men	26	21	26	27	23	19	24
Women	18	16	14	13	7	12	13
All persons	23	18	21	22	18	17	20
Improved method							
Drank more than 4/3 units on at least one day[2]							
Men	47	42	48	51	49	54	49
Women	42	43	44	48	51	58	46
All persons	44	42	46	50	50	55	47
Drank more than 8/6 units on at least one day[2]							
Men	30	23	30	34	31	28	29
Women	24	24	23	25	24	28	24
All persons	27	23	27	31	29	28	27
Weighted base (000's) =100%							
Men	*834*	*1,394*	*2,083*	*2,932*	*1,344*	*1,644*	*11,195*
Women	*606*	*1,581*	*1,427*	*1,476*	*635*	*458*	*6,649*
All persons	*1,440*	*2,974*	*3,511*	*4,410*	*1,979*	*2,101*	*17,844*
Unweighted sample							
Men	*301*	*492*	*742*	*1084*	*517*	*656*	*4172*
Women	*222*	*580*	*544*	*557*	*255*	*180*	*2522*
All persons	*523*	*1072*	*1286*	*1641*	*772*	*836*	*6694*

1 Usual gross weekly earnings for the respondent
2 The first of each pair of figures shown relates to men, and the second, to women
3 Results for 2006 include longitudinal data (see Appendix B).

Smoking and drinking among adults 2006

Table 2.21 Drinking last week, by sex and Government Office Region

Persons aged 16 and over
Great Britain: 2006[1]

Government Office Region	Drinking last week		Weighted base (000's) =100%	Unweighted sample
	Drank last week	Drank on 5 or more days last week		
Men	*Percentages*			
North East	70	19	*802*	*310*
North West	76	20	*2,219*	*900*
Yorkshire and the Humber	77	21	*1,805*	*723*
East Midlands	72	24	*1,698*	*686*
West Midlands	68	21	*1,748*	*677*
East of England	73	20	*1,990*	*786*
London	62	19	*2,237*	*662*
South East	72	23	*2,858*	*1115*
South West	76	24	*1,824*	*745*
England	72	21	*17,182*	*6604*
Wales	69	23	*1,021*	*410*
Scotland	67	14	*1,732*	*667*
Great Britain	71	21	*19,935*	*7681*
Women				
North East	53	11	*919*	*368*
North West	60	10	*2,669*	*1110*
Yorkshire and the Humber	62	14	*1,980*	*819*
East Midlands	58	14	*1,749*	*741*
West Midlands	52	11	*1,933*	*773*
East of England	57	11	*2,201*	*907*
London	46	7	*2,656*	*796*
South East	59	14	*3,235*	*1301*
South West	59	15	*2,126*	*884*
England	57	12	*19,468*	*7699*
Wales	53	11	*1,151*	*476*
Scotland	52	9	*2,123*	*838*
Great Britain	56	11	*22,742*	*9013*
All persons				
North East	61	14	*1,722*	*678*
North West	67	15	*4,888*	*2010*
Yorkshire and the Humber	69	17	*3,785*	*1542*
East Midlands	65	19	*3,446*	*1427*
West Midlands	60	16	*3,681*	*1450*
East of England	65	15	*4,191*	*1693*
London	53	12	*4,893*	*1458*
South East	65	18	*6,093*	*2416*
South West	67	19	*3,949*	*1629*
England	64	16	*36,651*	*14303*
Wales	61	16	*2,172*	*886*
Scotland	58	11	*3,856*	*1505*
Great Britain	63	16	*42,679*	*16694*

1 Results for 2006 include longitudinal data (see Appendix B).

General Household Survey, 2006

Smoking and drinking among adults 2006

Table 2.22 Maximum drunk on any one day last week, by sex and Government Office Region

Persons aged 16 and over   Great Britain: 2006[2]

Government Office Region	Original method		Improved method		Weighted base (000's) =100%	Unweighted sample
	Drank more than 4/3 units on at least one day[1]	Drank more than 8/6 units on at least one day[1]	Drank more than 4/3 units on at least one day[1]	Drank more than 8/6 units on at least one day[1]		
			Percentages			
Men						
North East	37	18	43	21	803	310
North West	41	26	47	31	2,216	899
Yorkshire and the Humber	41	23	48	29	1,794	718
East Midlands	34	18	41	23	1,703	688
West Midlands	31	16	37	19	1,745	676
East of England	30	15	37	20	1,987	785
London	30	16	35	21	2,239	662
South East	29	14	37	20	2,858	1115
South West	32	17	39	21	1,819	743
England	33	18	40	23	17,162	6596
Wales	36	18	42	22	1,024	411
Scotland	33	18	40	23	1,732	667
Great Britain	33	18	40	23	19,918	7674
Women						
North East	20	7	33	11	915	367
North West	27	12	40	20	2,668	1110
Yorkshire and the Humber	28	14	40	23	1,977	818
East Midlands	18	9	32	14	1,749	741
West Midlands	17	7	29	13	1,932	773
East of England	17	5	30	12	2,200	907
London	14	5	27	11	2,656	796
South East	18	8	32	15	3,237	1302
South West	21	7	34	16	2,126	884
England	20	8	33	15	19,465	7698
Wales	19	7	34	12	1,152	477
Scotland	20	7	33	14	2,124	838
Great Britain	20	8	33	15	22,741	9013
All persons						
North East	28	12	37	16	1,719	677
North West	34	18	43	25	4,886	2009
Yorkshire and the Humber	34	18	44	26	3,772	1536
East Midlands	26	13	36	18	3,451	1429
West Midlands	24	11	33	16	3,677	1449
East of England	23	10	33	15	4,189	1692
London	21	10	31	15	4,896	1458
South East	23	11	34	17	6,095	2417
South West	26	12	36	18	3,943	1627
England	26	13	36	19	36,628	14294
Wales	27	12	38	17	2,175	888
Scotland	26	12	36	18	3,856	1505
Great Britain	26	13	36	18	42,659	16687

1 The first of each pair of figures shown relates to men, and the second, to women.
2 Results for 2006 include longitudinal data (see Appendix B).

Table 3.1 Trends in household size: 1971 to 2006

(a) Households and (b) Persons
Great Britain

Number of persons in household (all ages)	Unweighted								Weighted							
	1971	1975	1981	1985	1991	1995	1996	1998	1998	2000	2001	2002	2003	2004	2005[1]	2006[2]
Percentage of households of each size																
(a) Households	%	%	%	%	%	%	%	%	%	%	%	%	%	%	%	%
1	17	20	22	24	26	28	27	29	31	32	31	31	30	31	31	30
2	31	32	31	33	34	35	34	36	34	34	34	35	36	35	35	34
3	19	18	17	17	17	16	16	15	16	15	16	16	15	15	15	16
4	18	17	18	17	16	15	15	14	14	13	13	13	13	13	13	14
5	8	8	7	6	6	5	5	5	4	5	4	4	5	4	4	4
6 or more	6	5	4	2	2	2	2	2	2	2	2	2	2	2	2	2
Weighted base (000's) =100%[3]									24,450	24,845	24,592	24,529	24,423	24,688	24,829	24,815
Unweighted sample[3]	11988	12097	12006	9993	9955	9758	9158	8636		8221	8989	8620	10283	8700	12802	9731
Average (mean) household size	2.91	2.78	2.70	2.56	2.48	2.40	2.43	2.36	2.32	2.30	2.33	2.31	2.32	2.30	2.30	2.34
Percentage of persons in households of each size																
(b) Persons	%	%	%	%	%	%	%	%	%	%	%	%	%	%	%	%
1	6	7	8	10	11	12	11	12	13	14	13	13	13	14	14	13
2	22	23	23	26	27	29	28	30	30	30	29	30	31	30	31	29
3	20	19	19	20	20	20	20	19	20	20	20	20	20	19	19	20
4	25	25	27	27	25	24	25	24	23	22	23	22	22	23	22	23
5	15	14	14	12	11	10	10	10	9	10	9	9	10	9	9	10
6 or more	13	11	9	6	5	5	6	5	4	5	5	5	5	5	5	5
Weighted base (000's) =100%[3]									56,751	57,106	57,260	56,570	56,721	56,873	56,985	58,041
Unweighted sample[3]	34849	33579	32410	25555	24657	23385	22274	20396		19266	21180	20149	24489	20421	30069	22924

1 2005 data includes last quarter of 2004/5 data due to survey change from financial year to calendar year.
2 Results for 2006 include longitudinal data (see Appendix B).
3 Trend tables show unweighted and weighted figures for 1998 to give an indication of the effect of the weighting. For the weighted data (1998 and 2000 to 2006) the weighted base (000's) is the base for percentages. Unweighted data (up to 1998) are based on the unweighted sample.

Table 3.2 Trends in household type: 1971 to 2006

(a) Households and (b) Persons
Great Britain

Household type	Unweighted								Weighted							
	1971	1975	1981	1985	1991	1995	1996	1998	1998	2000	2001	2002	2003	2004	2005	2006
Percentage of households of each type																
(a) Households	%	%	%	%	%	%	%	%	%	%	%	%	%	%	%	%
1 adult aged 16-59	5	6	7	8	10	12	11	13	15	16	15	16	15	16	16	15
2 adults aged 16-59	14	14	13	15	16	17	15	16	16	17	17	16	16	16	16	16
Youngest person aged 0-4	18	15	13	13	14	13	13	13	12	11	11	11	11	11	11	11
Youngest person aged 5-15	21	22	22	18	16	16	17	16	16	15	16	16	16	16	16	16
3 or more adults	13	11	13	12	12	10	11	9	10	10	11	10	11	10	10	11
2 adults, 1 or both aged 60 or over	17	17	17	17	16	16	16	17	15	15	15	15	16	16	16	16
1 adult aged 60 or over	12	15	15	16	16	15	16	16	15	16	15	15	15	15	15	16
Weighted base (000's) =100%[3]									24,450	24,845	24,592	24,529	24,423	24,688	24,829	24,815
Unweighted sample[3]	11934	12090	12006	9993	9955	9758	9158	8636		8221	8989	8620	10283	8700	12802	9731
Percentage of persons in each type of household																
(b) Persons	%	%	%	%	%	%	%	%	%	%	%	%	%	%	%	%
1 adult aged 16-59	2	2	3	3	4	5	5	5	7	7	7	7	6	7	7	6
2 adults aged 16-59	10	10	10	12	13	14	13	14	14	14	14	14	14	14	14	14
Youngest person aged 0-4	27	23	21	21	22	20	21	21	19	19	18	18	18	17	17	18
Youngest person aged 5-15	31	34	33	28	25	26	27	25	26	25	25	26	25	25	25	25
3 or more adults	15	13	16	17	16	15	15	13	15	15	16	15	16	16	15	17
2 adults, 1 or both aged 60 or over	11	12	12	13	13	14	13	14	13	13	13	13	14	14	14	13
1 adult aged 60 or over	4	5	6	6	7	6	7	7	7	7	7	6	7	7	7	7
Weighted base (000's) =100%[3]									56,751	57,106	57,260	56,570	56,721	56,873	56,985	58,041
Unweighted sample[3]	34720	33561	32410	25555	24657	23385	22274	20396		19266	21180	20149	24489	20421	30069	22924

1 2005 data includes last quarter of 2004/5 data due to survey change from financial year to calendar year.
2 Results for 2006 include longitudinal data (see Appendix B).
3 Trend tables show unweighted and weighted figures for 1998 to give an indication of the effect of the weighting. For the weighted data (1998 and 2000 to 2006) the weighted base (000's) is the base for percentages. Unweighted data (up to 1998) are based on the unweighted sample.

Table 3.3 Percentage living alone, by age: 1973 to 2006

All persons aged 16 and over *Great Britain*

	Percentage who lived alone																	
	Unweighted									Weighted[1]								
	1973	1983	1987	1991	1993	1995	1996	1998		1998	2000	2001	2002	2003	2004	2005[2]	2006[3]
16-24	2	2	3	3	4	5	4	4		4	5	5	5	4	4	5	3
25-44	2	4	6	7	8	9	8	10		12	12	12	12	12	12	12	11
45-64	8	9	10	11	11	12	11	14		15	16	15	15	14	16	16	15
65-74	26	28	28	29	28	27	31	27		28	29	27	27	27	26	26	26
75 and over	40	47	50	50	50	51	47	48		48	50	49	48	48	47	48	50
All aged 16 and over	9	11	12	14	14	15	14	16		17	17	16	17	16	17	17	16
Unweighted sample[4]																	
16-24	3811	3498	3558	2819	2574	2318	2233	1885		1870	2064	2023	2427	2069	2990	2160
25-44	8169	7017	7418	7118	6875	6761	6489	5861		5393	6118	5579	6843	5710	8234	6061
45-64	7949	5947	5802	5493	5360	5615	5114	4892		4803	5147	5169	5952	5198	7646	6068
65-74	2847	2494	2389	2196	2303	2129	1943	1862		1672	1882	1766	2273	1794	2754	2194
75 and over	1432	1490	1596	1603	1581	1451	1485	1374		1344	1474	1435	1683	1404	2126	1731
All aged 16 and over	24208	20446	20763	19229	18693	18274	17264	15874		15082	16685	15972	19178	16175	23750	18214

1 Weighted bases are shown in Table 3.21.
2 2005 data includes last quarter of 2004/5 data due to survey change from financial year to calendar year.
3 Results for 2006 include longitudinal data (see Appendix B).
4 Trend tables show unweighted and weighted figures for 1998 to give an indication of the effect of the weighting. For the weighted data (1998 and 2000 to 2006) the weighted base (000's) is the base for percentages. Unweighted data (up to 1998) are based on the unweighted sample.

Table 3.4   Percentage of men and women living alone by age

All persons aged 16 and over   *Great Britain: 2006[1]*

	Percentage living alone		
	Men	Women	Total
16-24	4	2	3
25-44	14	8	11
45-64	16	14	15
65-74	21	31	26
75 and over	32	61	50
All aged 16 and over	15	17	16
All persons[2]	12	14	13
Weighted bases (000's) =100%			
16-24	*3,487*	*3,369*	*6,856*
25-44	*8,194*	*8,347*	*16,541*
45-64	*7,142*	*7,361*	*14,503*
65-74	*2,294*	*2,553*	*4,847*
75 and over	*1,662*	*2,499*	*4,162*
All aged 16 and over	*22,779*	*24,130*	*46,909*
All persons[2]	*28,382*	*29,659*	*58,041*
Unweighted sample			
16-24	*1048*	*1112*	*2160*
25-44	*2873*	*3188*	*6061*
45-64	*2943*	*3125*	*6068*
65-74	*1041*	*1153*	*2194*
75 and over	*776*	*955*	*1731*
All aged 16 and over	*8681*	*9533*	*18214*
All persons[2]	*11060*	*11864*	*22924*

1   Results for 2006 include longitudinal data (see Appendix B)
2   Including children.

Table 3.5 Type of household: 1979 to 2006

(a) Households and (b) Persons   Great Britain

Household type	Unweighted						Weighted							
	1979	1985	1991	1995	1996	1998	1998	2000	2001	2002	2003	2004	2005[1]	2006[2]
	\multicolumn{14}{c}{Percentage of households of each type}													
(a) Households	%	%	%	%	%	%	%	%	%	%	%	%	%	%
1 person only	23	24	26	28	27	29	31	32	31	31	30	31	31	31
2 or more unrelated adults	3	4	3	2	3	2	3	3	3	2	3	3	3	2
Married/cohabiting couple														
with dependent children	31	28	25	24	25	23	22	21	22	21	21	21	21	22
with non-dependent children only	7	8	8	6	6	6	6	6	6	6	6	6	6	6
no children	27	27	28	29	28	30	28	28	29	28	29	29	29	28
Married couple														
with dependent children	..	..	..	..	23	20	19	18	18	18	18	18	17	18
with non-dependent children	..	..	..	..	6	5	6	6	6	6	6	6	5	6
no children	..	..	..	..	25	26	24	24	23	23	23	23	23	23
Cohabiting couple														
with dependent children	..	..	..	..	3	3	3	3	3	3	4	3	3	4
with non-dependent children	..	..	..	..	0	0	0	0	0	0	0	0	0	0
no children	..	..	..	..	4	4	5	5	5	5	5	5	5	5
Lone parent														
with dependent children	4	4	6	7	7	7	7	7	7	8	7	7	7	7
with non-dependent children only	4	4	4	3	3	3	3	2	2	2	3	3	3	3
Two or more families	1	1	1	1	1	1	1	1	1	1	1	1	1	1
Weighted base (000's) =100%[3]							24,389	24,787	24,493	24,449	24,326	24,613	24,751	24,727
Unweighted sample[3,4]	11454	9993	9955	9738	9138	8617		8204	8955	8594	10248	8675	12763	9705
	\multicolumn{14}{c}{Percentage of persons in each type of household}													
(b) Persons	%	%	%	%	%	%	%	%	%	%	%	%	%	%
1 person only	9	10	11	12	11	12	13	14	13	13	13	14	14	13
2 or more unrelated adults	2	3	2	2	3	2	3	3	3	3	3	4	3	3
Married/cohabiting couple														
with dependent children	49	45	41	40	42	39	38	36	37	36	37	36	36	37
with non-dependent children only	9	11	11	9	9	8	9	9	9	9	9	9	8	9
no children	20	21	23	25	24	26	25	25	25	25	25	25	26	24
Married couple														
with dependent children	..	..	..	..	37	34	33	32	32	32	31	31	30	31
with non-dependent children	..	..	..	..	9	8	8	9	9	8	8	8	8	9
no children	..	..	..	..	21	22	21	21	20	21	21	21	21	20
Cohabiting couple														
with dependent children	..	..	..	..	4	5	5	5	5	5	6	5	5	5
with non-dependent children	..	..	..	..	0	0	0	0	0	0	1	0	0	0
no children	..	..	..	..	3	4	4	4	5	4	5	5	5	5
Lone parent														
with dependent children	5	5	7	8	8	9	8	9	8	9	9	8	9	9
with non-dependent children only	3	4	3	3	3	3	3	2	2	2	3	3	3	3
Two or more families	2	1	2	1	1	2	2	2	2	2	2	2	2	2
Weighted base (000's) =100%[3]							56,605	56,955	56,921	56,245	56,319	56,565	56,663	57,684
Unweighted sample[3,4]	30546	25454	24657	23325	22190	20350		19220	21065	20045	24346	20319	29907	22819

1 2005 data includes last quarter of 2004/5 data due to survey change from financial year to calendar year.
2 Results for 2006 include longitudinal data (see Appendix B).
3 Trend tables show unweighted and weighted figures for 1998 to give an indication of the effect of the weighting. For the weighted data (1998 and 2000 to 2006) the weighted base (000's) is the base for percentages. Unweighted data (up to 1998) are based on the unweighted sample.
4 Total includes a very small number of same sex cohabitees.
.. Data are not available (not collected in these years).

Table 3.6 Family type, and marital status of lone parents: 1971 to 2006

Families with dependent children [1] *Great Britain*

Family type	Unweighted								Weighted							
	1971	1975	1981	1985	1991	1995	1996	1998	1998	2000	2001	2002	2003	2004	2005[2]	2006[3]
	%	%	%	%	%	%	%	%	%	%	%	%	%	%	%	%
Married/cohabiting couple[4]	92	90	87	86	81	78	79	75	76	74	75	73	74	75	74	75
Lone mother	7	9	11	12	18	20	20	22	21	23	22	24	23	23	24	22
single	1	1	2	3	6	8	7	9	8	11	10	12	11	11	11	10
widowed	2	2	2	1	1	1	1	1	1	1	1	0	0	1	1	1
divorced	2	3	4	5	6	7	6	8	7	7	7	7	7	7	8	7
separated	2	2	2	3	4	5	5	5	5	5	4	5	5	4	4	4
Lone father	1	1	2	2	1	2	2	2	3	3	3	2	3	2	2	3
All lone parents	8	10	13	14	19	22	21	25	24	26	25	27	26	25	26	25
Weighted base (000's) =100%[5]									7,182	7,105	7,146	7,206	7,071	7,000	7,025	7,271
Unweighted sample[5]	4864	4776	4445	3348	3143	3022	2975	2659		2464	2700	2582	3166	2570	3831	2886

1 Dependent children are persons aged under 16, or aged 16-18 and in full-time education, in the family unit, and living in the household.
2 2005 data includes last quarter of 2004/5 data due to survey change from financial year to calendar year.
3 Results for 2006 include longitudinal data (see Appendix B).
4 Including married women whose husbands were not defined as resident in the household.
5 Trend tables show unweighted and weighted figures for 1998 to give an indication of the effect of the weighting. For the weighted data (1998 and 2000 to 2006) the weighted base (000's) is the base for percentages. Unweighted data (up to 1998) are based on the unweighted sample.

Table 3.7 Families with dependent children: 1972 to 2006

Dependent children [1] *Great Britain*

	Percentage of all dependent children in each family type															
	Unweighted								Weighted							
	1972	1975	1981	1985	1991	1995	1996	1998	1998	2000	2001	2002	2003	2004	2005[2]	2006[3]
	%	%	%	%	%	%	%	%	%	%	%	%	%	%	%	%
Married/cohabiting couple with																
1 dependent child	16	17	18	19	17	16	17	15	17	17	17	18	18	18	18	19
2 or more dependent children	76	74	70	69	66	64	63	62	61	58	60	57	58	60	57	57
Lone mother with																
1 dependent child	2	3	3	4	5	5	5	6	6	7	6	8	7	7	8	7
2 or more dependent children	5	6	7	7	12	14	13	15	13	15	15	15	14	14	15	15
Lone father with																
1 dependent child	0	0	1	1	0	1	0	1	1	1	1	1	1	1	1	1
2 or more dependent children	1	1	1	1	1	1	1	1	1	2	1	1	2	1	1	1
Weighted base (000's) =100%[4]									12,799	12,641	12,606	12,451	12,368	12,190	12,189	12,549
Unweighted sample[4]	9474	9293	8216	5966	5799	5559	5431	4897		4499	4846	4561	5743	4583	6828	5166

1 Dependent children are persons aged under 16, or aged 16-18 and in full-time education, in the family unit, and living in the household.
2 2005 data includes last quarter of 2004/5 data due to survey change from financial year to calendar year.
3 Results for 2006 include longitudinal data (see Appendix B).
4 Trend tables show unweighted and weighted figures for 1998 to give an indication of the effect of the weighting. For the weighted data (1998 and 2000 to 2006) the weighted base (000's) is the base for percentages. Unweighted data (up to 1998) are based on the unweighted sample.

Table 3.8 Average (mean) number of dependent children by family type: 1971 to 2006

Families with dependent children[1]   Great Britain

Family type	Average (mean) number of children															
	Unweighted								Weighted[2]							
	1971	1975	1981	1985	1991	1995	1996	1998	1998	2000	2001	2002	2003	2004	2005[3]	2006[4]
Married/cohabiting couple[5]	2.0	2.0	1.9	1.8	1.9	1.9	1.9	1.9	1.8	1.8	1.8	1.8	1.8	1.8	1.8	1.8
Married couple	..	..	..	..	..	..	1.9	1.9	1.8	1.8	1.9	1.8	1.8	1.8	1.8	1.8
Cohabiting couple	..	..	..	..	..	..	1.7	1.7	1.7	1.7	1.6	1.6	1.7	1.6	1.6	1.6
Lone parent	1.8	1.7	1.6	1.6	1.7	1.7	1.7	1.7	1.6	1.7	1.7	1.6	1.6	1.6	1.6	1.7
Total: all families with dependent children	2.0	1.9	1.8	1.8	1.8	1.8	1.8	1.8	1.8	1.8	1.8	1.7	1.8	1.8	1.7	1.7
Unweighted sample[6]																
Married/cohabiting couple	*4482*	*4299*	*3887*	*2890*	*2541*	*2358*	*2329*	*2004*		*1804*	*2004*	*1889*	*2369*	*1946*	*2878*	*2167*
Married couple	..	..	..	..	..	..	*2086*	*1753*		*1558*	*1720*	*1636*	*1969*	*1651*	*2418*	*1828*
Cohabiting couple	..	..	..	..	..	..	*243*	*251*		*246*	*284*	*253*	*400*	*295*	*460*	*339*
Lone parent	*382*	*477*	*558*	*458*	*595*	*658*	*635*	*652*		*660*	*682*	*679*	*781*	*608*	*931*	*705*
Total	*4864*	*4776*	*4445*	*3348*	*3136*	*3016*	*2964*	*2656*		*2464*	*2686*	*2568*	*3150*	*2554*	*3809*	*2872*

1 Dependent children are persons aged under 16, or aged 16-18 and in full-time education, in the family unit, and living in the household.
2 Weighted bases are shown in Table 3.21.
3 2005 data includes last quarter of 2004/5 data due to survey change from financial year to calendar year.
4 Results for 2006 include longitudinal data (see Appendix B).
5 Including married women whose husbands were not defined as resident in the household.
6 Trend tables show unweighted and weighted figures for 1998 to give an indication of the effect of the weighting. For the weighted data (1998 and 2000 to 2006) the weighted base (000's) is the base for percentages. Unweighted data (up to 1998) are based on the unweighted sample.

.. Data are not available (not collected in these years).

Table 3.9 Age of youngest dependent child by family type

Families with dependent children[1]   Great Britain: 2005[2] and 2006[3] combined

Family type		Age of youngest dependent child				Unweighted sample[4]	Total
		0-4	5-9	10-15	16 and over		
							%
Married/cohabiting couple[5]	%	40	24	28	8	*4591*	75
Lone mother	%	34	29	29	9	*1362*	23
Lone father	%	16	33	44	7	*121*	2
All lone parents	%	32	29	30	9	*1483*	25
Total	%	38	25	29	8	*6074*	100

1 Dependent children are persons aged under 16, or aged 16-18 and in full-time education, in the family unit, and living in the household.
2 2005 data includes last quarter of 2004/5 data due to survey change from financial year to calendar year.
3 Results for 2006 include longitudinal data (see Appendix B).
4 Weighted base not shown for combined data sets
5 Including married women whose husbands were not defined as resident in the household.

Table 3.10 Stepfamilies with dependent children by family type

Stepfamilies with dependent children[1]
(Family head aged 16-59)   Great Britain: 2006[2]

Type of stepfamily	%
Couple with child(ren) from the woman's previous marriage/ cohabitation	84
Couple with child(ren) from the man's previous marriage/ cohabitation	10
Couple with child(ren) from both partners' previous marriage/ cohabitation	6
Weighted base (000's) = 100%	545
Unweighted sample	209

1 Dependent children are persons under 16, or aged 16-18 and in full-time education, in the family unit, and living in the household.
2 Results for 2006 include longitudinal data (see Appendix B).

Table 3.11 Usual gross weekly household income of families with dependent children by family type

Families with dependent children[1]   Great Britain: 2006[2]

Family type		Usual gross weekly household income							Weighted base (000's) = 100%[3]	Unweighted sample
		£0.00-£100.00	£100.01-£200.00	£200.01-£300.00	£300.01-£400.00	£400.01-£500.00	£500.01-£700.00	£700.01 and over		
Married couple	%	3	4	4	5	7	18	59	4,146	1658
Cohabiting couple	%	4	5	8	10	15	20	38	804	314
Lone mother[4]	%	8	31	24	13	8	10	5	1,532	604
Single	%	11	40	24	11	5	5	5	707	272
Divorced	%	5	25	23	16	11	14	7	493	204
Separated	%	7	19	23	17	12	16	5	277	110
Lone father	%	5	20	17	11	11	23	14	169	65
All lone parents	%	8	30	23	13	8	11	6	1,701	669
All families with dependent children	%	5	11	9	8	8	16	43	6,651	2641

1 Dependent children are persons aged under 16, or aged 16-18 and in full-time education, in the family unit, and living in the household.
2 Results for 2006 include longitudinal data (see Appendix B).
3 Bases exclude cases where income is not known.
4 Includes nineteen widowed lone mothers.
Note: Shaded figures indicate the estimates are unreliable and any analysis using these figures may be invalid. Any use of these shaded figures must be accompanied by this disclaimer.

Table 3.12 The distribution of the population by sex and age: 1971 to 2006

All persons *Great Britain*

Age	Unweighted								Weighted							
	1971	1975	1981	1985	1991	1995	1996	1998	1998	2000	2001	2002	2003	2004	2005[1]	2006[2]
	%	%	%	%	%	%	%	%	%	%	%	%	%	%	%	%
Males																
0-4	9	8	7	7	8	7	7	8	6	6	6	6	6	6	6	6
5-15[3]	19	18	18	16	15	16	16	16	15	15	15	15	15	15	14	14
16-44[3]	39	40	41	42	41	39	39	38	42	42	42	41	40	40	40	41
45-64	24	23	22	22	22	24	23	24	23	24	24	25	25	25	25	25
65-74	7	8	8	9	8	9	8	9	8	8	8	8	8	8	8	8
75 and over	3	3	4	4	5	5	6	5	5	5	5	6	6	6	6	6
Weighted base (000's) =100%[4]									27,921	28,134	28,212	27,524	27,732	27,802	27,732	28,382
Unweighted sample[4]	16908	16242	15735	12551	11913	11376	10781	9831		9322	10166	9706	11924	9911	14580	11060
	%	%	%	%	%	%	%	%	%	%	%	%	%	%	%	%
Females																
0-4	8	6	6	6	7	6	7	7	6	6	6	5	5	5	5	6
5-15[1]	16	17	16	15	14	14	15	14	14	14	14	14	13	13	13	13
16-44[1]	37	38	39	41	39	39	39	38	40	40	40	40	40	40	39	40
45-64	24	24	22	21	22	24	23	24	23	24	24	24	24	25	25	25
65-74	9	10	10	10	10	9	9	9	9	9	9	9	9	9	9	9
75 and over	5	6	7	8	8	8	7	8	8	8	8	8	8	8	8	8
Weighted base (000's) =100%[4]									28,828	28,973	29,048	29,047	28,989	29,072	29,253	29,659
Unweighted sample[4]	17871	17328	16675	13522	12744	12009	11493	10564		9944	11014	10443	12565	10510	15489	11864
	%	%	%	%	%	%	%	%	%	%	%	%	%	%	%	%
Total																
0-4	8	7	6	6	7	7	7	7	6	6	6	6	6	6	6	6
5-15[1]	17	17	17	15	15	15	15	15	14	14	14	14	14	14	14	13
16-44[1]	38	39	40	42	40	39	39	38	41	41	41	40	40	40	40	40
45-64	24	23	22	21	22	24	23	24	23	24	24	24	25	25	25	25
65-74	8	9	9	9	9	9	9	9	8	8	8	8	9	9	9	8
75 and over	4	4	5	6	7	6	7	7	7	7	7	7	7	7	7	7
Weighted base (000's) =100%[4]									56,749	57,106	57,260	56,571	56,721	56,873	56,985	58,041
Unweighted sample[4]	34779	33570	32410	26073	24657	23385	22274	20395		19266	21180	20149	24489	20421	30069	22924

1 2005 data includes last quarter of 2004/5 data due to survey change from financial year to calendar year.
2 Results for 2006 include longitudinal data (see Appendix B).
3 These age-groups were 5-14 and 15-44 in 1971 and 1975
4 Trend tables show unweighted and weighted figures for 1998 to give an indication of the effect of the weighting. For the weighted data (1998 and 2000 to 2006) the weighted base (000's) is the base for percentages. Unweighted data (up to 1998) are based on the unweighted sample.

Table 3.13 Percentage of males and females by age

All persons *Great Britain: 2006[1]*

Age		Males	Females	Weighted base (000's) =100%	Unweighted sample
0-4	%	50	50	3,378	1364
5-15	%	50	50	7,755	3346
16-19	%	52	48	3,353	1061
20-24	%	50	50	3,503	1099
25-29	%	48	52	3,520	1231
30-34	%	50	50	3,949	1426
35-39	%	50	50	4,396	1678
40-44	%	49	51	4,677	1726
45-49	%	48	52	4,049	1593
50-54	%	48	52	3,511	1415
55-59	%	51	49	3,854	1658
60-64	%	50	50	3,089	1402
65-69	%	48	52	2,623	1179
70-74	%	47	53	2,224	1015
75 and over	%	40	60	4,162	1731
Total	%	49	51	58,041	22924

1 Results for 2006 include longitudinal data (see Appendix B).

Table 3.14   Socio-economic classification based on own current or last job by sex and age

All persons aged 16 and over   *Great Britain: 2006[1]*

Socio-economic classification[2]	Age group						
	16-24	25-34	35-44	45-54	55-64	65 and over	All
	%	%	%	%	%	%	%
Men							
Higher managerial and professional	4	20	23	22	18	17	18
Lower managerial and professional	9	26	23	22	21	19	21
Intermediate	9	7	4	6	5	6	6
Small employers and own account	3	9	12	15	15	12	12
Lower supervisory and technical	14	11	13	12	13	17	13
Semi-routine	21	11	11	10	11	11	12
Routine	20	12	11	13	16	17	14
Never worked and long-term unemployed	19	3	2	2	1	1	4
Weighted base (000's)=100%	*2,162*	*3,476*	*4,356*	*3,506*	*3,439*	*3,925*	*20,865*
Unweighted sample	*653*	*1169*	*1582*	*1384*	*1480*	*1804*	*8072*
Women							
Higher managerial and professional	2	13	11	8	6	2	7
Lower managerial and professional	13	29	30	28	22	18	24
Intermediate	19	19	18	17	18	21	19
Small employers and own account	1	4	5	6	7	6	5
Lower supervisory and technical	4	4	4	5	6	6	5
Semi-routine	28	18	19	22	26	25	23
Routine	15	8	9	11	14	18	12
Never worked and long-term unemployed	18	5	4	3	2	5	5
Weighted base (000's)=100%	*2,154*	*3,570*	*4,377*	*3,849*	*3,363*	*5,008*	*22,320*
Unweighted sample	*699*	*1338*	*1705*	*1543*	*1523*	*2092*	*8900*
Total							
Higher managerial and professional	3	16	17	15	12	9	13
Lower managerial and professional	11	27	26	25	21	18	22
Intermediate	14	13	11	12	12	14	13
Small employers and own account	2	6	9	10	11	9	8
Lower supervisory and technical	9	8	9	8	9	11	9
Semi-routine	24	14	15	16	18	19	17
Routine	18	10	10	12	15	17	13
Never worked and long-term unemployed	18	4	3	3	2	3	5
Weighted base (000's)=100%	*4,315*	*7,047*	*8,734*	*7,354*	*6,802*	*8,934*	*43,185*
Unweighted sample	*1352*	*2507*	*3287*	*2927*	*3003*	*3896*	*16972*

1 Results for 2006 include longitudinal data (see Appendix B).
2 Full-time students and persons in inadequately described occupations are excluded.

Table 3.15   Ethnic group of GHS respondents: 2001 to 2006

All persons — Great Britain

Ethnic group	2001	2002	2003	2004	2005[1]	2006[2]
	%	%	%	%	%	%
White British	89 ⎫ 92	88 ⎫ 91	87 ⎫ 91	87 ⎫ 90	86 ⎫ 90	87 ⎫ 91
Other White	3 ⎭	3 ⎭	3 ⎭	4 ⎭	3 ⎭	4 ⎭
Mixed background	1	1	1	1	1	1
Indian	2 ⎫	2 ⎫	2 ⎫	2 ⎫	2 ⎫	2 ⎫
Pakistani	2 ⎪ 4	1 ⎪ 4	2 ⎪ 4	1 ⎪ 4	1 ⎪ 5	2 ⎪ 6
Bangladeshi	2 ⎪	0 ⎪	1 ⎪	1 ⎪	1 ⎪	1 ⎪
Other Asian background	0 ⎭	1 ⎭	1 ⎭	1 ⎭	1 ⎭	1 ⎭
Chinese	0	0	0	1	0	1
Black Caribbean	1 ⎫ 2 [3]	1 ⎫ 2 [3]	1 ⎫ 3 [3]	1 ⎫ 3 [3]	1 ⎫ 3 [3]	1 ⎫ 2 [3]
Black African	1 ⎭	1 ⎭	1 ⎭	1 ⎭	2 ⎭	1 ⎭
Other ethnic group	1	1	1	1	1	0
Weighted base (000's) = 100%	57,034	56,302	56,581	56,722	56,847	57,886
Unweighted sample	21102	20053	24430	20374	30006	22866

1 2005 data includes last quarter of 2004/5 data due to survey change from financial year to calendar year.
2 Results for 2006 include longitudinal data (see Appendix B).
3 Including other Black groups not shown separately.

Table 3.16   GHS respondents: age by ethnic group

All persons — Great Britain: 2004, 2005[1] and 2006[2] combined

Age	White British	Other White	Mixed back-ground	Indian	Pakistani	Bangladeshi	Other Asian background	Chinese	Black Caribbean	Black African	Other Black background	Other ethnic group	All
	%	%	%	%	%	%	%	%	%	%	%	%	%
0-15	19	13	46	20	32	38	21	13	20	36	44	24	20
16-24	11	13	17	13	18	18	9	33	14	13	12	17	11
25-44	27	39	25	37	30	31	46	35	33	36	28	35	28
45-64	26	24	9	23	15	9	19	17	18	13	10	18	25
65 and over	17	11	3	7	5	5	5	4	15	2	5	6	16
Unweighted sample[3]	59929	2173	759	1218	959	276	470	222	714	966	67	530	68283

1 2005 data includes last quarter of 2004/5 data due to survey change from financial year to calendar year.
2 Results for 2006 include longitudinal data (see Appendix B).
3 Weighted bases not shown for combined data sets.
Note: Shaded figures indicate the estimates are unreliable and any analysis using these figures may be invalid. Any use of these shaded figures must be accompanied by this disclaimer.

Table 3.17 GHS respondents: sex by ethnic group

All persons
Great Britain: 2004, 2005[1] and 2006[2] combined

Ethnic group		Male	Female	Unweighted sample[3]
White British	%	49	51	59929
Other White	%	47	53	2173
Mixed background	%	50	50	759
Indian	%	51	49	1218
Pakistani	%	53	47	959
Bangladeshi	%	51	49	276
Other Asian background	%	49	51	470
Chinese	%	54	46	222
Black Caribbean	%	44	56	714
Black African	%	47	53	966
Other Black background	%	43	57	67
Other ethnic group	%	48	52	530

1 2005 data includes last quarter of 2004/5 data due to survey change from financial year to calendar year.
2 Results for 2006 include longitudinal data (see Appendix B).
3 Weighted bases not shown for combined data sets.
Note: Shaded figures indicate the estimates are unreliable and any analysis using these figures may be invalid. Any use of these shaded figures must be accompanied by this disclaimer.

Table 3.18 GHS respondents: Ethnic group by Government Office Region

All persons
Great Britain: 2004, 2005[1] and 2006[2] combined

Government Office Region	White British	Other White	Mixed back-ground	Indian	Pakistani	Bangladeshi	Other Asian background	Chinese	Black Caribbean	Black African	Other Black background	Other ethnic group	Total
	%	%	%	%	%	%	%	%	%	%	%	%	%
England													
North East	5	1	2	0	1	0	3	8	1	0	0	2	4
North West	12	4	7	5	12	6	3	10	4	3	4	8	11
Yorkshire and the Humber	9	4	7	3	13	3	7	7	3	3	6	5	9
East Midlands	8	8	12	13	6	10	4	2	5	3	1	3	8
West Midlands	8	5	9	15	26	20	7	14	15	3	25	11	9
East of England	10	10	11	4	10	1	4	13	4	5	0	6	9
London	8	37	30	42	19	43	54	22	58	73	55	49	13
South East	15	15	13	12	10	3	10	15	4	8	5	10	14
South West	10	4	5	1	0	1	3	2	4	1	4	3	9
Wales	6	5	2	1	1	7	1	3	1	0	1	1	5
Scotland	10	7	2	4	1	4	4	5	2	1	0	2	9
Unweighted sample[3]	*59929*	*2173*	*759*	*1218*	*959*	*276*	*470*	*222*	*714*	*966*	*67*	*530*	*68283*

1 2005 data includes last quarter of 2004/5 data due to survey change from financial year to calendar year.
2 Results for 2006 include longitudinal data (see Appendix B).
3 Weighted bases not shown for combined data sets.
Note: Shaded figures indicate the estimates are unreliable and any analysis using these figures may be invalid. Any use of these shaded figures must be accompanied by this disclaimer

Table 3.19 GHS respondents: average household size by ethnic group of household reference person

Households	Great Britain: 2004, 2005[1] and 2006[2] combined	
Ethnic group	Average (mean) household size	Unweighted sample[3]
White British	2.27	26094
Other White	2.29	922
Mixed background	2.42	178
Indian	2.97	408
Pakistani	3.99	241
Bangladeshi	4.05	65
Other Asian background	2.59	172
Chinese	2.72	90
Black Caribbean	2.24	341
Black African	3.00	323
Other black background	2.73	18
Other ethnic group	2.61	192
Total	2.32	29044

1 2005 data includes last quarter of 2004/5 data due to survey change from financial year to calendar year.
2 Results for 2006 include longitudinal data (see Appendix B).
3 Weighted bases not shown for combined data sets.

Note: Shaded figures indicate the estimates are unreliable and any analysis using these figures may be invalid. Any use of these shaded figures must be accompanied by this disclaimer.

Table 3.20 GHS respondents: percentage born in the UK by age and ethnic group

All persons				Great Britain: 2004, 2005[1] and 2006[2] combined		
Ethnic group	Percentage born in the United Kingdom			Unweighted sample[3]		
	Age			Age		
	Under 25	25 and over	Total	Under 25	25 and over	Total
White British	99	97	98	17529	42381	59910
Other White	51	29	35	559	1614	2173
Mixed background	93	61	81	491	268	759
Indian	87	23	44	413	805	1218
Pakistani	86	21	53	489	470	959
Bangladeshi	84	12	52	154	122	276
Other Asian Background	54	6	21	149	321	470
Chinese	39	11	24	95	127	222
Black Caribbean	89	42	58	241	473	714
Black African	60	12	36	476	489	965
Other Black background	80	37	61	38	29	67
Other ethnic group	46	25	34	215	315	530
Total	94	89	90	20849	47414	68263

1 2005 data includes last quarter of 2004/5 data due to survey change from financial year to calendar year.
2 Results for 2006 include longitudinal data (see Appendix B).
3 Weighted base not shown for combined data sets.
Note: Shaded figures indicate the estimates are unreliable and any analysis using these figures may be invalid. Any use of these shaded figures must be accompanied by this disclaimer.

Table 3.21   Weighted bases for Tables 3.3 and 3.8

(a) Persons aged 16 and over

Great Britain

	1998	2000	2001	2002	2003	2004	2005[1]	2006[2]	Table Reference
16-24	6,139	6,191	6,192	6,286	6,381	6,470	6,511	6,856	3.3
25-44	17,117	17,130	17,275	16,404	16,304	16,225	16,177	16,541	3.3
45-64	13,226	13,519	13,540	13,757	13,927	14,102	14,244	14,503	3.3
65-74	4,767	4,719	4,729	4,792	4,822	4,849	4,863	4,847	3.3
75 and over	3,836	3,888	3,898	3,993	4,014	4,044	4,067	4,162	3.3
All aged 16 and over	45,085	45,447	45,632	45,232	45,448	45,689	45,861	46,909	3.3

(b) Families with dependent children[3]

Great Britain

	1998	2000	2001	2002	2003	2004	2005[1]	2006[2]	Table Reference
Married/cohabiting couple[4]	5,465	5,232	5,366	5,244	5,207	5,237	5,174	5,422	3.8
Married couple	4,765	4,496	4,580	4,515	4,330	4,431	4,348	4,560	3.8
Cohabiting couple	700	736	785	729	877	806	827	862	3.8
Lone parent	1,717	1,861	1,739	1,922	1,821	1,715	1,806	1,798	3.8
Total	7,182	7,093	7,105	7,166	7,028	6,952	6,980	7,219	3.8

1 2005 data includes last quarter of 2004/5 data due to survey change from financial year to calendar year.
2 Results for 2006 include longitudinal data (see Appendix B).
3 Dependent children are persons aged under 16, or aged 16-18 and in full-time education, in the family unit, and living in the household.
4 Including married women whose husbands were not defined as resident in the household.

Table 4.1 Housing tenure: 1971 to 2006

Households — Great Britain

Tenure	Unweighted								Weighted							
	1971	1975	1981	1985	1991	1995	1996	1998	1998	2000	2001	2002	2003	2004	2005[1]	2006[2]
	%	%	%	%	%	%	%	%	%	%	%	%	%	%	%	%
Owner occupied, owned outright	22	22	23	24	25	25	26	28	26	27	27	29	29	30	30	30
Owner occupied, with mortgage	27	28	31	37	42	42	41	41	42	41	41	40	40	39	38	40
Rented from council[3]	31	33	34	28	24	18	19	16	17	16	15	14	13	13	12	12
Rented from housing association[4]	1	1	2	2	3	4	5	5	5	6	6	7	7	7	7	8
Rented with job or business	5	3	2	2	1	2	**	**	**	**	**	**	**	**	**	**
Rented privately, unfurnished[5]	12	10	6	5	4	5	7	7	7	7	7	8	8	8	8	8
Rented privately, furnished	3	3	2	2	2	3	3	2	3	3	3	3	3	3	3	3
Weighted base (000's) =100%[6]									24,436	24,838	24,592	24,508	24,418	24,677	24,826	24,815
Unweighted sample[6]	11936	11970	11939	9933	9922	9723	9155	8631		8219	8989	8613	10281	8696	12800	9731

1 2005 data includes last quarter of 2004/5 data due to survey change from financial year to calendar year.
2 Results for 2006 include longitudinal data (see Appendix B).
3 Council includes local authorities.
4 Since 1996 housing associations are more correctly described as Registered Social Landlords (RSLs).
5 Unfurnished includes the answer 'partly furnished'.
6 Trend tables show unweighted and weighted figures for 1998 to give an indication of the effect of the weighting. For the weighted data (1998 and 2000 to 2006) the weighted base (000's) is the base for percentages. Unweighted data (up to 1998) are based on the unweighted sample.
** From 1996 onwards, accommodation that is provided as part of the employment contract of a member of the household, has been allocated to "rented privately". Squatters are also included in this category.

Table 4.2 Type of accommodation: 1971 to 2006

Households   *Great Britain*

Type of accommodation[1]	Unweighted data								Weighted data							
	1971	1975	1981	1985	1991	1995	1996	1998	1998	2000	2001	2002	2003	2004	2005[2]	2006[3]
	%	%	%	%	%	%	%	%	%	%	%	%	%	%	%	%
Detached house	16	15	16	19	19	22	21	23	21	21	21	22	22	22	22	23
Semi-detached house	33	34	32	31	32	31	32	33	32	31	31	32	32	31	31	30
Terraced house	30	28	31	29	29	28	27	26	26	28	28	27	27	27	28	28
Purpose-built flat or maisonette	13	14	15	15	14	15	15	15	17	16	16	16	15	16	16	15
Converted flat or maisonette/rooms	6	8	5	5	4	4	5	4	4	4	4	4	4	4	4	3
With business premises/other	2	1	1	1	1	1	0	0	0	0	0	0	0	0	0	0
Weighted base (000's) =100%[4]									*24,398*	*24,806*	*24,520*	*24,421*	*24,378*	*24,644*	*24,780*	*24,780*
Unweighted sample[4]	*11846*	*12041*	*11978*	*9890*	*9917*	*9730*	*9128*	*8615*		*8207*	*8963*	*8581*	*10262*	*8684*	*12778*	*9714*

1 Tables for type of accommodation exclude households living in caravans.
2 2005 data includes last quarter of 2004/5 data due to survey change from financial year to calendar year.
3 Results for 2006 include longitudinal data (see Appendix B).
4 Trend tables show unweighted and weighted figures for 1998 to give an indication of the effect of the weighting. For the weighted data (1998 and 2000 to 2006) the weighted base (000's) is the base for percentages. Unweighted data (up to 1998) are based on the unweighted sample.

Table 4.3 Type of accommodation occupied by households renting from a council compared with other households: 1981 to 2006

Households *Great Britain*

Type of accommodation[1]	Unweighted						Weighted							
	1981	1987	1991	1995	1996	1998	1998	2000	2001	2002	2003	2004	2005[2]	2006[3]
	%	%	%	%	%	%	%	%	%	%	%	%	%	%
Renting from council														
Detached house	1	1	1	0	1	1	1	1	1	0	1	1	1	1
Semi-detached house	30	28	28	28	28	29	27	27	26	27	27	26	26	24
Terraced house	34	35	34	33	31	28	27	31	29	30	30	27	28	28
Purpose-built flat or maisonette	33	34	35	38	38	40	43	40	42	40	41	43	42	45
Converted flat or maisonette	2	2	3	1	2	2	2	2	2	2	2	3	2	1
Weighted base (000's) =100%[1]							*4,021*	*3,870*	*3,695*	*3,404*	*3,075*	*3,312*	*3,044*	*2,862*
Unweighted sample[1]	*4007*	*2600*	*2339*	*1770*	*1748*	*1410*		*1240*	*1325*	*1138*	*1233*	*1101*	*1476*	*1028*
	%	%	%	%	%	%	%	%	%	%	%	%	%	%
Other households														
Detached house	24	25	25	26	25	27	26	24	25	25	25	25	25	26
Semi-detached house	33	33	33	32	33	33	32	32	32	32	33	32	31	31
Terraced house	29	28	28	27	27	26	26	27	27	26	27	27	28	28
Purpose-built flat or maisonette	6	7	8	10	9	10	11	11	11	12	11	12	12	11
Converted flat or maisonette	7	5	5	4	5	4	5	5	4	4	4	4	4	4
Weighted base (000's) =100%[1]							*20,328*	*###*	*20,808*	*20,980*	*21,282*	*21,304*	*21,717*	*21,894*
Unweighted sample[1]	*7904*	*7511*	*7578*	*7953*	*7379*	*7189*		*6954*	*7632*	*7431*	*9020*	*7573*	*11292*	*8675*
	%	%	%	%	%	%	%	%	%	%	%	%	%	%
All households														
Detached house	16	18	19	22	21	23	22	21	21	22	22	22	22	23
Semi-detached house	32	32	32	31	32	33	32	31	31	32	32	31	31	30
Terraced house	31	30	29	28	27	26	26	28	28	27	27	27	28	28
Purpose-built flat or maisonette	15	14	14	15	15	15	17	16	16	16	15	16	16	15
Converted flat or maisonette	5	5	4	4	5	4	4	4	4	4	4	4	4	3
Weighted base (000's) =100%[4]							*24,349*	*###*	*24,503*	*24,384*	*24,357*	*24,616*	*24,761*	*24,755*
Unweighted sample[4]	*11911*	*10111*	*9917*	*9723*	*9127*	*8599*		*8194*	*8957*	*8569*	*10253*	*8674*	*12768*	*9703*

1 Tables for type of accommodation exclude households living in caravans.
2 2005 data includes last quarter of 2004/5 data due to survey change from financial year to calendar year.
3 Results for 2006 include longitudinal data (see Appendix B).
4 Trend tables show unweighted and weighted figures for 1998 to give an indication of the effect of the weighting. For the weighted data (1998 and 2000 to 2006) the weighted base (000's) is the base for percentages. Unweighted data (up to 1998) are based on the unweighted sample.

Table 4.4 (a) Type of accommodation by tenure
(b) Tenure by type of accommodation

Households Great Britain: 2006[1]

Tenure		Type of accommodation[2]							Weighted base (000's) =100%	Unweighted sample
		Detached house	Semi-detached house	Terraced house	All houses	Purpose-built flat or maisonette	Converted flat or maisonette/ rooms	All flats/ rooms		
(a)										
Owner occupied, owned outright	%	36	34	21	91	7	2	9	7,479	3258
Owner occupied, with mortgage	%	27	34	31	91	7	2	9	9,820	3807
All owners	%	31	34	27	91	7	2	9	17,298	7065
Rented from council[3]	%	1	24	28	54	45	1	46	2,862	1028
Rented from housing association[4]	%	1	19	34	54	41	5	46	2,014	737
Social sector tenants	%	1	22	30	54	43	3	46	4,876	1765
Rented privately, unfurnished[5]	%	14	24	34	71	16	13	29	1,885	661
Rented privately, furnished	%	8	13	29	50	30	20	50	697	212
Private renters[6]	%	12	21	32	65	20	15	35	2,582	873
Total	%	23	30	28	81	15	3	19	24,755	9703
(b)		%	%	%	%	%	%	%	Total %	
Owner occupied, owned outright		47	34	23	34	14	15	14	30	
Owner occupied, with mortgage		46	44	44	45	17	24	19	40	
All owners		94	78	66	79	31	38	32	70	
Rented from council[3]		1	9	12	8	34	5	29	12	
Rented from housing association[4]		0	5	10	5	22	12	20	8	
Social sector tenants		1	14	21	13	55	17	48	20	
Rented privately, unfurnished[5]		5	6	9	7	8	29	12	8	
Rented privately, furnished		1	1	3	2	6	16	7	3	
Private renters[6]		5	7	12	8	14	45	19	10	
Weighted base (000's) =100%		5,673	7,500	6,935	20,108	3,791	856	4,647	24,755	
Unweighted sample		2513	3039	2688	8240	1199	264	1463	9703	

1 Results for 2006 include longitudinal data (see Appendix B).
2 Tables for type of accommodation exclude households living in caravans.
3 Council includes local authorities.
4 Since 1996, housing associations are more correctly described as Registered Social Landlords (RSLs).
5 Unfurnished includes the answer 'partly furnished'.
6 Tenants whose accommodation goes with the job of someone in the household have been allocated to 'rented privately'. Squatters are also included

Table 4.5 (a) **Household type by tenure**
(b) **Tenure by household type**

Households *Great Britain: 2006[1]*

Tenure		Household type							Weighted base (000's) =100%	Unweighted sample
		1 adult aged 16-59	2 adults aged 16-59	Small family	Large family	Large adult household	2 adults, 1 or both aged 60 or over	1 adult aged 60 or over		
(a)										
Owner occupied, owned outright	%	7	8	4	1	12	37	30	*7,513*	*3275*
Owner occupied, with mortgage	%	15	23	27	7	21	5	2	*9,828*	*3811*
All owners	%	12	17	17	4	17	19	14	*17,341*	*7086*
Rented from council[2]	%	16	10	21	7	10	11	26	*2,865*	*1029*
Rented from housing association[3]	%	21	8	20	7	9	9	25	*2,014*	*737*
Social sector tenants	%	18	9	21	7	10	10	25	*4,878*	*1766*
Rented privately, unfurnished[4]	%	26	23	19	6	12	5	10	*1,898*	*667*
Rented privately, furnished	%	31	26	11	4	22	2	4	*697*	*212*
Private renters[5]	%	27	23	17	5	15	4	8	*2,595*	*879*
Total	%	15	16	18	5	15	16	16	*24,815*	*9731*
(b)									Total	
		%	%	%	%	%	%	%	%	
Owner occupied, owned outright		15	15	7	9	25	71	58	30	
Owner occupied, with mortgage		41	58	60	52	53	14	5	40	
All owners		56	73	67	61	78	85	63	70	
Rented from council[2]		12	7	14	15	7	8	19	12	
Rented from housing association[3]		12	4	9	12	5	4	13	8	
Social sector tenants		24	11	23	27	12	13	32	20	
Rented privately, unfurnished[4]		14	11	8	9	6	2	5	8	
Rented privately, furnished		6	5	2	2	4	0	1	3	
Private renters[5]		20	15	10	12	10	3	6	10	
Weighted base (000's) =100%		*3,610*	*3,940*	*4,407*	*1,231*	*3,822*	*3,877*	*3,927*	*24,815*	
Unweighted sample		*1151*	*1547*	*1834*	*531*	*1256*	*1844*	*1568*	*9731*	

1 Results for 2006 include longitudinal data (see Appendix B).
2 Council includes local authorities.
3 Since 1996, housing associations are more correctly described as Registered Social Landlords (RSLs).
4 Unfurnished includes the answer 'partly furnished'.
5 Tenants whose accommodation goes with the job of someone in the household have been allocated to 'rented privately'. Squatters are also included in the privately rented category.

Table 4.6   Housing profile by family type: lone-parent families compared with other families

Families with dependent children[1]   *Great Britain: 2005[2] and 2006[3] combined*

	Lone-parent families	Other families
	%	%
Tenure		
Owner occupied, owned outright	5	9
Owner occupied, with mortgage	29	70
Rented from council	32	8
Rented from housing association	21	5
Rented privately unfurnished	11	7
Rented privately furnished	2	1
Central heating	%	%
Yes	97	97
No	3	3
Type of accommodation	%	%
Detached house	7	29
Semi-detached house	30	36
Terraced house	40	29
Purpose-built flat or maisonette	21	6
Converted flat or maisonette/rooms	3	1
Bedroom standard	%	%
2 or more below standard	1	0
1 below standard	8	4
Equals standard	56	32
1 above standard	31	42
2 or more above standard	5	21
Persons per room	%	%
Under 0.5	23	10
0.5-0.99	70	76
1.0-1.49	6	14
1.5 or above	0	0
Unweighted sample[4]	*1533*	*5006*

1   Dependent children are persons aged under 16, or aged 16-18 and in full-time education, and living in the household.
2   2005 data includes last quarter of 2004/5 data due to survey change from financial year to calendar year.
3   Results for 2006 include longitudinal data.
4   Weighted base not shown for combined data sets.

Table 4.7 Type of accommodation by household type

Households *Great Britain: 2006[1]*

Household type		Type of accommodation[2]							Weighted base (000's) =100%	Unweighted sample
		Detached house	Semi-detached house	Terraced house	All houses	Purpose-built flat or maisonette	Converted flat or maisonette/ rooms	All flats/ rooms		
One adult aged 16-59	%	10	19	31	61	30	10	39	3,599	1146
Two adults aged 16-59	%	21	29	30	81	14	5	19	3,934	1544
Small family	%	21	33	33	87	12	1	13	4,400	1831
Large family	%	23	32	35	90	8	2	10	1,231	531
Large adult household	%	30	36	27	93	6	1	7	3,820	1255
Two adults, one or both aged 60 or over	%	36	34	21	91	8	1	9	3,860	1835
One adult aged 60 or over	%	19	28	24	71	26	3	29	3,911	1561
Total	%	23	30	28	81	15	3	19	24,755	9703

1 Results for 2006 include longitudinal data (see Appendix B).
2 Tables for type of accommodation exclude households living in caravans.

Table 4.8 Usual gross weekly income by tenure

Households *Great Britain: 2006[1]*

Usual gross weekly income (£)	Tenure									
	Owners			Social sector tenants			Private renters			Total
	Owned outright	With mortgage	All owners	Council[2]	Housing association[3]	Social sector tenants	Unfurnished private[4]	Furnished private	Private renters[5]	
Income of household reference person										
Mean	370	641	524	185	187	186	456	475	461	449
Lower quartile	131	339	196	101	92	98	160	138	157	148
Median	228	500	387	149	152	150	316	320	316	312
Upper quartile	413	719	621	239	250	244	476	523	480	538
Income of household reference person and partner										
Mean	461	869	693	228	225	227	549	561	552	584
Lower quartile	157	439	256	113	104	110	196	138	180	181
Median	293	689	508	168	175	171	370	358	369	392
Upper quartile	517	1010	846	290	295	290	632	654	633	727
Total household income										
Mean	498	918	737	256	251	254	596	605	599	624
Lower quartile	168	471	280	115	110	114	165	212	200	198
Median	319	733	549	180	189	184	410	383	398	431
Upper quartile	575	1088	921	320	348	323	729	688	700	785
Weighted base (000's) =100%	7,513	9,828	17,341	2,865	2,014	4,878	1,898	697	2,595	24,815
Unweighted sample	2996	3498	6494	990	705	1695	624	186	810	8999

1 Results for 2006 include longitudinal data (see Appendix B).
2 Council includes local authorities.
3 Since 1996, housing associations are more correctly described as Registered Social Landlords (RSLs).
4 Unfurnished includes the answer 'partly furnished'.
5 Tenants whose accommodation goes with the job of someone in the household have been allocated to 'rented privately'. Squatters are also included in the privately rented category.

Table 4.9 (a) Age of household reference person by tenure
(b) Tenure by age of household reference person

Household reference persons *Great Britain: 2006[1]*

Tenure		Age of household reference person[2]								Weighted base (000's) =100%	Unweighted sample
		Under 25	25-29	30-44	45-59	60-64	65-69	70-79	80 and over		
(a)											
Owner occupied, owned outright	%	0	0	5	24	13	15	27	15	7,513	3275
Owner occupied, with mortgage	%	1	8	48	36	4	1	1	0	9,828	3811
All owners	%	1	4	29	31	8	7	12	7	17,341	7086
Rented from council[3]	%	8	6	27	21	7	8	13	10	2,865	1029
Rented from housing association[4]	%	6	7	28	25	5	7	12	12	2,014	737
Social sector tenants	%	7	6	27	22	6	7	13	11	4,878	1766
Rented privately, unfurnished[5]	%	10	16	40	19	4	2	6	4	1,898	667
Rented privately, furnished	%	29	19	32	13	1	2	1	2	697	212
Private renters[6]	%	15	17	38	17	3	2	4	3	2,595	879
Total	%	4	6	30	28	7	7	12	7	24,815	9731
(b)										Total	
		%	%	%	%	%	%	%	%	%	
Owner occupied, owned outright		1	2	5	26	55	67	70	63	30	
Owner occupied, with mortgage		16	49	64	51	23	9	4	2	40	
All owners		17	51	69	78	78	76	75	66	70	
Rented from council[3]		26	11	10	9	11	13	13	16	12	
Rented from housing association[4]		13	9	8	7	5	8	8	13	8	
Social sector tenants		39	20	18	16	17	21	21	30	20	
Rented privately, unfurnished[5]		21	20	10	5	4	2	4	4	8	
Rented privately, furnished		23	9	3	1	0	1	0	1	3	
Private renters[6]		44	29	13	6	5	3	4	5	10	
Weighted base (000's) =100%		886	1,518	7,372	6,871	1,801	1,661	2,875	1,830	24,815	
Unweighted sample		271	517	2683	2747	789	744	1250	730	9731	

1 Results for 2006 include longitudinal data (see Appendix B).

2 Boxed figures indicate median age-groups.

3 Council includes local authorities.

4 Since 1996, housing associations are more correctly described as Registered Social Landlords (RSLs).

5 Unfurnished includes the answer 'partly furnished'.

6 Tenants whose accommodation goes with the job of someone in the household have been allocated to 'rented privately'. Squatters are also included in the privately rented category.

Table 4.10 Tenure by sex and marital status of household reference person

Household reference persons   *Great Britain: 2006[1]*

Tenure	Males						Females						Total
	Married	Cohabiting	Single	Widowed	Divorced/ separated	All males	Married	Cohabiting	Single	Widowed	Divorced/ separated	All females	
	%	%	%	%	%	%	%	%	%	%	%	%	%
Owner occupied, owned outright	36	7	18	66	19	31	28	12	13	61	20	30	30
Owner occupied, with mortgage	49	66	33	6	35	45	52	47	27	5	32	30	40
All owners	85	72	51	72	54	76	80	59	41	66	52	60	70
Rented from council[2]	5	7	11	15	17	8	8	15	26	17	23	18	12
Rented from housing association[3]	3	4	13	10	15	6	6	10	15	12	15	12	8
Social sector tenants	8	11	24	25	32	13	14	25	41	29	38	30	20
Rented privately, unfurnished[4]	5	14	14	3	12	7	4	13	12	5	9	8	8
Rented privately, furnished	1	3	12	1	2	3	1	3	6	1	2	2	3
Private renters[5]	6	17	25	4	14	10	6	17	19	6	11	11	10
Weighted base (000's) =100%	*9,867*	*1,493*	*2,120*	*748*	*1,268*	*15,496*	*1,995*	*817*	*1,989*	*2,304*	*2,192*	*9,298*	*24,794*
Unweighted sample	*4131*	*572*	*658*	*327*	*446*	*6134*	*828*	*312*	*708*	*884*	*857*	*3589*	*9723*

1 Results for 2006 include longitudinal data (see Appendix B).
2 Council includes local authorities.
3 Since 1996, housing associations are more correctly described as Registered Social Landlords (RSLs).
4 Unfurnished includes the answer 'partly furnished'.
5 Tenants whose accommodation goes with the job of someone in the household have been allocated to 'rented privately'. Squatters are also included in the privately rented category.

Table 4.11 Housing tenure by ethnic group of household reference person

Households

Great Britain: 2004, 2005[1], 2006[2] combined

Tenure	Ethnic group											
	White British	Other White	Mixed background	Indian	Pakistani	Bangladeshi	Chinese	Other Asian background	Black Caribbean	Black African	Other Black background	Other ethnic group	Total
	%	%	%	%	%	%	%	%	%	%	%	%	%
Owner occupied, owned outright	32	20	10	23	26	9	18	14	14	2	0	17	30
Owner occupied, with mortgage	40	31	30	48	41	35	26	33	32	29	6	26	39
Rented from council	12	11	26	8	13	31	1	10	34	28	43	18	12
Rented from housing association	7	9	14	4	6	12	4	11	11	16	27	10	8
Rented privately unfurnished	7	17	13	8	6	7	16	9	6	16	14	13	8
Rented privately furnished	2	12	6	11	8	5	35	23	3	9	10	15	3
Unweighted sample[3]	26094	922	178	408	241	65	90	172	341	323	18	192	29044

1 2005 data includes last quarter of 2004/5 data due to survey change from financial year to calendar year
2 Results for 2006 include longitudinal data (see Appendix B).
3 Weighted bases not shown for combined data sets.

Note: Shaded figures indicate the estimates are unreliable and any analysis using these figures may be invalid. Any use of these shaded figures must be accompanied by this disclaimer

Table 4.12 (a) **Socio-economic classification and economic activity status of household reference person by tenure**
             (b) **Tenure by socio-economic classification and economic activity status of household reference person**

Household reference persons *Great Britain: 2006[1]*

Socio-economic classification and economic activity status of household reference person[2]		Tenure										
		Owners			Social sector tenants			Private renters			Total	
		Owned outright	With mortgage	All owners	Council[3]	Housing association[4]	Social sector tenants	Unfurnished private[5]	Furnished private	Private Renters[6]		
(a)		%	%	%	%	%	%	%	%	%	%	
Economically active HRP:												
Large employers and higher managerial		3	12	8	0	1	0	3	6	4	6	
Higher professional		5	12	9	0	1	0	9	13	10	7	
Lower managerial and professional		8	29	20	5	7	5	21	16	20	17	
Intermediate		3	8	6	3	4	4	5	10	6	6	
Small employers and own account		6	9	8	2	3	3	10	6	9	7	
Lower supervisory and technical		3	10	7	3	5	4	8	3	7	6	
Semi-routine		4	7	5	11	10	11	11	10	11	7	
Routine		3	7	5	9	7	8	10	7	9	6	
Never worked and long-term unemployed		0	0	0	2	2	2	1	2	1	1	
Economically inactive HRP		66	7	33	63	60	62	23	26	23	38	
Weighted base (000's) =100%		7,441	9,585	17,027	2,826	1,968	4,794	1,819	591	2,410	24,230	
Unweighted sample		3245	3721	6966	1016	724	1740	640	181	821	9527	
(b)											Weighted base (000's)	Unweighted sample
Economically active HRP:												
Large employers and higher manageri	%	14	77	92	1	1	2	4	2	6	1,477	598
Higher professional	%	20	65	85	0	1	1	9	5	14	1,711	675
Lower managerial and professional	%	14	68	82	3	3	6	9	2	12	4,090	1583
Intermediate	%	17	59	76	7	6	13	6	5	11	1,350	510
Small employers and own account	%	26	54	79	4	4	7	11	2	13	1,638	620
Lower supervisory and technical	%	14	62	76	6	7	13	10	1	11	1,508	558
Semi-routine	%	17	38	54	18	12	30	12	4	16	1,681	639
Routine	%	16	43	59	18	10	27	12	3	14	1,501	538
Never worked and long-term unemployed	%	12	2	14	44	29	73	6	7	13	152	49
Economically inactive HRP	%	54	7	61	20	13	33	4	2	6	9,122	3757
Total	%	31	40	70	12	8	20	8	2	10	24,230	9527

1 Results for 2006 include longitudinal data (see Appendix B).
2 Full-time students are classified as economically inactive.
3 Council includes local authorities.
4 Since 1996, housing associations are more correctly described as Registered Social Landlords (RSLs).
5 Unfurnished includes the answer 'partly furnished'.
6 Tenants whose accommodation goes with the job of someone in the household have been allocated to 'rented privately'. Squatters are also included in the privately rented category.

Note: Shaded figures indicate the estimates are unreliable and any analysis using these figures may be invalid. Any use of these shaded figures must be accompanied by this disclaimer.

Table 4.13 (a) **Length of residence of household reference person by tenure**
(b) **Tenure by length of residence of household reference person**

Household reference persons   *Great Britain: 2006[1]*

Length of residence[2] (years)	Tenure									Total	
	Owners			Social sector tenants			Private renters				
	Owned outright	With mortgage	All owners	Council[3]	Housing association[4]	Social sector tenants	Unfurnished private[5]	Furnished private	Private renters[6]		
(a)	%	%	%	%	%	%	%	%	%	%	
Less than 12 months	1	3	2	3	3	3	11	☐18	13	3	
12 months but less than 2 years	2	7	5	9	10	9	27	☐38	30	8	
2 years but less than 3 years	2	8	6	8	8	8	☐18	17	☐18	8	
3 years but less than 5 years	5	17	12	13	14	14	18	17	18	13	
5 years but less than 10 years	14	☐27	21	☐23	☐26	☐24	12	5	10	☐21	
10 years or more	☐76	37	☐54	43	40	42	14	5	12	47	
Weighted base (000's) =100%	*7,499*	*9,826*	*17,325*	*2,859*	*2,012*	*4,872*	*1,889*	*691*	*2,580*	*24,776*	
Unweighted sample	*3269*	*3810*	*7079*	*1027*	*736*	*1763*	*664*	*209*	*873*	*9715*	
(b)										*Weighted base (000's) =100%*	*Unweighted sample*
Less than 12 months	% 8	36	44	11	7	18	24	14	38	*862*	*335*
12 months but less than 2 years	% 6	35	41	12	10	22	24	13	37	*2,060*	*720*
2 years but less than 3 years	% 10	45	55	12	8	21	19	6	25	*1,873*	*663*
3 years but less than 5 years	% 12	52	65	12	9	21	11	4	14	*3,165*	*1194*
5 years but less than 10 years	% 20	52	72	13	10	23	4	1	5	*5,156*	*2014*
10 years or more	% 49	31	80	11	7	17	2	0	3	*11,660*	*4789*
Total	% 30	40	70	12	8	20	8	3	10	*24,776*	*9715*

1 Results for 2006 include longitudinal data (see Appendix B).
2 Boxed figures indicate median length of residence.
3 Council includes local authorities.
4 Since 1996, housing associations are more correctly described as Registered Social Landlords (RSLs).
5 Unfurnished includes the answer 'partly furnished'.
6 Tenants whose accommodation goes with the job of someone in the household have been allocated to 'rented privately'. Squatters are also included in the privately rented category.
Note: Derivation changed in 2005 to reflect the year the accommodation was purchased/commencement of rental agreement

Table 4.14 Persons per room: 1971 to 2006[1]

Households   *Great Britain*

Persons per room	Unweighted									Weighted							
	1971	1975	1981	1985	1991	1995	1996	1998		1998	2000	2001	2002	2003	2004	2005	2006
	%	%	%	%	%	%	%	%		%	%	%	%	%	%	%	%
Under 0.5	37	39	42	45	50	52	51	55		55	57	57	57	58	58	58	58
0.5 to 0.65	25	25	25	26	24	25	24	23		23	22	22	23	22	22	23	23
0.66 to 0.99	24	23	23	21	19	18	19	18		18	16	16	16	16	16	15	15
1	9	8	7	6	5	5	5	4		4	4	4	3	3	3	3	3
Over 1 to 1.5	4	3	2	1	1	1	1	1		1	1	1	1	1	1	1	1
Over 1.5	1	0	0	0	0	0	0	0		0	0	0	0	0	0	0	0
Weighted base (000's) =100%[2]										*24,450*	*24,845*	*24,592*	*24,529*	*24,423*	*24,688*	*24,829*	*24,813*
Unweighted sample[2]	*11990*	*12096*	*12002*	*9982*	*9646*	*9754*	*9154*	*8636*			*8221*	*8989*	*8620*	*10283*	*8700*	*12802*	*9730*
Mean persons per room	..	0.57	0.56	0.52	0.50	0.48	0.49	0.47		0.46	0.45	0.46	0.45	0.45	0.45	0.45	0.45

1 Results for 2006 include longitudinal data (see Appendix B).
2 Trend tables show unweighted and weighted figures for 1998 to give an indication of the effect of the weighting. For the weighted data (1998 and 2000 to 2006) the weighted base (000's) is the base for percentages. Unweighted data (up to 1998) are based on the unweighted sample.

Note: 2005 data includes last quarter of 2004/5 data due to survey change from financial year to calendar year.

Table 4.15 Persons per room and mean household size by tenure

Households Great Britain: 2006[1]

Persons per room[1]	Tenure									Total
	Owners			Social sector tenants			Private renters			
	Owned outright	With mortgage	All owners	Council[2]	Housing association[3]	Social sector tenants	Unfurnished private[4]	Furnished private	Private renters[5]	
	%	%	%	%	%	%	%	%	%	%
Under 0.5	80	46	61	51	56	53	51	40	48	58
0.5 to 0.65	14	29	22	23	21	22	25	27	26	23
0.66 to 0.99	5	21	14	18	14	17	18	24	19	15
1	1	3	2	5	7	6	5	6	5	3
Over 1	0	1	1	3	2	3	2	2	2	1
Weighted base (000's) =100%	7,513	9,828	17,341	2,865	2,014	4,878	1,896	697	2,593	24,813
Unweighted sample	3275	3811	7086	1029	737	1766	666	212	878	9730
Mean persons per room	0.34	0.49	0.43	0.49	0.49	0.49	0.49	0.52	0.50	0.45
Mean household size	1.93	2.78	2.41	2.14	2.10	2.12	2.24	2.34	2.27	2.34

1 Results for 2006 include longitudinal data (see Appendix B).
2 Boxed figures indicate median density of occupation.
3 Council includes local authorities.
4 Since 1996, housing associations are more correctly described as Registered Social Landlords (RSLs).
5 Unfurnished includes the answer 'partly furnished'.
6 Tenants whose accommodation goes with the job of someone in the household have been allocated to 'rented privately'. Squatters are also included in the privately rented category.

Table 4.16 Closeness of fit relative to the bedroom standard by tenure

Households Great Britain: 2006[1]

Difference from bedroom standard (bedrooms)	Tenure									Total
	Owners			Social sector tenants			Private renters			
	Owned outright	With mortgage	All owners	Council[2]	Housing association[3]	Social sector tenants	Unfurnished private[4]	Furnished private	Private renters[5]	
	%	%	%	%	%	%	%	%	%	%
1 or more below standard	1	2	1	5	5	5	2	6	3	2
Equals standard	10	19	15	52	58	54	39	48	41	25
1 above standard	32	42	38	30	28	29	41	25	37	36
2 or more above standard	57	37	46	13	10	12	18	20	18	36
Weighted base (000's) =100%	7,513	9,828	17,341	2,865	2,014	4,878	1,898	697	2,595	24,815
Unweighted sample	3275	3811	7086	1029	737	1766	667	212	879	9731

1 Results for 2006 include longitudinal data (see Appendix B).
2 Council includes local authorities.
3 Since 1996, housing associations are more correctly described as Registered Social Landlords (RSLs).
4 Unfurnished includes the answer 'partly furnished'.
5 Tenants whose accommodation goes with the job of someone in the household have been allocated to 'rented privately'. Squatters are also included in the privately rented category.

Table 4.17 Cars or vans: 1972 to 2006

Households Great Britain

Cars or vans	Unweighted								Weighted							
	1972	1975	1981	1985	1991	1995	1996	1998	1998	2000	2001	2002	2003	2004	2005[1]	2006[2]
	%	%	%	%	%	%	%	%	%	%	%	%	%	%	%	%
Households with:																
no car or van	48	44	41	38	32	29	30	28	28	27	28	27	26	27	25	23
one car or van	43	45	44	45	44	45	46	44	45	45	44	45	45	45	45	44
two cars or vans	8	10	12	14	19	22	21	23	22	22	23	22	24	24	25	26
three or more cars or vans	1	1	2	3	4	4	4	6	6	6	5	5	5	5	6	7
Weighted base (000's) =100%[3]									24,450	24,845	24,592	24,529	24,423	24,688	24,829	24,815
Unweighted sample[3]	11624	11929	11989	9963	9910	9758	9158	8636		8221	8989	8620	10283	8700	12802	9731

1 2005 data includes last quarter of 2004/5 data due to survey change from financial year to calendar year.
2 Results for 2006 include longitudinal data (see Appendix B).
3 Trend tables show unweighted and weighted figures for 1998 to give an indication of the effect of the weighting. For the weighted data (1998 and 2000 to 2006) the weighted base (000's) is the base for percentages. Unweighted data (up to 1998) are based on the unweighted sample.

Table 4.18 Availability of a car or van by socio-economic classification of household reference person

Households
Great Britain: 2006[1]

Socio-economic classification of household reference person[2]		Number of cars or vans available to household			Weighted base (000's) =100%	Unweighted sample
		None	1	2 or more		
Economically active HRP						
Large employers and higher managerial	%	3	36	61	1,477	598
Higher professional	%	7	38	55	1,711	675
Lower managerial and professional	%	8	45	47	4,090	1583
Intermediate	%	14	50	36	1,350	510
Small employers and own account	%	4	37	59	1,638	620
Lower supervisory and technical	%	9	45	45	1,508	558
Semi-routine	%	25	47	29	1,681	639
Routine	%	23	48	29	1,501	538
Never worked and long-term unemployed	%	69	27	4	152	49
Economically inactive HRP	%	43	45	12	9,122	3757
Total	%	23	44	33	24,230	9527

1 Results for 2006 include longitudinal data (see Appendix B).
2 Full-time students are classed as economically inactive.

Note: Shaded figures indicate the estimates are unreliable and any analysis using these figures may be invalid. Any use of these shaded figures must be accompanied by this disclaimer.

Table 4.19 Consumer durables, central heating and cars: 1975 to 2006

Households
Great Britain

										Weighted						
	1972	1975	1981	1985	1991	1993	1995	1996	1998	1998	2000	2001	2002	2003	2004	2005[1]	2006[2]
Percentage of households with:																	
Colour television	93	96	74	86	95	95	97	97	98	97	98	98	99	*	*	99	99
Home computer	*	*	*	13	21	24	25	27	34	34	45	50	54	58	60	63	69
Washing machine	66	71	78	81	87	88	90	90	92	91	93	92	93	94	95	95	95
Telephone	42	54	75	81	88	90	93	94	96	96	98	98	99	99	99	99	99
fixed telephone	**	**	**	**	**	**	**	**	**	**	93	93	92	92	92	92	92
mobile telephone	**	**	**	**	**	**	**	**	**	**	58	70	75	76	79	80	83
Central heating	37	43	59	69	82	83	86	88	90	90	92	92	93	93	94	95	95
Car or van	52	56	59	62	67	68	71	70	72	72	73	72	73	74	73	75	77
one	43	45	44	45	44	45	45	46	44	45	45	44	45	45	45	45	44
more than one	9	11	14	17	23	23	26	24	28	27	28	28	28	29	29	30	33
Weighted base (000's) = 100%[3]										24,450	24,575	24,592	24,529	24,423	24,688	24,829	24,815
Unweighted sample[1]	11663	11929	11718	9993	9955	9850	9757	9156	8636		8213	8989	8618	10283	8700	12802	9731

1 2005 data includes last quarter of 2004/5 data due to survey change from financial year to calendar year.
2 Results for 2006 include longitudinal data (see Appendix B).
3 Trend tables show unweighted and weighted figures for 1998 to give an indication of the effect of the weighting. For the weighted data (1998 and 2000 to 2006) the weighted base (000's) is the base for percentages. Unweighted data (up to 1998) are based on the unweighted sample.

* Data are not available (consumer durable item not on GHS list that year).
** Data only available for 2000 to 2006. Percentages for fixed and mobile telephones sum to greater than 100 because many households owned both.

Table 4.20 Consumer durables, central heating and cars by socio-economic classification of household reference person

Household reference persons
Great Britain: 2006[1]

Consumer durables	Socio-economic classification of household reference person[2]								Total
	Economically active							Economically inactive	
	Large employers and higher managerial	Higher professional	Lower managerial and professional	Intermediate	Small employers and own account	Lower supervisory and technical	Semi-routine	Routine	
Percentage of households with:									
Television	99	98	99	99	99	99	99	99	99	99
Home computer	94	95	90	85	87	82	70	67	43	69
Washing machine	99	99	99	99	97	98	96	96	91	95
Telephone (fixed or mobile)	100	100	100	100	100	100	99	99	99	99
fixed telephone[3]	97	96	96	93	94	94	87	83	91	92
mobile telephone[3]	94	95	94	91	94	95	90	87	67	83
Central heating	98	97	97	96	94	95	94	93	95	95
Car or van - more than one	61	55	47	36	59	45	29	29	12	33
Weighted base (000's) = 100%	*1,477*	*1,711*	*4,090*	*1,350*	*1,638*	*1,508*	*1,681*	*1,501*	*9,274*	*24,230*
Unweighted sample	*598*	*675*	*1583*	*510*	*620*	*558*	*639*	*538*	*3806*	*9527*

1 Results for 2006 include longitudinal data (see Appendix B).
2 Where the household reference person had never worked or was long-term unemployed these are not shown as a separate category, but are included in the total. Full-time students are classified as economically inactive.
3 Percentages for fixed and mobile telephones sum to greater than 100 because many households owned both.

Table 4.21 Consumer durables, central heating and cars by usual gross weekly household income

Households   *Great Britain: 2006[1]*

Consumer durables	Usual gross weekly household income (£)							
	0.00-100	100.01-200	200.01-300	300.01-400	400.01-500	500.01-700	700.01 or more	Total
Percentage of households with:								
Television	96	99	99	99	99	99	99	99
Home computer	43	35	53	63	73	84	94	68
Washing machine	85	88	95	95	98	99	100	95
Telephone	97	98	99	100	100	100	100	99
fixed telephone[2]	80	87	91	91	93	94	98	92
mobile telephone[2]	65	61	77	85	90	93	95	83
Central heating	92	93	94	94	95	96	98	95
Car or van - more than one	10	7	10	16	28	37	64	32
Weighted base (000's) =100%	*2,183*	*3,635*	*2,743*	*2,250*	*1,994*	*3,163*	*6,904*	*22,871*
Unweighted sample	*795*	*1428*	*1109*	*897*	*787*	*1235*	*2692*	*8943*

1 Results for 2006 include longitudinal data (see Appendix B).
2 Percentages for fixed and mobile telephones sum to greater than 100 because many households owned both.

Table 4.22 Consumer durables, central heating and cars by household type

Households   *Great Britain: 2006[1]*

Consumer durables	Household type							
	1 adult aged 16-59	2 adults aged 16-59	Small family	Large family	Large adult household	2 adults, 1 or both aged 60 or over	1 adult aged 60 or over	Total
Percentage of households with:								
Television	96	99	99	100	100	100	99	99
Home computer	66	85	86	88	92	56	22	69
Washing machine	90	99	100	99	99	98	84	95
Telephone (fixed or mobile)	98	100	100	100	100	100	99	99
fixed telephone[2]	80	91	90	92	96	99	96	92
mobile telephone[2]	88	92	94	93	93	80	47	83
Central heating	91	96	97	98	97	96	94	95
Car or van - more than one	7	50	41	44	63	27	2	33
Weighted base (000's) =100%	*3,610*	*3,940*	*4,407*	*1,231*	*3,822*	*3,877*	*3,927*	*24,815*
Unweighted sample	*1151*	*1547*	*1834*	*531*	*1256*	*1844*	*1568*	*9731*

1 Results for 2006 include longitudinal data (see Appendix B).
2 Percentages for fixed and mobile telephones sum to greater than 100 because many households owned both.

Table 4.23 Consumer durables, central heating and cars: lone-parent families compared with other families

Families with dependent children[1]   *Great Britain: 2005[2] and 2006[3] combined*

Consumer durables	Lone-parent families	Other families
Percentage of households with:		
Television	99	99
Home computer	71	91
Washing machine	98	100
Telephone (fixed or mobile)	99	100
fixed telephone[4]	80	96
mobile telephone[4]	91	94
Central heating	97	97
Car or van - one or more	58	94
Unweighted sample[5]	*1533*	*5006*

1 Dependent children are persons aged under 16, or aged 16-18 and in full time education, and living in the household.
2 2005 data includes last quarter of 2004/5 data due to survey change from financial year to calendar year.
3 Results for 2006 include longitudinal data (see Appendix B).
4 Percentages for fixed and mobile telephones sum to greater than 100 because many households owned both.
5 Weighted base not shown for combined data sets.

Table 5.1 Sex by marital status

All persons aged 16 and over *Great Britain: 2006[1]*

Marital status[2]	Men	Women
	%	%
Married	52	50
Civil Partnership[3]	0	0
Cohabiting	10	10
Single	27	21
Widowed	3	10
Divorced	5 ⎤ 6	8 ⎤ 10
Separated	2 ⎦	2 ⎦
Weighted base	*22,779*	*24,130*
Unweighted s	*8681*	*9533*

1 Results for 2006 include longitudinal data (see Appendix B).
2 Marital status as recorded at the beginning of the interview.
3 Since December 2005 same sex couples have been able to obtain legal recognition of their partnership by registering as Civil Partners
4 Total includes a very small number of same sex cohabitees.

Table 5.2 (a) **Age by sex and marital status**
(b) **Marital status by sex and age**

Persons aged 16 and over   *Great Britain: 2006[1]*

Age	Marital status[2]						
	Married[3]	Cohabiting	Single	Widowed	Divorced	Separated	Total[4]
(a)	%	%	%	%	%	%	%
Men							
16-24	1	10	51	0	0	2	15
25-34	11	39	22	1	5	12	16
35-44	23	28	13	1	19	22	20
45-54	21	12	6	1	31	24	16
55-64	22	7	5	14	28	25	15
65-74	14	2	2	27	13	11	10
75 and over	9	1	1	56	5	4	7
Weighted base (000's) =100%	*11,949*	*2,348*	*6,201*	*795*	*1,037*	*354*	*22,779*
Unweighted sample	*4995*	*896*	*1916*	*347*	*365*	*133*	*8681*
	%	%	%	%	%	%	%
Women							
16-24	2	20	54	0	0	3	14
25-34	14	39	20	0	6	13	16
35-44	23	22	12	2	24	32	19
45-54	22	11	5	3	28	29	16
55-64	20	5	3	9	23	13	14
65-74	13	2	3	25	12	6	11
75 and over	6	1	3	61	6	3	10
Weighted base (000's) =100%	*11,921*	*2,348*	*4,977*	*2,465*	*1,822*	*528*	*24,130*
Unweighted sample	*4989*	*896*	*1756*	*949*	*715*	*205*	*9533*
	%	%	%	%	%	%	%
Total							
16-24	1	15	52	0	0	2	15
25-34	12	39	21	0	5	13	16
35-44	23	25	12	2	22	28	19
45-54	22	12	6	3	29	27	16
55-64	21	6	4	11	25	18	15
65-74	13	2	2	26	13	8	10
75 and over	7	1	2	60	6	4	9
Weighted base (000's) =100%	*23,870*	*4,696*	*11,178*	*3,261*	*2,859*	*882*	*46,909*
Unweighted sample	*9984*	*1792*	*3672*	*1296*	*1080*	*338*	*18214*

(b)									Weighted base (000's) =100%[3]	Unweighted sample[3]
Men										
16-24	%	2	7	90	0	0	0		*3,487*	*1048*
25-34	%	34	25	37	0	1	1		*3,684*	*1241*
35-44	%	61	15	18	0	4	2		*4,510*	*1632*
45-54	%	70	8	11	0	9	2		*3,614*	*1427*
55-64	%	73	5	8	3	8	2		*3,528*	*1516*
65-74	%	75	3	5	9	6	2		*2,294*	*1041*
75 and over	%	62	2	6	27	3	1		*1,662*	*776*
Total	%	52	10	27	3	5	2		*22,779*	*8681*

Table 5.2   (a)   **Age by sex and marital status**
                (b)   **Marital status by sex and age**

Persons aged 16 and over   *Great Britain: 2006*[1]

Age	Marital status[2]								
		Married[3]	Cohabiting	Single	Widowed	Divorced	Separated	Total[4]	
Women									
16-24	%	6	14	80	0	0	0	*3,369*	*1112*
25-34	%	44	24	27	0	3	2	*3,784*	*1416*
35-44	%	61	11	13	1	10	4	*4,557*	*1770*
45-54	%	68	7	7	2	13	4	*3,949*	*1582*
55-64	%	71	3	4	7	12	2	*3,418*	*1545*
65-74	%	59	2	5	24	9	1	*2,553*	*1153*
75 and over	%	28	1	6	60	4	1	*2,499*	*955*
Total	%	50	10	21	10	8	2	*24,130*	*9533*
Total									
16-24	%	4	10	85	0	0	0	*6,856*	*2160*
25-34	%	40	25	32	0	2	1	*7,469*	*2657*
35-44	%	61	13	15	1	7	3	*9,067*	*3402*
45-54	%	69	7	9	1	11	3	*7,563*	*3009*
55-64	%	72	4	6	5	10	2	*6,946*	*3061*
65-74	%	66	2	5	17	7	1	*4,847*	*2194*
75 and over	%	41	1	6	47	4	1	*4,162*	*1731*
Total	%	51	10	24	7	6	2	*46,909*	*18214*

1 Results for 2006 include longitudinal data (see Appendix B).
2 Marital status as recorded at the beginning of the interview.
3 Married includes persons in a legally recognised Civil Partnership.
4 Total includes a very small number of same sex cohabitees.

Table 5.3 Percentage currently cohabiting by sex and age

Men and women aged 16-59 Great Britain: 2006[1]

Age	All	Non-married[1]	Weighted base (000's) =100% All	Non-married[2]	Unweighted sample All	Non-married[2]
Men		Percentage cohabiting				
16-19	1	1	1,345	1,343	406	405
20-24	17	15	1,244	1,142	369	335
25-29	32	37	1,381	968	454	313
30-34	22	39	1,717	833	588	270
35-39	17	37	1,888	766	683	252
40-44	11	28	1,972	643	711	201
45-49	9	22	1,684	565	663	203
50-54	6	23	1,525	371	607	137
55-59	5	16	1,753	446	744	165
Total	13	23	14,508	7,077	5225	2281
Women						
16-19	3	2	1,317	1,302	437	432
20-24	27	27	1,543	1,330	505	435
25-29	29	41	1,633	996	602	374
30-34	21	46	1,831	782	699	302
35-39	12	29	2,060	750	824	287
40-44	11	26	2,252	813	847	302
45-49	8	22	1,998	626	788	242
50-54	5	16	1,746	514	712	203
55-59	4	12	1,782	496	781	200
Total	13	25	16,163	7,609	6195	2777

1 Results for 2006 include longitudinal data (see Appendix B).
2 Men and women describing themselves as 'separated' were in a legal sense still married. However, because the separated can cohabit, they have been included in the 'non-married' category.

Table 5.4 Percentage currently cohabiting by legal marital status and age

Men and women aged 16-59 *Great Britain: 2005[1] and 2006[2] combined*

Legal marital status[3]	16-24	25-34	35-49	50-59	Total	Unweighted sample[4]				
						16-24	25-34	35-49	50-59	Total
Men		Percentage cohabiting								
Married[5]	-	-	-	-	-	56	936	2854	2091	5937
Non-married										
Single	9	38	29	12	22	1584	1247	868	255	3954
Widowed	..	26	0	10	10	..	3	4	32	39
Divorced	100	49	37	31	36	1	57	420	303	781
Separated	58	24	28	11	22	2	32	118	73	225
Non-married total	9	38	31	20	24					
Total	8	23	11	5	12	1643	2275	4264	2754	10936
Women										
Married[5]	-	-	-	-	-	145	1309	3278	2180	6912
Non-married										
Single	17	43	32	16	27	1502	1241	824	146	4016
Widowed	..	0	12	5	7	..	3	51	138	192
Divorced	23	42	27	18	25	9	129	685	463	1286
Separated	0	9	12	7	10	10	86	209	109	414
Non-married total	17	41	27	14	25					
Total	15	22	10	4	12	1969	2768	5047	3036	12820

1 Note: 2005 data includes last quarter of 2004/05 data due to survey change from financial year to calendar year.
2 Results for 2006 include longitudinal data (see Appendix B).
3 Men and women describing themselves as 'separated' were in a legal sense still married. However, because the separated can cohabit they have been included in the 'non-married' category.
4 Weighted bases not shown for combined data sets.
5 Married includes persons in a legally recognised Civil Partnership.
.. Data are not available (base = 0).
Note: Shaded figures indicate the estimates are unreliable and any analysis using these figures may be invalid. Any use of these shaded figures must be accompanied by this disclaimer.

Table 5.5 Cohabitees: age by legal marital status

Cohabiting persons aged 16-59 *Great Britain: 2005[1] and 2006[2] combined*

Legal marital status[3]	16-24	25-34	35-49	50-59	Total
	%	%	%	%	%
Men					
Married[4]	-	-	-	-	-
Non-married					
Single	99	93	59	23	74
Widowed	0	0	0	2	0
Divorced	0	5	34	68	21
Separated	1	1	7	6	4
Unweighted sample[5]	147	558	508	163	1376
Women					
Married	-	-	-	-	-
Non-married					
Single	99	91	55	19	76
Widowed	0	0	1	5	1
Divorced	1	8	38	69	21
Separated	0	1	5	6	3
Unweighted sample[5]	304	600	510	141	1555

1 2005 data includes last quarter of 2004/05 data due to survey change from financial year to calendar year.
2 Results for 2006 include longitudinal data (see Appendix B).
3 Men and women describing themselves as 'separated' were in a legal sense still married. However, because the separated can cohabit they have been included in the 'non-married' category.
4 Married includes persons in a legally recognised Civil Partnership.
5 Weighted bases not shown for combined data sets.

Table 5.6 Cohabitees: age by sex

Cohabiting persons aged 16-59   *Great Britain: 2006[1]*

Age	Men	Women
	%	%
16-19	1	2
20-24	11	19
25-29	23	22
30-34	20	18
35-39	17	12
40-44	11	11
45-49	8	8
50-54	5	4
55-59	5	3
Weighted base (000's) =100%	1,913	2,114
Unweighted sample	722	799

1 Results for 2006 include longitudinal data (see Appendix B).

Table 5.7 Legal marital status of women aged 18-49: 1979 to 2006

Women aged 18-49   *Great Britain*

Legal marital status[1]	Unweighted									Weighted							
	1979	1983	1985	1989	1991	1993	1995	1996	1998	1998	2000	2001	2002	2003	2004	2005[2]	2006[3]
	%	%	%	%	%	%	%	%	%	%	%	%	%	%	%	%	%
Married[4]	74	70	68	63	61	59	58	57	53	53	51	50	49	47	48	47	50
Non-married																	
Single	18	21	22	26	26	28	28	29	30	32	35	36	38	38	39	39	38
Widowed	1	1	1	1	1	1	1	1	1	1	0	1	0	1	1	1	1
Divorced	4	6	6	7	8	9	9	9	11	10	9	9	9	10	9	10	8
Separated	3	2	3	3	3	4	4	4	5	4	5	4	4	4	3	3	3
Weighted base (000's) =100%[5]										11,827	11,946	11,689	11,752	11,466	11,630	11,541	11,909
Unweighted sample[5]	6006	5285	5364	5483	5359	5171	4953	4695	4181		3979	4325	4092	4821	4139	5874	4453

1 Women describing themselves as 'separated' were in a legal sense still married. However, because the separated can cohabit they have been included in the 'non-married' category.
2 2005 data includes last quarter of 2004/05 data due to survey change from financial year to calendar year.
3 Results for 2006 include longitudinal data (see Appendix B).
4 Married includes persons in a legally recognised Civil Partnership.
5 Trend tables show unweighted and weighted figures for 1998 to give an indication of the effect of the weighting. For the weighted data (1998 and 2000 to 2006) the weighted base (000's) is the base for percentages. Unweighted data (up to 1998) are based on the unweighted sample.

Table 5.8 Percentage of women aged 18-49 cohabiting by legal marital status: 1979 to 2006

Women aged 18-49   Great Britain

Legal marital status	Unweighted								Weighted								
	1979	1985	1991	1993	1995	1996	1998		1998	2000	2001	2002	2003	2004	2005[2]	2006[3]
						Percentage cohabiting											
Married[4]	-	-	-	-	-	-	-		-	-	-	-	-	-	-	-
Non-married																	
Single	8	14	23	23	26	28	31		32	31	35	31	33	29	31	31
Widowed	0	5	2	8	8	5	8		11	15	11	13	6	13	10	10
Divorced	20	21	30	25	27	31	31		32	35	33	32	35	32	31	28
Separated	17	20	13	11	11	7	12		12	11	14	15	14	9	10	10
All non-married	11	16	23	22	25	26	29		30	30	32	29	31	28	29	29
Total	3	5	9	9	10	11	13		14	15	16	15	16	15	16	14
weighted bases (000's) =100%[5]																	
Married									6,212	6,051	5,899	5,727	5,445	5,540	5,430	5,986
Non-married																	
Single									3,760	4,176	4,155	4,458	4,331	4,580	4,539	4,571
Widowed									99	55	99	52	63	98	77	84
Divorced									1,229	1,120	1,023	1,029	1,171	1,060	1,097	933
Separated									528	544	513	486	456	353	397	334
Total									11,828	11,946	11,689	11,752	11,466	11,630	11,541	11,909
Unweighted sample[5]																	
Married[4]	4461	3653	3265	3053	2864	2683	2234		2032	2176	2050	2398	2040	2919	2311
Non-married																	
Single	1061	1175	1416	1431	1405	1361	1268		1342	1523	1490	1720	1561	2166	1614
Widowed	61	55	55	49	40	44	36		20	37	18	28	34	34	31
Divorced	256	338	448	453	437	421	443		393	389	363	484	376	551	366
Separated	167	143	175	185	206	186	200		192	200	171	191	128	204	131
Total	6006	5364	5359	5171	4952	4695	4181		3979	4325	4092	4821	4139	5874	4450

1 Men and women describing themselves as 'separated' were in a legal sense still married. However, because the separated can cohabit they have been included in the 'non-married' category.
2 2005 data includes last quarter of 2004/5 data due to survey change from financial year to calendar year.
3 Results for 2006 include longitudinal data (see Appendix B)
4 Married includes persons in a legally recognised Civil Partnership.
5 Trend tables show unweighted and weighted figures for 1998 to give an indication of the effect of the weighting. For the weighted data (1998 and 2000 to 2006) the weighted base (000's) is the base for percentages. Unweighted data (up to 1998) are based on the unweighted sample.

Table 5.9   Women aged 16-59:
(a) Whether has dependent children in the household by marital status
(b) Marital status by whether has dependent children in the household

Women aged 16-59   *Great Britain: 2006[1]*

Marital status		Children			Weighted base (000's) =100%	Unweighted sample
		Dependent children	Non-dependent children only	No children		
(a)						
Married[2]	%	53	15	32	8,166	3278
Non-married						
Cohabiting	%	39	2	59	2,216	836
Single	%	19	1	79	3,937	1378
Widowed	%	25	19	56	207	82
Divorced	%	39	18	43	1,146	438
Separated	%	60	12	28	416	158
Total	%	42	10	48	16,089	6170
(b)						
Married[2]		65	75	34	51	
Non-married						
Cohabiting		13	3	17	14	
Single		11	3	40	24	
Widowed		1	2	1	1	
Divorced		7	13	6	7	
Separated		4	3	1	3	
Weighted base (000's) =100%		6,718	1,586	7,784	16,089	
Unweighted sample		2666	515	2989	6170	

1 Results for 2006 include longitudinal data (see Appendix B).
2 Married includes persons in a legally recognised Civil Partnership.
Note: Shaded figures indicate the estimates are unreliable and any analysis using these figures may be invalid. Any use of these shaded figures must be accompanied by this disclaimer.

Table 5.10

Women aged 16-59   *Great Britain: 2006[1]*

Legal marital status	Has dependent children	No dependent children	Total	Weighted bases (000's) =100%			Unweighted sample		
				Has dependent children	No dependent children	Total[2]	Has dependent children	No dependent children	Total[2]
	Percentage cohabiting								
Married[3]	-	-	-	4,422	3,996	8,491	1768	1604	3396
Non-married									
Single	45	21	27	1,311	4,054	5,416	503	1395	1917
Widowed	13	5	7	60	163	223	21	69	90
Divorced	26	34	23	21	24	25	608	899	1,513	251	339	592
Separated	9	8	9	275	181	456	109	69	178
Total	12	12	12	6,676	9,293	16,100	2652	3476	6173

1 Results for 2006 include longitudinal data (see Appendix B).
2 Totals include a small number of children for whom dependency could not be established.
3 Married includes persons in a legally recognised Civil Partnership.
Note: Shaded figures indicate the estimates are unreliable and any analysis using these figures may be invalid. Any use of these shaded figures must be accompanied by this disclaimer.

Table 5.11 Cohabiting women aged 16-59: whether has dependent children in the household by legal marital status

Cohabiting women aged 16-59   *Great Britain: 2006[1]*

Legal marital status	Has dependent children	No dependent children	Total[2]
	%	%	%
Non-married			
Single	75	81	78
Widowed	1	1	1
Divorced	20	17	19
Separated	3	1	2
Weighted base (000's) = 100%	*776*	*1,052*	*1,865*
Unweighted sample	*307*	*386*	*707*

1 Results for 2006 include longitudinal data (see Appendix B).
2 Totals include a small number of children for whom dependency could not be established.

Table 5.12   Number of past cohabitations not ending in marriage by sex and age

Men and women aged 16-59   *Great Britain: 2006[1]*

Age		Number of completed cohabitations[2]					Weighted base (000's) =100%	Unweighted sample
		None	One	Two	Three or more	Total at least one		
Men								
16-19	%	98	2	0	0	2	*999*	*302*
20-24	%	91	9	1	0	9	*1,100*	*327*
25-29	%	82	13	4	2	18	*1,215*	*404*
30-34	%	76	15	5	4	24	*1,553*	*537*
35-39	%	77	14	7	2	23	*1,721*	*632*
40-44	%	78	15	5	3	22	*1,798*	*659*
45-49	%	86	10	2	2	14	*1,594*	*633*
50-54	%	90	7	2	1	10	*1,427*	*574*
55-59	%	93	5	2	0	7	*1,680*	*719*
Total	%	85	10	3	2	15	*13,087*	*4787*
Women								
16-19	%	98	2	0	0	2	*1,054*	*352*
20-24	%	85	12	2	1	15	*1,366*	*446*
25-29	%	77	17	4	1	23	*1,477*	*554*
30-34	%	76	18	4	2	24	*1,733*	*667*
35-39	%	74	19	6	1	26	*1,962*	*786*
40-44	%	82	13	4	1	18	*2,133*	*806*
45-49	%	88	9	3	0	12	*1,941*	*768*
50-54	%	90	8	1	1	10	*1,688*	*693*
55-59	%	95	4	1	0	5	*1,720*	*761*
Total	%	85	12	3	1	16	*15,074*	*5833*
All								
16-19	%	98	2	0	0	2	*2,053*	*654*
20-24	%	87	11	1	0	13	*2,467*	*773*
25-29	%	79	15	4	1	21	*2,692*	*958*
30-34	%	76	17	5	3	24	*3,286*	*1204*
35-39	%	75	17	6	2	25	*3,683*	*1418*
40-44	%	80	14	4	2	20	*3,931*	*1465*
45-49	%	87	9	3	1	13	*3,535*	*1401*
50-54	%	90	8	2	1	10	*3,115*	*1267*
55-59	%	94	5	1	0	6	*3,401*	*1480*
Total	%	84	11	3	1	16	*28,161*	*10620*

1 Results for 2006 include longitudinal data (see Appendix B).
2 Excludes current cohabitations.

Table 5.13   Number of past cohabitations not ending in marriage by current marital status and sex

Men and women aged 16-59   *Great Britain: 2006[1]*

Number of cohabitations[2]	Marital status						
	Married[3]	Non-married					
		Cohabiting	Single	Widowed	Divorced	Separated	Total[4]
	%	%	%	%	%	%	%
Men							
None	91	79	78	94	73	86	85
One	7	15	12	0	17	13	10
Two	2	4	6	0	6	1	3
Three or more	1	2	4	6	3	0	2
Total at least one	9	21	22	6	27	14	15
Weighted base (000's) =100%	*6,590*	*1,748*	*3,860*	*64*	*653*	*212*	*13,207*
Unweighted sample	*2666*	*668*	*1167*	*22*	*214*	*75*	*4835*
Women							
None	91	78	74	97	77	85	84
One	8	15	19	2	17	12	12
Two	1	6	5	1	4	2	3
Three or more	0	1	2	0	1	1	1
Total at least one	9	22	26	3	23	15	16
Weighted base (000's) =100%	*7,874*	*1,981*	*3,424*	*220*	*1,256*	*427*	*15,218*
Unweighted sample	*3177*	*755*	*1211*	*89*	*484*	*163*	*5893*

1   Results for 2006 include longitudinal data (see Appendix B).
2   Excludes current cohabitations.
3   Married includes persons in a legally recognised Civil Partnership.
4   Total includes a small number of same sex cohabitees.

Table 5.14
Age at first cohabitation which did not end in marriage by year cohabitation began and sex

Persons aged 16-59 who have cohabited   *Great Britain: 2006[1]*

Age at first cohabitation	Year first cohabitation began			
	1960-1979	1980-1989	1990-2006	All
	%	%	%	%
Men				
16-19	27	29	19	22
20-24	46	45	35	38
25-29	20	14	24	21
30-34	6	9	8	8
35-59	0	4	14	10
Weighted base (000's) =100%	*137*	*429*	*1,130*	*1,696*
Unweighted sample	*51*	*157*	*386*	*594*
Women				
16-19	41	37	30	32
20-24	47	40	36	38
25-29	10	15	17	16
30-34	2	5	8	7
35-59	0	3	8	6
Weighted base (000's) =100%	*152*	*535*	*1,404*	*2,092*
Unweighted sample	*59*	*212*	*541*	*812*
All				
16-19	34	33	25	28
20-24	46	42	35	38
25-29	15	14	20	18
30-34	4	7	8	8
35-59	0	3	11	8
Weighted base (000's) =100%	*289*	*965*	*2,534*	*3,788*
Unweighted sample	*110*	*369*	*927*	*1406*

1 Results for 2006 include longitudinal data (see Appendix B).

Note: Shaded figures indicate the estimates are unreliable and any analysis using these figures may be invalid. Any use of these shaded figures must be accompanied by this disclaimer.

Table 5.15 Duration of past cohabitations which did not end in marriage by number of past cohabitations and sex

Persons aged 16-59 who have cohabited[1] Great Britain: 2006[2]

Duration of cohabitation	First cohabitation			Second cohabitation
	One only	One of two or more	All	All
	%	%	%	%
Men				
Less than 1 year	25	24	24	35
1 year, less than 2	22	20	22	20
2 years, less than 3	19	22	20	17
3 years, less than 5	14	14	14	16
5 years or more	20	20	20	12
Mean length in months	39	34	38	
Weighted base (000's) =100%	*1,169*	*550*	*1,719*	*549*
Unweighted sample	*417*	*185*	*602*	*182*
Women				
Less than 1 year	17	18	17	26
1 year, less than 2	20	22	20	24
2 years, less than 3	16	21	17	16
3 years, less than 5	19	19	19	20
5 years or more	28	21	27	15
Mean length in months	47	37	45	
Weighted base (000's) =100%	*1,663*	*471*	*2,134*	*494*
Unweighted sample	*643*	*183*	*826*	*192*
All persons				
Less than 1 year	20	21	21	31
1 year, less than 2	21	21	21	22
2 years, less than 3	17	21	18	16
3 years, less than 5	17	16	17	18
5 years or more	25	20	24	13
Mean length in months	44	35	42	
Weighted base (000's) =100%	*2,832*	*1,021*	*3,853*	*1,043*
Unweighted sample	*1060*	*368*	*1428*	*374*

1 Includes current cohabitations.
2 Results for 2006 include longitudinal data (see Appendix B).

Table 6.1 Current pension scheme membership by age and sex

Employees aged 16 and over excluding YT and ET   *Great Britain: 2006[1]*

Pension scheme members	Age						
	16-17	18-24	25-34	35-44	45-54	55 and over	Total
				Percentages			
Men full time							
Occupational pension[2]	6	20	49	60	65	55	53
Personal pension	0	6	15	25	28	28	21
Any pension	6	22	58	73	77	69	63
Women full time							
Occupational pension[2]	7	30	57	65	67	60	58
Personal pension	0	5	12	16	18	12	13
Any pension	7	32	63	73	73	67	64
Women part time							
Occupational pension[2]	0	6	39	53	44	28	36
Personal pension	0	1	9	14	11	11	10
Any pension	0	7	43	59	50	33	40
Weighted base (000's) =100%							
Men full time	*106*	*1,217*	*2,671*	*3,229*	*2,365*	*1,600*	*11,187*
Women full time	*33*	*944*	*1,883*	*1,925*	*1,599*	*795*	*7,180*
Women part time	*268*	*624*	*729*	*1,330*	*1,031*	*870*	*4,852*
Unweighted sample							
Men full time	32	364	898	1185	945	*687*	*4,111*
Women full time	*11*	*308*	*666*	*728*	*653*	*350*	*2,716*
Women part time	*93*	*206*	*293*	*540*	*409*	*388*	*1,929*

1 Results for 2006 include longitudinal data (see Appendix B).
2 Including a few people who were not sure if they were in a scheme but thought it possible.
Note: Shaded figures indicate the estimates are unreliable and any analysis using these figures may be invalid. Any use of these shaded figures must be accompanied by this disclaimer.

Table 6.2 Membership of current employer's pension scheme by sex and whether working full time or part time

Employees aged 16 and over excluding YT and ET   *Great Britain: 2006*

Pension scheme coverage	Men			Women		
	Working full time	Working part time	Total	Working full time	Working part time	Total
	%	%	%	%	%	%
Current employer has a pension scheme						
Member[2]	53	13	49	58	36	49
Eligible but not a member	12	15	13	14	16	15
Not eligible to belong	5	19	7	5	9	7
Does not know if a member	0	0	0	0	0	0
(subtotal)	*71*	*47*	*68*	*77*	*60*	*70*
Current employer does not have a pension scheme	28	48	30	22	38	28
Not known if employer has a pension scheme	1	5	1	1	2	1
Weighted base (000's) =100%	*11,148*	*1,263*	*12,411*	*7,152*	*4,798*	*11,949*
Unweighted sample	*4094*	*441*	*4535*	*2705*	*1905*	*4610*

1 Results for 2006 include longitudinal data (see Appendix B).
2 Including a few people who were not sure if they were in a scheme but thought it possible.

Table 6.3 Membership of current employer's pension scheme by sex: 1983 to 2006[1]

Employees aged 16 and over excluding YT and ET[2]

Great Britain

Pension scheme coverage[3]	Unweighted									Weighted							
	1983	1987	1989	1991	1993	1995	1996	1998	1998	2000	2001	2002	2003	2004	2005[4]	2006
	%	%	%	%	%	%	%	%	%	%	%	%	%	%	%	%
Men full time																
Current employer has a pension scheme																
Member[3]	66	63	64	61	60	58	58	57	55	54	54	55	55	53	54	53
Not a member	10 77	12 74	14 79	16 77	16 76	16 74	16 76	15 72	15 71	16 70	19 73	18 73	16 72	18 71	17 71	18 71
Does not know if a member	1	0	0	1	0	0	0	0	0	0	0	0	0	0	0	0
Current employer does not have a pension scheme	22	22	19	21	22	25	25	28	29	29	26	26	28	28	28	28
Not known if employer has a pension scheme	2	3	2	2	2	1	1	1	1	1	1	1	1	1	1	1
Weighted base (000's) =100%[5]									11,009	11,323	11,220	10,820	10,708	10,422	9,325	11,148
Unweighted sample[5]	5087	5129	4906	4563	3976	4062	3937	3697		3558	3881	3709	4427	3649	4680	4094
	%	%	%	%	%	%	%	%	%	%	%	%	%	%	%	%
Women full time																
Current employer has a pension scheme																
Member[3]	55	52	55	55	54	55	53	56	55	58	58	60	56	56	58	58
Not a member	17 72	16 68	21 76	21 77	22 77	20 76	20 73	17 73	18 73	17 75	20 78	18 78	20 76	19 76	20 78	19 77
Does not know if a member	0	1	0	0	0	0	0	0	0	0	0	0	0	0	0	0
Current employer does not have a pension scheme	24	28	21	20	22	24	26	26	27	25	21	21	24	23	21	22
Not known if employer has a pension scheme	4	4	3	3	2	1	1	0	0	1	1	1	1	1	1	1
Weighted base (000's) =100%[5]									6,429	6,353	6,465	6,362	6,554	6,601	6,099	7,152
Unweighted sample[5]	2256	2562	2602	2484	2239	2331	2143	2244		2089	2384	2236	2685	2345	3026	2705
	%	%	%	%	%	%	%	%	%	%	%	%	%	%	%	%
Women part time																
Current employer has a pension scheme																
Member[3]	13	11	15	17	19	24	26	27	26	31	33	33	33	34	37	36
Not a member	39 53	34 46	37 52	34 52	35 55	32 55	28 53	26 52	26 52	25 56	29 63	25 58	27 60	26 61	25 62	24 60
Does not know if a member	0	0	0	1	0	0	0	0	0	0	0	0	0	0	0	0
Current employer does not have a pension scheme	40	44	40	39	38	42	44	45	46	42	36	40	37	37	35	38
Not known if employer has a pension scheme	7	10	7	8	7	3	2	2	3	2	2	2	3	2	2	2
Weighted base (000's) =100%[5]									4,628	5,059	4,990	4,963	5,044	5,182	4,930	4,798
Unweighted sample[5]	1638	2126	2102	1977	1938	2038	1908	1674		1732	1878	1795	2179	1897	2597	1905

1 Results for 2006 include longitudinal data (Appendix B).
2 Prior to 1985 full-time students are excluded. Figures since 1987 include full-time students who were working but exclude those on Government schemes.
3 Including a few people who were not sure if they were in a scheme but thought it possible.
4 2005 data includes last quarter of 2004/5 data due to survey change from financial year to calendar year.
5 Trend tables show unweighted and weighted figures for 1998 to give an indication of the effect of the weighting. For the weighted data (1998 and 2000 to 2006) the weighted base (000's) is the base for percentages. Unweighted data (up to 1998) are based on the unweighted sample.

Table 6.4 Current pension scheme membership by sex and socio-economic classification

Employees aged 16 and over excluding YT and ET   *Great Britain: 2006[1]*

Pension scheme members	Socio-economic classification			
	Managerial and professional	Intermediate	Routine and manual	Total[2]
	Percentages			
Men full time				
Occupational pension[3]	65	60	37	53
Personal pension	24	18	18	21
Any pension	76	69	48	63
Women full time				
Occupational pension[3]	69	60	34	58
Personal pension	16	12	11	13
Any pension	75	65	41	64
Women part time				
Occupational pension[3]	64	47	24	36
Personal pension	15	14	7	10
Any pension	70	52	28	40
Weighted base (000's) =100%				
Men full time	*5,552*	*817*	*4,611*	*11,187*
Women full time	*3,592*	*1,536*	*1,820*	*7,180*
Women part time	*1,057*	*947*	*2,327*	*4,852*
Unweighted sample				
Men full time	*2,109*	*287*	*1,648*	*4,111*
Women full time	*1,391*	*576*	*667*	*2,716*
Women part time	*447*	*390*	*909*	*1,929*

1 Results for 2006 include longitudinal data (see Appendix B).
2 Total includes a small number of employees for whom socio-economic classification could not be derived.
3 Including a few people who were not sure if they were in a scheme but thought it possible.

Table 6.5 Membership of current employer's pension scheme by sex and socio-economic classification

Employees aged 16 and over excluding YT and ET *Great Britain: 2006[1]*

Pension scheme coverage	Socio-economic classification			
	Managerial and professional	Intermediate	Routine and manual	Total[2]
Men full time	%	%	%	%
Current employer has a pension scheme				
Member[3]	65 ⎤ 77	60 ⎤ 79	37 ⎤ 61	53 ⎤ 71
Not a member[4]	12 ⎦	19 ⎦	23 ⎦	18 ⎦
Current employer does not have a pension scheme	22	19	38	28
Not known if employer has a pension scheme	0	1	2	1
Weighted base (000's) =100%	5,552	817	4,611	11,148
Unweighted sample	2109	287	1648	4094
Women full time	%	%	%	%
Current employer has a pension scheme				
Member[3]	69 ⎤ 84	60 ⎤ 80	34 ⎤ 62	58 ⎤ 77
Not a member[4]	15 ⎦	20 ⎦	28 ⎦	19 ⎦
Current employer does not have a pension scheme	16	20	37	22
Not known if employer has a pension scheme	0	1	1	1
Weighted base (000's) =100%	3,592	1,536	1,820	7,152
Unweighted sample	1391	576	667	2705
Women part time	%	%	%	%
Current employer has a pension scheme				
Member[3]	64 ⎤ 79	47 ⎤ 67	24 ⎤ 52	36 ⎤ 60
Not a member[4]	15 ⎦	21 ⎦	28 ⎦	24 ⎦
Current employer does not have a pension scheme	21	32	45	38
Not known if employer has a pension scheme	0	1	2	2
Weighted base (000's) =100%	1,057	947	2,327	4,798
Unweighted sample	447	390	909	1905

1 Results for 2006 include longitudinal data (see Appendix B).
2 Total includes a small number of employees for whom socio-economic classification could not be derived.
3 Including a few people who were not sure if they were in a scheme but thought it possible.
4 Including people who were not eligible and a few people who did not know if they were a member.

Table 6.6   Current pension scheme membership by sex and usual gross weekly earnings

Employees aged 16 and over excluding YT and ET   *Great Britain: 2006[1]*

Pension scheme members	Usual gross weekly earnings (£)							
	0.01–100.00	100.01–200.00	200.01–300.00	300.01–400.00	400.01–500.00	500.01–600.00	600.01 or more	Total[2]
	Percentages							
Men full time								
Occupational pension[3]	19	14	29	46	56	68	72	53
Personal pension	14	8	11	18	27	24	32	21
Any pension	28	18	37	58	70	79	87	63
Women full time								
Occupational pension[3]	28	25	40	61	68	75	82	58
Personal pension	14	9	11	14	15	19	22	13
Any pension	41	31	47	68	76	82	89	64
Women part time								
Occupational pension[3]	12	36	60	79	83	73	65	36
Personal pension	5	10	12	13	28	30	36	10
Any pension	15	42	64	85	89	81	80	40
Weighted base (000's) =100%								
Men full time	*102*	*342*	*1,234*	*1,917*	*1,472*	*1,260*	*2,775*	*11,187*
Women full time	*105*	*572*	*1,600*	*1,458*	*874*	*626*	*1,066*	*7,180*
Women part time	*1,531*	*1,688*	*649*	*277*	*122*	*54*	*77*	*4,852*
Unweighted sample								
Men full time	*37*	*106*	*431*	679	543	467	1085	*4,111*
Women full time	38	212	591	553	324	242	422	*2,716*
Women part time	595	664	258	117	52	24	31	*1,929*

1 Results for 2006 include longitudinal data (see Appendix B).
2 Totals include no answers to income.
3 Including a few people who were not sure if they were in a scheme but thought it possible.
Note: Shaded figures indicate the estimates are unreliable and any analysis using these figures may be invalid. Any use of these shaded figures must be accompanied by this disclaimer.

Table 6.7 Current pension scheme membership by sex and length of time with current employer

Employees aged 16 and over excluding YT and ET *Great Britain: 2006[1]*

Pension scheme members	Length of time with current employer			
	Less than 2 years	2 years, but less than 5 years	5 years or more	Total[2]
	Percentages			
Men full time				
Occupational pension[3]	29	46	70	53
Personal pension	17	23	26	21
Any pension	40	58	83	63
Women full time				
Occupational pension2	38	52	73	58
Personal pension	10	16	16	13
Any pension	43	59	80	64
Women part time				
Occupational pension[3]	16	29	54	36
Personal pension	6	9	13	10
Any pension	20	35	59	40
Weighted base (000's) =100%				
Men full time	*2,484*	*2,178*	*5,141*	*11,187*
Women full time	*1,951*	*1,602*	*3,167*	*7,180*
Women part time	*1,488*	*1,093*	*2,070*	*4,852*
Unweighted sample				
Men full time	867	790	1970	*4,111*
Women full time	682	594	1268	*2,716*
Women part time	572	427	852	*1,929*

1 Results for 2006 include longitudinal data (see Appendix B).
2 Including a few where length of time in job was not known.
3 Including a few people who were not sure if they were in a scheme but thought it possible.

Table 6.8 Whether or not current employer has a pension scheme by sex and length of time with current employer

Employees aged 16 and over excluding YT and ET *Great Britain: 2006[1]*

Pension scheme coverage	Length of time with current employer			
	Less than 2 years	2 years, but less than 5 years	5 years or more	Total[2]
	%	%	%	%
Men full time				
Current employer has a pension scheme				
Member[3]	29	46	70	53
Not a member[4]	20 60	18 68	7 80	12 71
Not eligible to belong	12	4	3	5
Current employer does not have a pension scheme	37	31	20	28
Not known if employer has a pension scheme	3	1	0	1
Weighted base (000's) =100%	*2,469*	*2,176*	*5,140*	*11,148*
Unweighted sample	*861*	*789*	*1969*	*4094*
	%	%	%	%
Women full time				
Current employer has a pension scheme				
Member[3]	38	52	73	58
Not a member[4]	18 67	21 76	9 85	14 77
Not eligible to belong	11	3	2	5
Current employer does not have a pension scheme	31	24	15	22
Not known if employer has a pension scheme	2	0	0	1
Weighted base (000's) =100%	*1,949*	*1,602*	*3,167*	*7,152*
Unweighted sample	*681*	*594*	*1268*	*2705*
	%	%	%	%
Women part time				
Current employer has a pension scheme				
Member[3]	16	29	54	36
Not a member[4]	15 45	17 55	15 74	16 60
Not eligible to belong	14	9	5	9
Current employer does not have a pension scheme	51	43	26	38
Not known if employer has a pension scheme	4	1	0	2
Weighted base (000's) =100%	*1,468*	*1,093*	*2,070*	*4,798*
Unweighted sample	*565*	*427*	*852*	*1905*

1 Results for 2006 include longitudinal data (see Appendix B).
2 Including a few whose length of time in job was not known.
3 Including a few people who were not sure if they were in a scheme but thought it possible.
4 Including a few people who did not know if they were a member.

Table 6.9 Current pension scheme membership by sex and number of employees in the establishment

Employees aged 16 and over excluding YT and ET *Great Britain: 2006[1]*

Pension scheme members	Number of employees at establishment					
	1-2	3-24	25-99	100-999	1000 or more	Total[2]
			Percentages			
Men full time						
Occupational pension[3]	33	28	49	67	82	53
Personal pension	29	25	23	18	16	21
Any pension	54	46	61	73	85	63
Women full time						
Occupational pension[3]	26	36	57	68	81	58
Personal pension	20	14	13	14	10	13
Any pension	42	44	63	73	83	64
Women part time						
Occupational pension[3]	10	21	40	53	73	36
Personal pension	9	9	9	11	13	10
Any pension	17	27	44	58	74	40
Weighted base (000's) =100%						
Men full time	*396*	*2,747*	*2,824*	*3,699*	*1,351*	*11,187*
Women full time	*167*	*1,819*	*1,849*	*2,248*	*1,054*	*7,180*
Women part time	*251*	*1,971*	*1,187*	*1,065*	*317*	*4,852*
Unweighted sample						
Men full time	*156*	*1005*	*1029*	*1360*	*502*	*4,111*
Women full time	*66*	*694*	*698*	*840*	*400*	*2,716*
Women part time	*104*	*773*	*478*	*419*	*132*	*1,929*

1 Results for 2006 include longitudinal data (see Appendix B).
2 Includes a few people for whom the number of employees at establishment was not known.
3 Including a few people who were not sure if they were in a scheme but thought it possible.
Note: Shaded figures indicate the estimates are unreliable and any analysis using these figures may be invalid. Any use of these shaded figures must be accompanied by this disclaimer.

Table 6.10   Membership of current employer's pension scheme by sex and number of employees at the establishment

Employees aged 16 and over excluding YT and ET   *Great Britain: 2006[1]*

Pension scheme coverage	Number of employees at establishment					
	1-2	3-24	25-99	100-999	1000 or more	Total[2]
	%	%	%	%	%	%
Men full time						
Current employer has a pension scheme						
Member[3]	33 } 39	28 } 45	49 } 68	67 } 87	82 } 95	53 } 71
Not a member[4]	7	17	19	20	13	18
Current employer does not have a pension scheme	59	53	31	12	4	28
Not known if employer has a pension scheme	2	1	1	0	0	1
Weighted base (000's) =100%	*394*	*2,740*	*2,815*	*3,694*	*1,345*	*11,148*
Unweighted sample	*155*	*1002*	*1025*	*1358*	*499*	*4094*
	%	%	%	%	%	%
Women full time						
Current employer has a pension scheme						
Member[3]	26 } 45	36 } 57	57 } 77	68 } 88	81 } 95	58 } 77
Not a member[4]	19	21	20	20	14	19
Current employer does not have a pension scheme	55	42	22	11	5	22
Not known if employer has a pension scheme	0	1	1	1	0	1
Weighted base (000's) =100%	*167*	*1,815*	*1,846*	*2,240*	*1,044*	*7,152*
Unweighted sample	*66*	*692*	*697*	*837*	*396*	*2705*
	%	%	%	%	%	%
Women part time						
Current employer has a pension scheme						
Member[3]	10 } 23	21 } 42	40 } 69	53 } 85	73 } 91	36 } 60
Not a member[4]	13	21	28	32	19	25
Current employer does not have a pension scheme	77	56	29	13	9	38
Not known if employer has a pension scheme	0	2	2	2	0	2
Weighted base (000's) =100%	*251*	*1,946*	*1,173*	*1,059*	*317*	*4,798*
Unweighted sample	*104*	*762*	*472*	*416*	*132*	*1905*

1 Results for 2006 include longitudinal data (see Appendix B).
2 Includes a few people for whom the number of employees at establishment was not known
3 Including a few people who were not sure if they were in a scheme but thought it possible.
4 Including people who were not eligible and a few people who did not know if they were a member.

Table 6.11 Membership of current employer's pension scheme by sex and industry group

Employees aged 16 and over excluding YT and ET *Great Britain: 2004, 2005[1] and 2006[2] combined*

Pension scheme members	Industry group[3]										
	Agriculture, forestry, fishing	Coal mining, energy and water supply	Mining (excl coal), manufacture of metals, minerals and chemicals	Metal goods, engineering and vehicle	Other manufacturing	Construction	Distribution, hotels, catering repairs	Transport and communications	Banking, finance, insurance business services	Public and other personal services	Total
	Percentage with an occupational pension[4]										
Men full time	26	72	61	56	50	37	32	55	50	78	53
Women full time	24	59	92	57	45	50	30	49	51	72	58
Women part time	29	40	*	35	26	18	16	36	36	50	36
Unweighted sample[5]											
Men full time	137	189	34	1409	1181	1144	1754	1143	1952	2566	11509
Women full time	18	40	6	314	454	154	1135	281	1347	3739	7488
Women part time	31	7	*	107	174	97	1753	143	666	2943	5925

1 2005 data includes last quarter of 2004/5 data due to survey change from financial year to calendar year.
2 Results for 2006 include longitudinal data (see Appendix b).
3 Standard Industrial Classification, 1992.
4 Including a few people who were not sure if they were in a scheme but thought it possible.
5 Weighted bases not shown for combined data sets.
* Information is suppressed for unweighted sample sizes below 5 as a measure of disclosure control

Table 6.12 Membership of personal pension scheme by sex and whether working full time or part time: self-employed persons

Self-employed persons aged 16 and over *Great Britain: 2006[1]*

Pension scheme coverage	Men			Women		
	Working full time	Working part time	Total	Working full time	Working part time	Total
	%	%	%	%	%	%
Informant belongs to a personal pension scheme	45	11	40	31	23	27
Informant no longer contributes to a personal pension scheme	21	38	24	22	16	19
Informant has never had a personal pension scheme	34	51	37	48	61	54
Weighted base (000's) =100%	1,854	337	2,190	489	428	918
Unweighted sample	723	135	858	192	185	377

1 Results for 2006 include longitudinal data (see Appendix B).

Table 6.13 Membership of a personal pension scheme for self-employed men working full time: 1991 to 2006[1]

Self-employed persons aged 16 and over   *Great Britain*

Pension scheme coverage	Unweighted								Weighted							
	1991	1992	1993	1994	1995	1996	1998	1998	2000	2001	2002	2003	2004	2005[2]	2006
							Percentages								
Men working full time															
Informant belongs to a personal pension scheme	66	65	67	60	61	64	65	64	54	54	52	49	49	45	45
Informant no longer contributes to a personal pension scheme	7	9	9	10	11	11	10	10	12	15	14	15	19	20	21
Informant has never had a personal pension scheme	27	26	24	30	28	25	26	27	34	32	34	36	33	35	34
Weighted base (000's) =100%[3]								1,958	1,904	1,942	1,926	1,871	1,834	1,827	1,854
Unweighted sample[3]	929	869	852	842	879	696	683		625	700	695	784	642	958	723

1 Results for 2006 include longitudinal data (see Appendix B).
2 2005 data includes last quarter of 2004/5 data due to survey change from financial year to calendar year.
3 Trend tables show unweighted and weighted figures for 1998 to give an indication of the effect of the weighting. For the weighted data (1998 and 2000 to 2006) the weighted base (000's) is the base for percentages. Unweighted data (up to 1998) are based on the unweighted sample.

Table 6.14 Membership of personal pension scheme by sex and length of time in self-employment

Self-employed persons aged 16 and over *Great Britain: 2004, 2005[1] and 2006[2] combined*

	Length of time in self-employment			
	Less than 2 years	2 years, but less than 5 years	5 years or more	Total
	Percentage of self-employed who belong to a personal pension scheme			
Men full time	20	32	55	46
Women full time	8	26	38	30
Women part time	17	12	29	23
Unweighted sample [1]				
Men full time	*305*	*341*	*1462*	*2108*
Women full time	*101*	*106*	*337*	*544*
Women part time	*132*	*93*	*300*	*525*

1 2005 data includes last quarter of 2004/5 data due to survey change from financial year to calendar year.
2 Results for 2006 include longitudinal data (see Appendix B).
3 Weighted bases not shown for combined data sets.

Table 7.1   Self perception of general health during the last 12 months: 1977 to 2006

Persons aged 16 and over[1]   *Great Britain*

	Unweighted data									Weighted data							
	1977	1981	1985	1987	1991	1995	1996	1998		1998	2000	2001	2002	2003	2004	2005[2]	2006[3]
	%	%	%	%	%	%	%	%		%	%	%	%	%	%	%	%
Percentage who reported their general health was:																	
Good	58	62	63	60	63	60	55	59		60	59	59	56	59	60	59	62
Fairly good	30	26	25	28	25	26	33	27		27	27	27	30	27	27	27	26
Not good	12	12	12	12	12	14	12	14		14	13	14	14	14	14	14	12
Weighted base (000's) =100%[4]										40,884	42,467	41,990	41,876	41,111	42,015	41,914	42,744
Unweighted sample[4]	23125	23242	18575	19477	18174	16724	15684	14410			14113	15385	14815	17451	14917	21717	16717

1 This question is not asked of proxies.
2 2005 data includes last quarter of 2004/5 data due to survey change from financial year to calendar year
3 Results for 2006 include longitudinal data (see Appendix B).
4 Trend tables show unweighted and weighted figures for 1998 to give an indication of the effect of the weighting. For the weighted data (1998 and 2000 to 2005) the weighted base (000's) is the base for percentages. Unweighted data (up to 1998) are based on the unweighted sample.

All persons

Great Britain

Table 7.2 Trends in self-reported sickness by sex and age, 1972 to 2006: percentage of persons who reported
(a) longstanding illness
(b) limiting longstanding illness
(c) restricted activity in the 14 days before interview

	Unweighted data								Weighted data									Weighted base 2006 (000's) =100%[3]	Unweighted sample[3] 2006
	1972	1975	1981	1985	1991	1995	1996	1998	1998	2000	2001	2002	2003	2004	2005[1]	2006[2]		

Percentage who reported:

(a) Longstanding illness

Males
0-4	5	8	12	11	13	14	14	15	15	14	17	17	14	15	14	11	1,686	682
5-15[4]	9	11	17	18	17	20	19	21	21	23	20	21	18	19	19	17	3,887	1683
16-44[4]	14	17	22	21	23	23	27	24	24	23	22	23	20	20	22	21	9,607	3228
45-64	29	35	40	42	42	43	46	44	44	45	44	46	41	43	43	45	6,424	2658
65-74	48	50	51	55	61	55	61	59	59	61	58	65	62	57	58	63	2,230	1016
75 and over	54	63	60	58	63	56	64	68	68	63	64	71	61	63	65	70	1,604	750
Total	20	23	28	29	31	31	34	33	33	33	32	34	31	31	32	33	25,437	10017

Females
0-4	3	6	7	9	10	11	13	15	15	13	12	12	10	11	10	10	1,673	674
5-15[4]	6	9	13	13	15	17	16	19	19	18	16	19	17	15	16	15	3,819	1641
16-44[4]	13	16	21	22	23	22	27	23	23	22	21	25	22	22	24	23	10,750	3955
45-64	31	33	41	43	41	39	47	43	43	42	42	44	41	42	43	44	7,090	3012
65-74	48	54	58	56	55	54	58	59	59	54	56	61	59	55	61	63	2,477	1121
75 and over	65	61	70	65	65	66	68	65	65	64	63	72	65	63	64	70	2,428	926
Total	21	25	30	31	32	31	35	34	34	32	31	35	32	32	33	34	28,236	11329

All persons
0-4	4	7	10	10	12	13	13	15	15	14	14	15	12	13	12	11	3,358	1356
5-15[4]	8	10	15	16	16	19	18	20	20	20	18	20	18	17	18	16	7,705	3324
16-44[4]	13	16	21	22	23	23	27	24	24	22	22	24	21	21	23	22	20,357	7183
45-64	30	34	41	43	41	41	47	43	43	44	43	45	41	43	43	45	13,514	5670
65-74	48	52	55	56	58	55	58	59	59	57	57	63	60	56	60	63	4,707	2137
75 and over	62	62	67	63	65	63	66	66	66	64	63	72	64	63	64	70	1,676
Total	21	24	29	30	31	31	35	33	33	32	32	35	31	32	33	33	53,673	21346

Percentage who reported:

(b) Limiting longstanding illness

Males
0-4	..	3	3	4	4	5	4	4	4	4	5	5	4	4	5	3	1,685	682
5-15[4]	..	6	8	8	7	8	8	8	8	9	9	8	7	8	7	7	3,887	1683
16-44[4]	..	9	10	10	10	12	14	12	12	11	10	12	10	10	11	10	9,607	3228
45-64	..	24	26	27	25	28	31	28	28	27	28	28	24	26	26	23	6,421	2657
65-74	..	36	35	38	40	37	42	36	36	38	36	43	37	33	36	37	2,231	1016
75 and over	..	46	44	43	46	41	50	48	48	44	47	52	41	43	44	47	1,602	749
Total	..	14	16	16	17	18	21	19	19	18	18	20	17	17	18	17	25,432	10015

Females
0-4	..	2	3	3	3	3	4	5	5	4	4	3	4	4	3	3	1,673	674
5-15[4]	..	4	6	6	5	8	8	8	8	8	8	9	7	7	7	6	3,819	1641
16-44[4]	..	9	11	11	11	13	16	13	13	11	12	14	11	12	13	12	10,750	3955
45-64	..	22	26	26	25	26	32	29	29	27	26	28	25	24	26	27	7,090	3012
65-74	..	39	41	38	34	37	40	39	39	35	37	39	37	33	39	39	2,477	1121
75 and over	..	49	56	51	51	52	53	51	51	48	45	53	46	48	48	51	2,428	926
Total	..	16	19	18	18	20	23	21	21	19	19	22	19	19	20	20	28,236	11329

All persons
0-4	..	2	3	3	4	4	4	4	4	4	4	4	4	4	4	3	3,358	1356
5-15[4]	..	5	7	7	6	8	8	8	8	8	8	8	7	8	7	7	77,005	3324
16-44[4]	..	9	11	10	10	12	15	13	13	11	11	13	11	11	12	11	20,357	7183
45-64	..	23	26	26	25	27	32	28	28	27	27	28	24	25	26	25	13,510	5669
65-74	..	38	38	38	37	37	41	37	37	37	36	41	37	33	37	38	4,707	2137
75 and over	..	48	52	48	49	48	52	50	50	47	46	53	44	46	47	50	4,030	1675
Total	..	15	17	17	18	19	22	20	20	19	19	21	18	18	19	19	53,668	21344

1 2005 data includes last quarter of 2004/5 data due to survey change from financial year to calendar year
2 Results for 2006 include longitudinal data (see Appendix B).
3 Trend tables show unweighted and weighted figures for 1998 to give an indication of the effect of the weighting. Bases for earlier years can be found in GHS reports for each year.
4 These age-groups were 5-14 and 15-44 in 1972 to 1975
.. Data are not available.

Table 7.2 - continued

All persons

Great Britain

Percentage who reported:

(c) Restricted activity in the 14 days before interview

	Unweighted data								Weighted data									Weighted base[3] 2006 (000's) =100%*	Unweighted sample[3] 2006
	1972	1975	1981	1985	1991	1993	1995	1996	1998	1998	2000	2001	2002	2003	2004	2005[1]	2006[2]		
Males																			
0-4	5	10	13	13	11	13	11	12	10	10	11	9	10	11	10	9	8	1,686	682
5-15[2]	6	9	12	11	11	11	10	10	9	9	10	9	9	8	10	9	7	3,889	1684
16-44[2]	7	7	8	9	9	11	10	13	11	11	10	10	11	11	8	9	9	9,697	3258
45-64	9	10	12	11	12	15	15	18	18	18	17	17	17	15	15	15	14	6,453	2671
65-74	10	8	11	13	14	16	17	19	18	18	20	16	19	18	16	15	16	2,229	1015
75 and over	10	12	15	17	18	17	20	23	24	24	23	23	24	20	20	18	19	1,606	751
Total	7	9	11	11	11	13	13	15	14	14	13	13	14	12	12	11	11	25,559	10,061
Females																			
0-4	6	8	12	13	10	10	11	9	8	8	7	8	10	9	11	9	7	1,673	674
5-15[2]	5	7	11	12	9	11	10	9	11	11	11	10	9	9	7	8	8	3,820	1642
16-44[2]	8	10	11	13	12	13	13	15	13	13	12	12	13	12	12	13	11	10,774	3964
45-64	9	10	13	14	13	17	17	22	20	20	19	18	19	19	18	17	17	7,091	3012
65-74	10	12	17	18	16	19	20	21	23	23	21	21	20	22	20	21	21	2,480	1123
75 and over	14	13	21	23	21	23	26	25	27	27	27	26	28	25	26	24	24	2,428	926
Total	8	10	13	14	13	15	15	17	16	16	15	15	16	15	14	15	14	28,267	11,341
All persons																			
0-4	6	9	13	13	11	11	11	10	9	9	9	9	10	10	10	9	8	3,358	1,356
5-15[2]	6	8	12	11	10	11	10	10	10	10	10	10	9	9	9	8	7	7,709	3,326
16-44[2]	8	9	10	11	10	12	12	14	12	12	11	11	12	11	10	11	10	20,471	7,222
45-64	9	10	12	12	13	16	16	20	19	19	18	17	18	17	16	16	15	13,544	5,683
65-74	10	11	14	16	15	18	19	20	21	21	21	19	20	20	18	18	19	4,709	2,138
75 and over	13	13	19	21	20	21	24	24	26	26	25	25	26	23	23	22	22	4,034	1,677
Total	8	9	12	12	12	14	14	16	15	15	14	14	15	14	13	13	13	53,827	21,402

1 Note: 2005 data includes last quarter of 2004/5 data due to survey change from financial year to calendar year.
2 Results for 2006 include longitudinal data (see Appendix B).
3 Trend tables show unweighted and weighted figures for 1998 to give an indication of the effect of the weighting. Bases for earlier years can be found in GHS reports for each year.
4 These age-groups were 5-14 and 15-44 in 1972 to 1975

Table 7.3 Acute sickness: average number of restricted activity days per person per year, by sex and age

All persons Great Britain: 2006[1]

	Number of days			Weighted bases (000's) =100%			Unweighted sample		
	Males	Females	Total	Males	Females	Total	Males	Females	Total
Age									
0-4	9	10	10	1,684	1,673	3,356	681	674	1,355
5-15	10	11	10	3,889	3,820	7,709	1,684	1,642	3,326
16-44	16	21	19	9,697	10,774	20,471	3,258	3,964	7,222
45-64	35	40	37	6,453	7,091	13,544	2,671	3,012	5,683
65-74	41	55	48	2,226	2,478	4,705	1,014	1,122	2,136
75 and over	56	70	65	1,604	2,426	4,030	750	925	1,675
Total	24	31	28	25,552	28,263	53,815	10,058	11,339	21,397

1 Results for 2006 include longitudinal data (see Appendix B).

Table 7.4 Chronic sickness: prevalence of reported longstanding illness by sex, age and socio-economic classification of household reference person

All persons Great Britain: 2006[1]

Socio-economic classification of household reference person[2]	Males					Females				
	Age					Age				
	0-15	16-44	45-64	65 and over	Total	0-15	16-44	45-64	65 and over	Total
	Percentage who reported longstanding illness									
Large employers and higher managerial	11	19	35	68	28	13	21	36	62	27
Higher professional	14 (15)	17 (19)	32 (38)	67 (64)	27 (29)	10 (13)	18 (22)	33 (38)	59 (62)	24 (29)
Lower managerial and professional	16	20	41	60	30	14	24	40	63	32
Intermediate	15 (14)	21 (19)	45 (45)	68 (65)	33 (32)	16 (16)	27 (21)	45 (43)	66 (65)	40 (35)
Small employers and own account	13	17	45	63	32	16	16	42	63	30
Lower supervisory and technical	17	20	49	67	36	9	22	50	66	34
Semi-routine	17 (16)	26 (25)	55 (54)	66 (68)	36 (38)	15 (14)	27 (25)	47 (51)	69 (71)	40 (40)
Routine	14	28	57	72	42	16	26	54	76	43
Never worked and long-term unemployed	23	*35*	*68*	*58*	38	9	26	*66*	*48*	29
All persons	16	21	45	66	33	14	23	44	67	34
Weighted bases (000's) =100%										
Large employers and higher managerial	526	886	577	305	2294	496	1014	586	226	2322
Higher professional	493	1024	706	327	2551	554	1007	665	240	2466
Lower managerial and professional	1253	2321	1509	753	5837	1303	2616	1698	959	6577
Intermediate	416	657	484	247	1805	353	916	639	711	2619
Small employers and own account	530	932	777	416	2655	505	926	754	363	2548
Lower supervisory and technical	550	1050	688	626	2914	541	1044	718	470	2772
Semi-routine	804	1076	684	466	3029	718	1332	918	936	3904
Routine	627	1038	824	623	3112	625	1129	873	827	3455
Never worked and long-term unemployed	205	197	82	40	525	247	282	88	136	753
All persons	5572	9607	6424	3834	25437	5491	10750	7090	4905	28236
Unweighted sample										
Large employers and higher managerial	236	323	255	154	968	226	398	264	110	998
Higher professional	225	356	310	164	1055	251	387	303	113	1054
Lower managerial and professional	554	807	653	362	2376	569	982	744	428	2723
Intermediate	178	215	195	114	702	146	329	264	283	1022
Small employers and own account	215	306	312	190	1023	211	333	321	154	1019
Lower supervisory and technical	221	344	271	278	1114	210	370	298	197	1075
Semi-routine	335	357	275	203	1170	301	498	376	376	1551
Routine	260	333	322	272	1187	261	400	347	325	1333
Never worked and long-term unemployed	76	55	29	16	176	82	87	30	48	247
All persons	2365	3228	2658	1766	10017	2315	3955	3012	2047	11329

1 Results for 2006 include longitudinal data (see Appendix B).
2 Where the household reference person was a full-time student or had an inadequately described occupation these are not shown as separate categories but are included in the figure for all persons (see Appendix A).

Note: Shaded figures indicate the estimates are unreliable and any analysis using these figures may be invalid. Any use of these shaded figures must be accompanied by this disclaimer.

Table 7.5 Chronic sickness: prevalence of reported limiting longstanding illness by sex, age and socio-economic classification of household reference person

All persons
Great Britain: 2006[1]

Socio-economic classification of household reference person [2]	Males					Females				
	Age					Age				
	0-15	16-44	45-64	65 and over	Total	0-15	16-44	45-64	65 and over	Total

Percentage who reported limiting longstanding illness

Large employers and higher managerial	5		8		15		40		13		5		9		20		36		13	
Higher professional	5	4	6	7	11	15	34	36	11	12	5	5	8	10	18	22	36	40	13	16
Lower managerial and professional	4		7		18		35		13		5		12		24		42		18	
Intermediate	6	6	13	11	25	24	43	43	19	18	6	5	14	10	27	26	47	44	25	21
Small employers and own account	5		10		23		42		18		4		6		26		39		16	
Lower supervisory and technical	7		8		25		44		20		6		8		32		43		20	
Semi-routine	8	7	16	13	33	32	44	45	22	22	4	6	18	14	32	33	46	49	25	25
Routine	6		15		37		46		25		8		16		36		56		29	
Never worked and long-term unemployed	9		27		54		33		25		6		20		54		44		24	
All persons	6		10		23		41		17		5		12		27		45		20	

Weighted bases (000's) =100%										
Large employers and higher managerial	526	886	577	305	2,294	496	1,014	586	226	2,322
Higher professional	493	1,024	706	325	2,549	554	1,007	665	240	2,466
Lower managerial and professional	1,253	2,321	1,506	753	5,833	1,303	2,616	1,698	959	6,577
Intermediate	416	657	484	247	1,805	353	916	639	711	2,619
Small employers and own account	530	932	778	416	2,655	505	926	754	363	2,548
Lower supervisory and technical	550	1,050	688	626	2,914	541	1,044	718	470	2,772
Semi-routine	804	1,076	684	466	3,029	718	1,332	918	936	3,904
Routine	627	1,038	824	623	3,112	625	1,129	873	827	3,455
Never worked and long-term unemployed	205	197	82	40	525	247	282	88	136	753
All persons	5,572	9,607	6,421	3,832	25,432	5,491	10,750	7,090	4,905	28,236

Unweighted sample										
Large employers and higher managerial	236	323	255	154	968	226	398	264	110	998
Higher professional	225	356	310	163	1054	251	387	303	113	1054
Lower managerial and professional	554	807	652	362	2375	569	982	744	428	2723
Intermediate	178	215	195	114	702	146	329	264	283	1022
Small employers and own account	215	306	312	190	1023	211	333	321	154	1019
Lower supervisory and technical	221	344	271	278	1114	210	370	298	197	1075
Semi-routine	335	357	275	203	1170	301	498	376	376	1551
Routine	260	333	322	272	1187	261	400	347	325	1333
Never worked and long-term unemployed	76	55	29	16	176	82	87	30	48	247
All persons	2365	3228	2657	1765	10015	2315	3955	3012	2047	11329

1 Results for 2006 include longitudinal data (see Appendix B).
2 Where the household reference person was a full-time student or had an inadequately described occupation these are not shown as separate categories but are included in the figure for all persons (see Appendix A).
Note: Shaded figures indicate the estimates are unreliable and any analysis using these figures may be invalid. Any use of these shaded figures must be accompanied by this disclaimer.

Table 7.6 Acute sickness
(a) Prevalence of reported restricted activity in the 14 days before interview, by sex, age, and socio-economic classification of household reference person
(b) Average number of restricted activity days per person per year, by sex, age, and socio-economic classification of household reference person

All persons   Great Britain: 2006[1]

Socio-economic classification of household reference person[2]	Males					Females				
	Age					Age				
	0-15	16-44	45-64	65 and over	Total	0-15	16-44	45-64	65 and over	Total

(a) Percentage who reported restricted activity in the 14 days before interview

Large employers and higher managerial	7	9	9	16	10	7	9	12	17	10
Higher professional	9	8	8	14	9	7	8	14	22	11
Lower managerial and professional	8	8	12	13	10	10	11	17	21	14
Intermediate	8	9	11	20	11	4	16	18	21	16
Small employers and own account	7	10	16	19	12	7	8	15	19	11
Lower supervisory and technical	9	9	13	21	12	9	11	19	20	14
Semi-routine	6	9	16	15	11	6	12	19	26	16
Routine	7	11	21	20	15	7	13	19	29	17
Never worked and long-term unemployed	4	10	22	5	9	7	12	16	24	13
All persons	7	9	14	17	11	8	11	17	23	14

(b) Average number of restricted activity days per person per year

Large employers and higher managerial	9	14	24	46	20	12	14	20	46	18
Higher professional	9	14	16	30	16	10	10	32	47	20
Lower managerial and professional	10	12	28	38	19	8	21	35	59	28
Intermediate	7	12	28	54	21	2	28	41	59	36
Small employers and own account	6	16	38	47	26	12	9	36	47	23
Lower supervisory and technical	9	19	32	64	30	18	23	47	67	36
Semi-routine	13	23	46	37	28	7	27	50	69	38
Routine	11	25	60	56	37	15	28	53	82	45
Never worked and long-term unemployed	6	24	52	19	21	8	31	51	52	29

Sub-totals (Total columns):

	0-15	16-44	45-64	65 and over	Total
(a) Males totals row –	11	13	17	20	12
	12	17	19	26	14
	15	19			16
(b) Males –	18	24	31	55	24
	24	34	38	50	30
	32	47	54	73	40

Table 7.6 Acute sickness
(a) Prevalence of reported restricted activity in the 14 days before interview, by sex, age, and socio-economic classification of household reference person
(b) Average number of restricted activity days per person per year, by sex, age, and socio-economic classification of household reference person

All persons
Great Britain: 2006[1]

Socio-economic classification of household reference person [2]	Males					Females					All persons
	Age					Age					
	0-15	16-44	45-64	65 and over	Total	0-15	16-44	45-64	65 and over	Total	
All persons	9	17	34	47	24	10	21	40	63	31	
Weighted bases (000's) =100%											
Large employers and higher managerial	526	888	583	305	2,302	496	1,014	586	226	2,322	
Higher professional	493	1,030	706	326	2,555	554	1,009	664	240	2,467	
Lower managerial and professional	1,253	2,351	1,509	753	5,866	1,303	2,623	1,702	961	6,589	
Intermediate	416	664	486	247	1,814	355	922	639	711	2,627	
Small employers and own account	530	939	781	418	2,669	505	928	754	364	2,551	
Lower supervisory and technical	550	1,063	699	626	2,938	541	1,044	719	470	2,774	
Semi-routine	804	1,089	688	466	3,046	718	1,335	918	936	3,907	
Routine	629	1,047	828	623	3,127	625	1,132	873	827	3,458	
Never worked and long-term unemployed	205	197	82	40	525	247	282	84	136	749	
All persons	5,575	9,697	6,453	3,835	25,559	5,493	10,774	7,091	4,909	28,267	
Unweighted sample											
Large employers and higher managerial	236	324	258	154	972	226	398	264	110	998	
Higher professional	225	358	310	163	1056	251	388	303	113	1055	
Lower managerial and professional	554	816	653	362	2385	569	984	745	429	2727	
Intermediate	178	218	196	114	706	147	331	264	283	1025	
Small employers and own account	215	309	314	191	1029	211	334	321	155	1021	
Lower supervisory and technical	221	348	275	278	1122	210	370	299	197	1076	
Semi-routine	335	361	276	203	1175	301	499	376	376	1552	
Routine	261	336	324	272	1193	261	401	347	325	1334	
Never worked and long-term unemployed	76	55	29	16	176	82	87	28	48	245	
All persons	2336	3258	2671	1766	10061	2316	3964	3012	2049	11341	

1 Results for 2006 include longitudinal data (see Appendix B)
2 Where the household reference person was a full-time student or had an inadequately described occupation these are not shown as separate categories but are included in the figure for all persons (see Appendix A).

Note: Shaded figures indicate the estimates are unreliable and any analysis using these figures may be invalid. Any use of these shaded figures must be accompanied by this disclaimer.

Table 7.7 Chronic sickness: prevalence of reported longstanding illness by sex, age, and economic activity status

Persons aged 16 and over   *Great Britain: 2006[1]*

Economic activity status	Age							
	Men				Women			
	16-44	45-64	65 and over	Total	16-44	45-64	65 and over	Total
	Percentage who reported longstanding illness							
Working	19	36	49	26	21	33	55	26
Unemployed	22	46	100	28	26	47	28	32
Economically inactive	35	76	68	63	28	62	67	55
Total	21	45	66	37	23	44	67	39
Weighted bases (000's) =100%								
Working	7,794	4,841	351	12,986	7,721	4,430	222	12,373
Unemployed	521	151	8	680	300	108	10	418
Economically inactive	1,284	1,432	3,472	6,188	2,721	2,552	4,673	9,946
Total	9,599	6,424	3,831	19,854	10,742	7,090	4,905	22,737
Unweighted sample								
Working	2,658	2,009	159	4,826	2,847	1,860	98	4,805
Unemployed	169	58	3	230	97	43	3	143
Economically inactive	398	591	1,603	2,592	1,008	1,109	1,946	4,063
Total	3,225	2,658	1,765	7,648	3,952	3,012	2,047	9,011

1 Results for 2006 include longitudinal data (see Appendix B)

Note: Shaded figures indicate the estimates are unreliable and any analysis using these figures may be invalid. Any use of these shaded figures must be accompanied by this disclaimer.

Table 7.8   Chronic sickness: prevalence of reported limiting longstanding illness by sex, age, and economic activity status

Persons aged 16 and over

Great Britain: 2006[1]

Economic activity status	Age							
	Men				Women			
	16-44	45-64	65 and over	Total	16-44	45-64	65 and over	Total
	Percentage who reported limiting longstanding illness							
Working	7	12	20	9	9	15	27	12
Unemployed	9	32	0	14	13	35	28	19
Economically inactive	26	59	43	43	19	48	46	39
Total	10	23	41	20	12	27	45	24
Weighted bases (000's) =100%								
Working	7,794	4,838	351	12,982	7,721	4,430	222	12,373
Unemployed	521	151	8	680	300	108	10	418
Economically inactive	1,284	1,432	3,470	6,186	2,721	2,552	4,673	9,946
Total	9,599	6,421	3,830	19,849	10,742	7,090	4,905	22,737
Unweighted sample								
Working	2,658	2,008	159	4,825	2,847	1,860	98	4,805
Unemployed	169	58	3	230	97	43	3	143
Economically inactive	398	591	1,602	2,591	1,008	1,109	1,946	4,063
Total	3,225	2,657	1,764	7,646	3,952	3,012	2,047	9,011

1 Results for 2006 include longitudinal data (see Appendix B).

Note: Shaded figures indicate the estimates are unreliable and any analysis using these figures may be invalid. Any use of these shaded figures must be accompanied by this disclaimer.

Table 7.9 Acute sickness
   (a) Prevalence of reported restricted activity in the 14 days before interview, by sex, age and economic activity status
   (b) Average number of restricted activity days per person per year, by sex, age, and economic activity status

Persons aged 16 and over *Great Britain: 2006[1]*

Economic activity status	Age							
	Men				Women			
	16-44	45-64	65 and over	Total	16-44	45-64	65 and over	Total
	(a) Percentage who reported restricted activity in the 14 days before interview							
Working	8	8	7	8	10	11	16	11
Unemployed	10	9	0	10	5	14	28	8
Economically inactive	14	33	18	21	16	27	23	22
Total	9	14	17	11	11	17	23	16
	(b) Average number of restricted activity days per person per year							
Working	13	18	13	15	16	23	32	19
Unemployed	15	19	0	16	7	33	103	16
Economically inactive	38	92	51	58	36	70	64	58
Total	16	35	47	28	21	40	63	36
Weighted bases (000's) =100%								
Working	7,875	4,871	351	13,097	7,738	4,438	222	12,397
Unemployed	526	151	8	686	300	108	10	418
Economically inactive	1,287	1,430	3,473	6,190	2,730	2,545	4,677	9,952
Total	9,688	6,453	3,832	19,973	10,767	7,091	4,909	22,767
Unweighted sample								
Working	2,685	2,023	159	4,867	2,853	1,863	98	4,814
Unemployed	171	58	3	232	97	43	3	143
Economically inactive	399	590	1,603	2,592	1,011	1,106	1,948	4,065
Total	3,255	2,671	1,765	7,689	3,961	3,012	2,049	9,020

1 Results for 2006 include longitudinal data (see Appendix B).

Note: Shaded figures indicate the estimates are unreliable and any analysis using these figures may be invalid. Any use of these shaded figures must be accompanied by this disclaimer.

Table 7.10 **Self-reported sickness by sex and Government Office Region: percentage of persons who reported**
(a) longstanding illness
(b) limiting longstanding illness
(c) restricted activity in the 14 days before interview

All persons *Great Britain: 2006[1]*

Government Office Region[2]		(a) Longstanding illness	(b) Limiting longstanding illness	(c) Restricted activity in the 14 days before interview	Weighted base (000's)	Unweighted sample
Males			Percentages			
England						
	North East	35	22	14	1,039	409
	North West	32	18	11	2,892	1203
	Yorkshire and the Humber	38	18	10	2,229	924
	East Midlands	33	18	12	2,062	845
	West Midlands	35	17	11	2,274	913
	East of England	32	16	10	2,522	1020
	London	25	14	9	3,108	952
	South East	34	16	12	3,610	1444
	South West	34	18	11	2,268	935
All England		33	17	11	22,003	8645
Wales		35	22	12	1,296	523
Scotland		31	15	10	2,138	849
Great Britain		33	17	11	25,437	10017
Females						
England						
	North East	39	23	19	1,069	435
	North West	35	22	13	3,376	1420
	Yorkshire and the Humber	38	22	15	2,458	1035
	East Midlands	38	23	15	2,136	906
	West Midlands	35	21	14	2,402	980
	East of England	31	16	13	2,710	1128
	London	28	18	15	3,505	1081
	South East	34	19	13	3,968	1624
	South West	36	21	13	2,672	1125
All England		34	20	14	24,296	9734
Wales		36	24	18	1,401	579
Scotland		32	20	12	2,538	1016
Great Britain		34	20	14	28,236	11329
All persons						
England						
	North East	37	22	16	2,109	844
	North West	34	20	12	6,268	2623
	Yorkshire and the Humber	38	20	12	4,687	1959
	East Midlands	36	20	14	4,197	1751
	West Midlands	35	19	13	4,677	1893
	East of England	32	16	12	5,231	2148
	London	27	16	12	6,613	2033
	South East	34	18	13	7,578	3068
	South West	35	20	12	4,939	2060
All England		33	19	13	46,300	18379
Wales		35	23	15	2,698	1102
Scotland		31	18	11	4,676	1865
Great Britain		33	19	13	53,673	21346

1 Results for 2006 include longitudinal data (see Appendix B).
2 The data have not been standardised to take account of age or socio-economic group.

Table 7.11 Chronic sickness: rate per 1000 reporting longstanding condition groups, by sex

Persons aged 16 and over *Great Britain: 2006[1]*

Condition group		Men	Women	Total
XIII	Musculoskeletal system	126	183	156
VII	Heart and circulatory system	120	115	118
VIII	Respiratory system	66	63	64
III	Endocrine and metabolic	48	69	58
IX	Digestive system	24	35	30
VI	Nervous system	27	33	30
V	Mental disorders	28	30	29
VI	Eye complaints	17	19	18
X	Genito-urinary system	14	17	16
VI	Ear complaints	15	15	15
II	Neoplasms and benign growths	11	16	14
XII	Skin complaints	9	8	9
IV	Blood and related organs	4	6	5
	Other complaints[2]	2	3	3
I	Infectious diseases	2	2	2
Average number of conditions reported by those with a longstanding illness		1.6	1.7	1.6
Weighted bases (000's) = 100%		*22,779*	*24,130*	*46,909*
Unweighted sample		*8681*	*9533*	*18214*

1 Results for 2006 include longitudinal data (see Appendix B).
2 Including general complaints such as insomnia, fainting, generally run down, old age and general infirmity and non-specific conditions such as war wounds or road accident injuries where no further details were given.

Table 7.12 Chronic sickness: rate per 1000 reporting longstanding condition groups, by age

Persons aged 16 and over *Great Britain: 2006[1]*

Condition group		16-44	45-64	65-74	75 and over
XIII	Musculoskeletal system	63	195	309	361
VII	Heart and circulatory system	18	143	321	353
VIII	Respiratory system	50	65	99	99
III	Endocrine and metabolic	19	80	123	132
IX	Digestive system	14	41	55	54
VI	Nervous system	20	39	41	38
V	Mental disorders	27	36	24	18
VI	Eye complaints	3	15	35	90
X	Genito-urinary system	10	14	28	42
VI	Ear complaints	5	14	33	57
II	Neoplasms and benign growths	4	16	35	38
XII	Skin complaints	8	8	8	11
IV	Blood and related organs	4	5	9	13
	Other complaints[2]	2	2	2	8
I	Infectious diseases	2	3	2	1
Average number of conditions reported by those with a longstanding illness		1.3	1.6	1.7	1.9
Weighted bases (000's) =100%		*23,397*	*14,503*	*4,847*	*4,162*
Unweighted sample		*8221*	*6068*	*2194*	*1731*

1 Results for 2006 include longitudinal data (see Appendix B).
2 Including general complaints such as insomnia, fainting, generally run down, old age and general infirmity and non-specific conditions such as war wounds or road accident injuries where no further details were given.

Table 7.13 Chronic sickness: rate per 1000 reporting selected longstanding condition groups, by age and sex

Persons aged 16 and over *Great Britain: 2006[1]*

Condition group			16-44	45-64	65-74	75 and over	All ages
XIII Musculoskeletal system		Men	60	164	251	261	126
		Women	66	226	361	428	183
VII Heart and circulatory system		Men	13	159	355	384	120
		Women	23	128	290	333	115
VIII Respiratory system		Men	53	60	106	121	66
		Women	48	71	93	84	63
III Endocrine and metabolic		Men	12	70	96	134	48
		Women	25	90	147	130	69
IX Digestive system		Men	10	34	47	54	24
		Women	18	48	62	54	35
VI Nervous system		Men	17	37	36	36	27
		Women	24	41	46	38	33
Weighted bases (000's) =100%		*Men*	*11,681*	*7,142*	*2,294*	*1,662*	*22,779*
		Women	*11,716*	*7,361*	*2,553*	*2,499*	*24,130*
Unweighted sample		*Men*	*3921*	*2943*	*1041*	*776*	*8681*
		Women	*4300*	*3125*	*1153*	*955*	*9533*

1 Results for 2006 include longitudinal data (see Appendix B).

Table 7.14 Chronic sickness: rate per 1000 reporting selected longstanding conditions, by sex and age

Persons aged 16 and over *Great Britain: 2006[1]*

Condition	Men					Women				
	16-44	45-64	65-74	75 and over	All ages	16-44	45-64	65-74	75 and over	All ages
Musculoskeletal (XIII)										
Arthritis and rheumatism	11	67	144	153	52	17	116	229	259	95
Back problems	24	53	44	30	36	30	57	45	30	40
Other bone and joint problems	25	45	63	78	39	19	53	87	139	49
Heart and circulatory (VII)										
Hypertension	5	69	128	110	45	10	75	145	141	58
Heart attack	2	24	64	72	20	1	12	39	65	15
Stroke	1	7	32	35	9	1	5	17	21	6
Other heart complaints	4	45	108	142	37	8	27	73	83	28
Other blood vessel/embolic disorders	1	11	19	22	7	3	6	13	19	7
Respiratory (VIII)										
Asthma	44	38	42	46	42	42	52	61	49	48
Bronchitis and emphysema	0	6	27	36	7	1	7	14	14	6
Hay fever	5	3	1	1	4	1	2	1	1	1
Other respiratory complaints	4	13	36	38	12	4	9	17	21	9
Weighted bases (000's) =100%	*11,681*	*7,142*	*2,294*	*1,662*	*22,779*	*11,716*	*7,361*	*2,553*	*2,499*	*24,130*
Unweighted sample	*3921*	*2943*	*1041*	*776*	*8681*	*4300*	*3125*	*1153*	*955*	*9533*

1 Results for 2006 include longitudinal data (see Appendix B).

Table 7.15 Chronic sickness: rate per 1000 reporting selected longstanding condition groups, by socio-economic classification of household reference person

Persons aged 16 and over   *Great Britain: 2006[1]*

Condition group		Managerial and professional	Intermediate	Small employers and own account	Lower supervisory and technical	Semi-routine and routine	Total[2]
XIII	Musculoskeletal system	117	177	152	177	215	156
VII	Heart and circulatory system	99	150	91	129	153	118
VIII	Respiratory system	56	69	65	75	75	64
III	Endocrine and metabolic	45	75	41	66	81	58
IX	Digestive system	23	29	27	28	46	30
VI	Nervous system	27	33	23	28	34	30
Average number of condition groups reported by those with a longstanding illness		1.47	1.68	1.59	1.67	1.74	1.54
Weighted bases (000's) =100%		18,950	3,955	4,706	4,996	11,734	46,909
Unweighted sample		7696	1502	1806	1907	4419	18214

1 Results for 2006 include longitudinal data (see Appendix B).
2 Where the household reference person was a full-time student, had an inadequately described occupation, had never worked or was long-term unemployed these are not shown as separate categories but are included in the figure for all persons (see Appendix A).

Table 7.16 Chronic sickness: rate per 1000 reporting selected longstanding condition groups, by sex and age and socio-economic classification of household reference person

Persons aged 16 and over   *Great Britain: 2006[1]*

Condition group	Men				Women				All aged 16 and over			
	16-44	45-64	65 and over	Total	16-44	45-64	65 and over	Total	16-44	45-64	65 and over	Total
XIII Musculoskeletal system												
Managerial and professional	51	109	203	93	63	171	349	142	57	140	277	117
Intermediate	56	192	283	140	54	215	379	184	55	203	342	163
Routine and manual	74	220	289	165	81	291	445	239	77	257	377	203
VII Heart and circulatory system												
Managerial and professional	15	128	398	109	20	96	309	89	17	112	353	99
Intermediate	13	162	349	121	24	101	301	116	19	131	320	118
Routine and manual	14	195	355	143	25	174	326	148	19	184	339	146
VIII Respiratory system												
Managerial and professional	56	54	76	58	45	60	70	54	50	57	73	56
Intermediate	48	60	108	62	58	82	82	71	53	71	92	67
Routine and manual	58	70	145	81	49	71	103	70	54	70	121	75
III Endocrine and metabolic												
Managerial and professional	12	54	98	39	24	62	125	52	18	58	111	45
Intermediate	11	68	109	47	27	76	120	65	19	72	116	57
Routine and manual	13	87	126	60	29	125	160	92	21	107	146	77
IX Digestive system												
Managerial and professional	9	30	35	20	13	35	61	27	11	32	48	23
Intermediate	6	34	53	23	22	35	43	32	14	35	47	28
Routine and manual	15	39	63	33	23	68	65	47	19	54	64	40
VI Nervous system												
Managerial and professional	16	30	24	22	26	38	46	33	21	34	35	27
Intermediate	13	40	39	26	14	29	55	29	13	34	49	28
Routine and manual	18	45	44	32	20	51	33	33	19	48	38	32
Weighted base (000's) =100%												
Managerial and professional	*4,979*	*3,030*	*1,435*	*9,445*	*4,977*	*3,053*	*1,475*	*9,505*	*9,956*	*6,083*	*2,911*	*18,950*
Intermediate	*2,016*	*1,400*	*701*	*4,116*	*2,000*	*1,442*	*1,102*	*4,545*	*4,016*	*2,841*	*1,804*	*8,661*
Routine and manual	*3,824*	*2,438*	*1,741*	*8,003*	*3,847*	*2,597*	*2,282*	*8,726*	*7,672*	*5,035*	*4,023*	*16,729*
Unweighted sample												
Managerial and professional	*1745*	*1321*	*702*	*3768*	*1898*	*1357*	*673*	*3928*	*3643*	*2678*	*1375*	*7696*
Intermediate	*659*	*561*	*319*	*1539*	*717*	*605*	*447*	*1769*	*1376*	*1166*	*766*	*3308*
Routine and manual	*1244*	*955*	*764*	*2963*	*1387*	*1056*	*920*	*3363*	*2631*	*2011*	*1684*	*6326*

1 Results for 2006 include longitudinal data (see Appendix B).

Table 7.17 Trends in consultations with an NHS GP in the 14 days before interview by sex and age: 1972 to 2006

All persons
Great Britain

	Unweighted								Weighted									Weighted base 2006 (000's)[3] =100%	Unweighted sample[3] 2006
	1972	1975	1981	1985	1991	1995	1996	1998	1998	2000	2001	2002	2003	2004	2005[1]	2006[2]		
								Percentage consulting GP										
Males																		
0-4	13	13	21	22	23	22	23	18	18	18	18	19	17	15	15	14	1,686	682
5-15[4]	7	7	8	9	10	9	9	8	8	8	7	8	7	7	7	6	3,891	1685
16-44[4]	8	7	7	7	9	10	10	9	9	8	8	9	8	8	8	8	11,337	3798
45-64	11	11	12	12	11	14	15	14	14	15	13	15	12	14	13	13	7,023	2893
65-74	12	12	13	15	17	17	19	17	17	20	18	22	18	17	17	18	2,294	1041
75 and over	19	20	17	19	21	22	21	21	21	20	22	21	21	21	21	24	1,652	772
Total	10	9	10	11	12	13	13	12	12	12	11	13	11	11	11	11	27,882	10871
Females																		
0-4	15	13	17	21	21	21	20	18	18	14	18	14	14	17	15	15	1,673	674
5-15[4]	6	7	9	11	11	13	9	10	10	9	9	8	7	7	6	6	3,820	1642
16-44[4]	15	13	15	17	17	18	20	17	17	16	15	18	15	16	17	15	11,495	4217
45-64	12	12	13	15	17	17	19	18	18	17	18	17	17	17	17	15	7,317	3107
65-74	15	16	16	17	19	23	21	19	19	22	18	21	22	21	21	22	2,549	1151
75 and over	20	17	20	20	19	23	23	20	20	22	20	27	20	22	21	20	2,494	953
Total	13	12	14	16	17	18	19	17	17	16	16	17	16	16	16	15	29,348	11744
All persons																		
0-4	14	13	19	21	22	21	22	18	18	16	18	17	16	16	15	15	3,358	1356
5-15[4]	7	7	9	10	10	11	9	9	9	8	8	8	7	7	7	6	7,711	3327
16-44[4]	12	10	11	12	13	14	15	13	13	12	11	14	12	12	12	12	22,831	8015
45-64	12	11	12	14	14	16	17	16	16	16	16	16	15	15	15	14	14,340	6000
65-74	14	14	15	16	18	20	20	18	18	21	18	22	20	19	19	20	4,843	2192
75 and over	20	18	19	20	19	23	22	21	21	21	21	25	20	22	21	22	4,146	1725
Total	12	11	12	14	14	16	16	14	14	14	13	15	13	14	14	13	57,230	22615

1 2005 data includes last quarter of 2004/5 data due to survey change from financial year to calendar year.
2 Results for 2006 include longitudinal data (see Appendix B).
3 Trend tables show unweighted and weighted figures for 1998 to give an indication of the effect of the weighting. Bases for earlier years can be found in GHS reports for each year.
4 These age-groups were 5-14 and 15-44 in 1972 to 1975.

Table 7.18 Average number of NHS GP consultations per person per year by sex and age: 1972 to 2006

All persons[1]
Great Britain

	Unweighted								Weighted								
	1972[2]	1975	1981	1985	1991	1995	1996	1998	1998	2000	2001	2002	2003	2004	2005[3]	2006[4]
Males																
0-4	4	4	7	7	7	7	8	6	6	6	6	7	6	5	5	4
5-15[5]	2	2	2	3	3	3	3	2	2	2	2	3	2	2	2	2
16-44[5]	3	2	2	2	3	3	3	3	3	3	3	3	3	3	3	2
45-64	4	4	4	4	4	4	5	4	4	5	4	5	4	4	4	4
65-74	4	4	4	5	5	5	6	5	5	6	5	7	6	5	5	6
75 and over	7	7	6	6	7	8	7	7	7	6	7	7	7	7	7	8
Total	3	3	3	3	4	4	4	4	4	4	4	4	3	3	4	3
Females																
0-4	5	4	5	7	7	7	6	6	6	4	6	4	5	6	5	5
5-15[5]	2	2	3	3	3	4	3	3	3	3	3	2	2	2	2	2
16-44[5]	5	4	5	5	5	6	7	5	5	5	5	6	5	5	6	5
45-64	4	4	4	5	5	5	6	6	6	5	6	5	5	6	5	5
65-74	5	5	5	5	6	7	7	6	6	7	5	7	7	7	7	7
75 and over	7	6	6	7	6	7	7	6	6	7	6	9	6	7	7	7
Total	4	4	4	5	5	6	6	5	5	5	5	6	5	5	5	5
All persons																
0-4	4	4	6	7	7	7	7	6	6	5	6	5	5	5	5	5
5-15[5]	2	2	3	3	3	3	3	3	3	2	3	2	2	2	2	2
16-44[5]	4	3	4	4	4	4	5	4	4	4	4	4	4	4	4	4
45-64	4	4	4	4	4	5	5	5	5	5	5	5	5	5	5	5
65-74	4	4	5	5	6	6	6	6	6	6	5	7	6	6	6	7
75 and over	7	7	6	6	6	7	7	6	6	7	6	8	7	7	7	7
Total	4	4	4	4	5	5	5	4	4	4	4	5	4	4	4	4

1 Trend tables show unweighted and weighted figures for 1998 to give an indication of the effect of weighting. Bases for 2006 are shown in table 7.18. Bases for earlier years can be found in GHS reports for each year.
2 1972 figures relate to England and Wales.
3 2005 data includes last quarter of 2004/5 data due to survey change from financial year to calendar year.
4 Results for 2006 include longitudinal data (see Appendix B).
5 These age-groups were 5-14 and 15-44 in 1972 to 1975.

Table 7.19 NHS GP consultations: trends in site of consultations: 1971 to 2006

Consultations in the 14 days before interview   Great Britain

Site of consultation	Unweighted								Weighted							
	1971	1975	1981	1985	1991	1995	1996	1998	1998	2000	2001	2002	2003	2004	2005[1]	2006[2]
	%	%	%	%	%	%	%	%	%	%	%	%	%	%	%	%
Surgery[3]	73	78	79	79	81	84	84	84	84	86	85	86	86	87	87	87
Home	22	19	14	14	11	9	8	6	6	5	5	5	4	4	3	4
Telephone	4	3	7	7	8	7	8	10	10	10	10	9	10	9	10	10
Weighted bases (000's) =100% [4]									9,658	9,744	9,161	10,284	9,165	9,379	9,316	9,265
Unweighted sample [4]	5031	4455	4704	4123	4228	4385	4341	3504	3504	3294	3418	3656	3980	3375	4941	3682

1 2005 data includes last quarter of 2004/5 data due to survey change from financial year to calendar year.
2 Results for 2006 include longitudinal data (see Appendix B).
3 Includes consultations with a GP at a health centre and those who had answered 'elsewhere'.
4 Trend tables show unweighted and weighted figures for 1998 to give an indication of the effect of the weighting. For the weighted data (1998 and 2000 to 2005) the weighted base (000's) is the base for percentages. Unweighted data (up to 1998) are based on the unweighted sample.

Table 7.20 Percentage of persons who consulted an NHS GP in the 14 days before interview by sex and site of consultation, and by age and site of consultation

Persons who consulted in the 14 days before interview
Great Britain: 2006[1]

Site of consultation[2]	Total	Males	Females	Age					
				0-4	5-15	16-44	45-64	65-74	75 and over
	%	%	%	%	%	%	%	%	%
Surgery	90	90	89	92	91	92	90	90	79
At home	3	3	3	0	1	1	1	4	15
Telephone	10	9	12	10	9	10	12	9	12
Weighted base (000's) =100%	*7,521*	*3,109*	*4,412*	*487*	*480*	*2,631*	*2,063*	*965*	*894*
Unweighted sample	*2993*	*1255*	*1738*	*188*	*206*	*946*	*854*	*423*	*376*

1 Results for 2006 include longitudinal data (see Appendix B).
2 Percentages add to more than 100 because some people consulted at more than one site during the reference period.

Table 7.21   NHS GP consultations
  (a) Percentage of persons who consulted a doctor in the 14 days before interview, by sex, age, and economic activity status
  (b) Average number of consultations per person per year, by sex, age, and economic activity status

Persons aged 16 and over *Great Britain: 2006[1]*

Economic activity status	Age							
	Men				Women			
	16-44	45-64	65 and over	Total	16-44	45-64	65 and over	Total
(a) Percentage who consulted a GP in the 14 days before interview								
Working	8	11	15	9	15	14	24	14
Unemployed	8	28	44	12	15	25	0	17
Economically inactive	11	23	21	19	16	18	21	19
Total	8	14	20	12	15	16	21	17
(b) Average number of consultations per person per year								
Working	2	3	4	3	5	4	8	5
Unemployed	2	8	11	4	5	7	0	6
Economically inactive	4	7	7	6	6	6	7	7
Total	3	4	7	4	5	5	7	5
Weighted bases (000's) =100%								
Working	*8,087*	*4,927*	*344*	*13,358*	*7,784*	*4,423*	*203*	*12,410*
Unemployed	*613*	*154*	*8*	*775*	*319*	*107*	*10*	*437*
Economically inactive	*1,576*	*1,483*	*3,569*	*6,628*	*2,960*	*2,633*	*4,806*	*10,398*
Total	*10,275*	*6,564*	*3,921*	*20,760*	*11,063*	*7,163*	*5,019*	*23,245*
Unweighted sample								
Working	*2753*	*2044*	*155*	*4952*	*2868*	*1855*	*90*	*4813*
Unemployed	*198*	*59*	*3*	*260*	*101*	*43*	*3*	*147*
Economically inactive	*488*	*612*	*1643*	*2743*	*1089*	*1143*	*2000*	*4232*
Total	*3439*	*2715*	*1801*	*7955*	*4058*	*3041*	*2093*	*9192*

1   Results for 2006 include longitudinal data (see Appendix B).

Note: Shaded figures indicate the estimates are unreliable and any analysis using these figures may be invalid. Any use of these shaded figures must be accompanied by this disclaimer.

Table 7.22 Percentage of persons consulting an NHS GP in the 14 days before interview who obtained a prescription from the doctor, by sex, age and socio-economic classification of household reference person

Persons who consulted in the 14 days before interview *Great Britain: 2006[1]*

Socio-economic classification of household reference person[2]	Males					Females				
	Age					Age				
	0-15	16-44	45-64	65 and over	Total	0-15	16-44	45-64	65 and over	Total
	Percentage consulting who obtained a prescription									
Managerial and professional	62	52	63	66	60	69	59	69	73	65
Intermediate	61	51	64	56	58	70	70	77	67	71
Routine and manual	68	67	60	66	65	59	69	74	79	72
All persons consulting	62	57	63	65	61	64	65	74	73	69
Weighted bases (000's) =100%										
Managerial and professional	*234*	*360*	*357*	*296*	*1,247*	*164*	*682*	*402*	*308*	*1,556*
Intermediate	*66*	*144*	*193*	*106*	*508*	*69*	*288*	*187*	*221*	*766*
Routine and manual	*153*	*312*	*353*	*371*	*1,190*	*200*	*640*	*466*	*487*	*1,793*
All persons consulting	*494*	*884*	*938*	*794*	*3,109*	*474*	*1,747*	*1,126*	*1,066*	*4,412*
Unweighted sample										
Managerial and professional	*99*	*128*	*159*	*141*	*527*	*69*	*260*	*178*	*136*	*643*
Intermediate	*27*	*50*	*74*	*49*	*200*	*29*	*102*	*77*	*90*	*298*
Routine and manual	*67*	*102*	*145*	*158*	*472*	*76*	*234*	*183*	*200*	*693*
All persons consulting	*206*	*302*	*391*	*356*	*1255*	*188*	*644*	*463*	*443*	*1738*

1 Results for 2006 include longitudinal data (see Appendix B).
2 Where the household reference person was a full-time student, had an inadequately described occupation, had never worked or was long term unemployed these are not shown as separate categories, but are included in the figure for all persons (see Appendix A).

Note: Shaded figures indicate the estimates are unreliable and any analysis using these figures may be invalid. Any use of these shaded figures must be accompanied by this disclaimer.

Table 7.23 GP consultations: consultations with a doctor in the 14 days before interview by sex of person consulting and whether consultation was NHS or private

Consultations in the 14 days before interview *Great Britain: 2006[1]*

Type of consultation	Consultations made by males	Consultations made by females	All persons
	%	%	%
NHS	96	98	97
Private	4	2	3
Weighted base (000's) =100%	*4,351*	*6,235*	*10,586*
Unweighted sample	*1758*	*2449*	*4207*

1 Results for 2006 include longitudinal data (see Appendix B).

Table 7.24 Trends in reported consultations with a practice nurse by sex and age: 2000 to 2006
(a) percentage consulting a practice nurse in the 14 days before interview
(b) average number of consultations with a practice nurse per person per year

All persons   Great Britain

	(a) percentage consulting a practice nurse							(b) average number of consultations with a practice nurse per person per year							Weighted base 2006 (000's) = 100%	Unweighted sample 2006
	2000	2001	2002	2003	2004	2005	2006	2000	2001	2002	2003	2004	2005[1]	2006[2]		
Males																
0-4	4	3	4	5	5	5	3	1	1	1	1	1	1	1	1,694	685
5-15	2	2	2	1	2	2	2	0	1	0	0	1	1	0	3,908	1693
16-44	2	2	3	2	2	2	2	1	1	1	1	1	1	1	11,374	3810
45-64	5	6	6	5	6	7	7	1	2	2	2	2	2	2	7,026	2895
65-74	10	12	13	13	11	13	11	3	4	4	4	3	4	4	2,294	1041
75 and over	8	13	11	10	14	15	15	3	4	3	3	4	5	5	1,652	772
Total	4	5	5	4	5	5	5	1	1	1	1	1	1	1	27,947	10896
Females																
0-4	5	3	5	6	3	4	3	1	1	1	2	1	1	1	1,684	679
5-15	1	1	1	1	1	2	2	0	0	0	0	0	0	1	3,845	1652
16-44	5	5	6	5	6	7	6	1	1	2	2	2	2	2	11,519	4224
45-64	6	7	7	7	7	7	6	2	2	2	2	2	2	2	7,324	3110
65-74	10	11	11	14	12	14	12	3	4	3	4	4	4	4	2,549	1151
75 and over	9	12	12	14	12	14	12	3	4	4	4	4	4	4	2,494	953
Total	5	6	7	7	6	7	6	2	2	2	2	2	2	2	29,415	11769
All persons																
0-4	4	3	4	6	4	5	3	1	1	1	2	1	1	1	3,378	1364
5-15	1	2	1	1	1	2	2	0	1	0	0	0	0	0	7,753	3345
16-44	3	4	4	4	4	5	4	1	1	1	1	1	1	1	22,893	8034
45-64	6	6	7	6	6	7	7	2	2	2	2	2	2	2	14,349	6005
65-74	10	12	12	14	12	13	12	3	4	4	4	3	4	4	4,843	2192
75 and over	9	12	12	12	13	14	13	3	4	4	4	4	5	4	4,146	1725
Total	5	5	6	6	6	6	6	1	2	2	2	2	2	2	57,362	22665

1 2005 data includes last quarter of 2004/5 data due to survey change from financial year to calendar year.
2 Results for 2006 include longitudinal data (see Appendix B).

Table 7.25 Percentage of children using health services other than a doctor in the 14 days before interview

All persons aged under 16 Great Britain: 2006[1]

	Male			Female			Total		
	0-4	5-15	Total	0-4	5-15	Total	0-4	5-15	Total
Percentage who reported:[2]									
Seeing a practice nurse at the GP surgery	4	2	2	4	2	3	4	2	2
Seeing a health visitor at the GP surgery	6	1	2	5	1	2	5	1	2
Going to a child health clinic	6	0	2	4	0	1	5	0	2
Going to a child welfare clinic	1	0	0	0	0	0	1	0	0
None of the above	85	97	93	89	97	95	87	97	94
Weighted base (000's) =100%	1,686	3,889	5,575	1,670	3,820	5,490	3,356	7,709	11,065
Unweighted sample	682	1684	2366	673	1642	2315	1355	3326	4681

1 Results for 2006 include longitudinal data (see Appendix B).
2 Percentages may sum to greater than 100 as respondents could give more than one answer

Table 7.26 Trends in percentages of persons who reported attending an outpatient or casualty department in the 3 months before interview by sex and age: 1972 to 2006

All persons[1] Great Britain

	Unweighted								Weighted								
	1972[2]	1975	1981	1985	1991	1995	1996	1998	1998	2000	2001	2002	2003	2004	2005[3]	2006[4]
						Percentages										
Males																
0-4	8	9	12	13	14	12	13	16	16	14	16	17	13	14	15	15
5-15[5]	9	8	11	12	11	11	12	12	12	11	10	11	10	11	10	11
16-44[5]	11	9	11	12	11	12	13	13	13	12	11	11	11	10	10	10
45-64	11	10	12	16	15	16	16	17	17	16	16	15	16	17	16	15
65-74	10	11	14	16	18	21	20	25	25	24	22	24	21	24	22	21
75 and over	10	12	14	15	22	26	25	29	29	26	31	26	24	26	26	27
Total	10	10	11	13	13	14	15	16	16	15	14	14	14	14	14	13
Females																
0-4	6	8	9	11	11	12	9	13	13	10	11	12	11	11	12	12
5-15[5]	6	6	8	9	8	9	10	11	11	8	8	9	9	8	9	8
16-44[5]	9	9	11	12	12	12	13	13	13	13	12	12	13	12	12	12
45-64	11	10	13	15	16	17	18	18	18	16	18	16	18	18	16	18
65-74	12	12	16	17	18	21	22	21	21	21	21	20	20	21	21	22
75 and over	13	10	16	17	20	22	24	26	26	24	23	25	22	22	24	27
Total	10	9	12	13	14	14	15	16	16	15	14	14	15	15	14	15
All persons																
0-4	7	9	10	12	13	12	11	14	15	12	13	14	12	13	13	13
5-15[5]	8	7	10	10	10	10	11	11	11	10	9	10	10	10	9	10
16-44[5]	10	9	11	12	12	12	13	13	13	13	12	12	12	11	11	11
45-64	11	10	13	15	16	16	17	18	18	16	17	15	17	18	16	17
65-74	11	11	15	17	18	21	21	23	23	22	21	22	21	22	21	22
75 and over	12	10	15	16	21	24	24	27	27	25	26	25	23	24	25	27
Total	10	9	12	13	13	14	15	16	16	15	14	14	14	14	14	14

1 Trend tables show unweighted and weighted figures for 1998 to give an indication of the effect of the weighting. Bases for 2004 are shown in Table 7.2. Bases for earlier years can be found in GHS reports for each year.
2 1972 figures relate to England and Wales.
3 2005 data includes last quarter of 2004/5 data due to survey change from financial year to calendar year.
4 Results for 2006 include longitudinal data (see Appendix B).
5 These age groups were 5-14 and 15-44 in 1972 to 1975.

Table 7.27 Trends in day-patient treatment in the 12 months before interview by sex and age, 1992 to 2006

All persons   Great Britain

	Unweighted					Weighted								Weighted base 2006 (000's) =100%[3]	Unweighted sample[3] 2006
	1992	1994	1995	1996	1998	1998	2000	2001	2002	2003	2004	2005[1]	2006[2]		

Percentage receiving day-patient treatment

Males
0-4	4	4	4	5	6	6	6	7	8	7	6	6	6	1,686	682
5-15	2	3	3	3	4	4	5	4	5	4	4	4	4	3,889	1684
16-44	4	5	6	5	6	6	6	7	7	7	7	6	5	11,357	3804
45-64	4	5	7	6	7	7	8	8	8	8	9	10	7	7,019	2892
65-74	5	6	6	7	6	6	10	8	11	10	12	12	9	2,294	1041
75 and over	4	5	5	6	12	11	7	10	12	12	13	10	8	1,652	772
Total	4	5	5	5	6	6	7	7	8	7	8	7	6	27,897	10875

Females
0-4	2	3	3	3	5	4	6	4	5	6	5	5	4	1,673	674
5-15	2	3	2	4	4	4	3	3	4	4	4	4	3	3,820	1642
16-44	5	7	6	7	8	8	8	8	9	8	8	8	7	11,497	4217
45-64	5	5	7	8	8	8	9	8	8	9	10	9	8	7,319	3108
65-74	4	5	5	6	6	6	12	8	10	9	10	10	9	2,549	1151
75 and over	3	5	5	7	8	8	8	8	10	11	9	9	9	2,494	953
Total	4	5	5	6	7	7	8	7	8	8	8	8	7	29,352	11745

All persons
0-4	3	3	3	4	5	5	6	5	6	6	6	6	5	3,358	1356
5-15	2	3	3	3	4	4	4	4	4	4	4	4	3	7,709	3326
16-44	4	6	6	6	7	7	7	7	8	7	7	7	6	22,854	8021
45-64	5	5	7	7	8	8	8	8	8	8	9	9	7	14,338	6000
65-74	4	5	6	7	6	6	11	8	10	9	11	11	9	4,843	2192
75 and over	3	5	5	6	9	9	8	9	11	12	11	10	9	4,146	1725
Total	4	5	5	6	7	7	7	7	8	8	8	8	6	57,248	22620

1 2005 data includes last quarter of 2004/5 data due to survey change from financial year to calendar year
2 Results for 2006 include longitudinal data (see Appendix B).
3 Trend tables show unweighted and weighted figures for 1998 to give an indication of the effect of the weighting. Bases for earlier years can be found in GHS reports for each year.

Table 7.28 Trends in inpatient stays in the 12 months before interview by sex and age, 1982 to 2006

All persons
Great Britain

	Unweighted							Weighted							Weighted base 2006 (000's) =100% [3]	Unweighted sample [3] 2006
	1982	1985	1991	1995	1996	1998		1998	2000	2001	2002	2003	2004	2005[1]	2006[2]	
	Percentage with inpatient stay															
Males																
0-4	14	12	10	9	9	9		9	8	11	11	8	8	8	7	1,686	682
5-15	6	8	6	5	5	5		5	5	4	4	3	3	3	2	3,889	1684
16-44	5	6	6	5	5	5		5	4	5	5	4	4	4	4	11,366	3807
45-64	8	8	8	9	8	8		8	8	8	8	8	7	8	7	7,019	2892
65-74	12	13	13	15	13	15		15	13	12	15	13	11	12	11	2,294	1041
75 and over	14	17	20	21	18	21		21	18	19	18	16	17	19	17	1,652	772
Total	7	8	8	8	7	8		8	7	7	7	7	6	7	6	27,906	10878
Females																
0-4	12	8	8	8	7	10		10	6	6	7	5	6	6	5	1,673	674
5-15	4	5	4	4	4	4		4	3	3	2	4	3	3	3	3,820	1642
16-44	15	16	15	12	12	11		11	10	9	9	9	10	9	8	11,497	4217
45-64	8	8	9	8	10	8		9	7	8	8	7	9	8	6	7,319	3108
65-74	8	18	11	11	12	10		10	13	10	8	10	11	11	11	2,549	1151
75 and over	12	13	16	20	16	15		15	18	15	16	14	15	17	15	2,492	952
Total	11	11	11	10	10	10		10	9	8	8	8	9	9	8	29,350	11744
All persons																
0-4	13	10	9	9	8	9		9	7	9	9	7	7	7	6	3,358	1356
5-15	5	6	5	4	4	5		5	4	4	3	3	3	3	3	7,709	3326
16-44	10	11	10	8	9	8		8	7	7	7	7	7	7	6	22,863	8024
45-64	8	8	8	8	9	8		9	8	8	8	7	8	8	7	14,338	6000
65-74	10	10	12	13	12	12		12	13	11	11	12	11	12	11	4,843	2192
75 and over	13	15	18	20	17	17		17	18	17	17	15	16	17	16	4,144	1724
Total	9	10	10	9	9	9		9	8	8	8	7	8	8	7	57,256	22622

1 2005 data includes last quarter of 2004/5 data due to survey change from financial year to calendar year.
2 Results for 2006 include longitudinal data (see Appendix B).
3 Trend tables show unweighted and weighted figures for 1998 to give an indication of the effect of the weighting. Bases for earlier years can be found in GHS reports for each year.

Table 7.29 Average number of nights spent in hospital as an inpatient during the 12 months before interview, by sex and age

All inpatients *Great Britain: 2006[1]*

Age	Male	Female	Total	Male	Female	Total	Male	Female	Total
	Average number of nights			Weighted base (all inpatients) (000's) =100%			Unweighted sample		
0-4	5	4	5	125	83	208	52	34	86
5-15	3	4	4	80	117	198	38	53	91
16-44	5	3	4	414	927	1,341	136	345	481
45-64	9	9	9	494	461	955	212	196	408
65-74	9	9	9	253	277	530	115	120	235
75 and over	10	12	12	274	377	651	127	151	278
All persons	7	7	7	1,641	2,242	3,883	680	899	1579

1 Results for 2006 include longitudinal data (see Appendix B).
Note: Shaded figures indicate the estimates are unreliable and any analysis using these figures may be invalid. Any use of these shaded figures must be accompanied by this disclaimer.

Table 7.30 Inpatient stays and outpatient attendances
(a) Average number of inpatient stays per 100 persons in a 12 month reference period, by sex and age
(b) Average number of outpatient attendances per 100 persons per year, by sex and age

All persons *Great Britain: 2006[1]*

Age	(a) Average number of inpatient stays per 100 persons during the 12 months before interview			(b) Average number of outpatient attendances per 100 persons per year			Weighted bases (000's) =100%			Unweighted sample		
	Males	Females	Total	Males	Females	Total	Males	Females	Total	Males	Females	Total
0-4	8	6	7	91	76	83	1,686	1,673	3,358	682	674	1356
5-15	3	4	3	60	52	56	3,889	3,820	7,709	1684	1642	3326
16-44	4	6	5	82	105	94	11,366	11,497	22,863	3807	4217	8024
45-64	9	8	9	126	160	143	7,019	7,319	14,338	2892	3108	6000
65-74	16	13	14	213	213	213	2,294	2,549	4,843	1041	1151	2192
75 and over	21	19	20	261	276	270	1,652	2,492	4,144	772	952	1724
Total	7	8	8	112	134	123	27,906	29,350	57,256	10878	11744	22622

1 Results for 2006 include longitudinal data (see Appendix B).